Microwave and RF Engineering

A Simulation Approach with Keysight Genesys Software

Microwave and RF Engineering

A Simulation Approach with Keysight Genesys Software

Ali A. Behagi

and

Stephen D. Turner

BT Microwave LLC

State College, PA 16803

Microwave and RF Engineering
A Simulation Approach with Keysight Genesys Software

ISBN 978-09835460-3-0

Copyright © 2015 by Ali A. Behagi

Published in USA
BT Microwave LLC
State College, PA 16803

All rights reserved. Printed and bound in the United States of America. No part of this book may be reproduced, stored in a retrieval system, or transmitted in any form or by any means, electronic, photocopying, recording, or otherwise, without prior permission in writing from the publishers.

Table of Contents

Foreword — xv

Preface — xvii

Chapter 1 RF and Microwave Concepts and Components — 1

1.1 Introduction — 1

1.2 Straight Wire, Flat Ribbon, and Skin Depth — 3
    1.2.1 Calculation of Straight Wire Inductance — 3
    1.2.2 Analysis of Straight Wire Inductor in Genesys — 5
    1.2.3 Skin Depth in Conductors — 8
    1.2.4 Wire Resistance at Different Frequencies — 10

1.3 Physical Resistors — 12
    1.3.1 Chip Resistors — 14

1.4 Physical Inductors — 16
    1.4.1 Air Core inductors — 18
    1.4.2 Modeling the Air Core Inductor in Genesys — 22
    1.4.3 Inductor Q Factor — 27
    1.4.4 Chip Inductors — 28
    1.4.5 Chip Inductor Simulation in Genesys — 29
    1.4.6 Magnetic Core Inductors — 32

1.5 Physical Capacitors — 40
    1.5.1 Single Layer Capacitor — 41
    1.5.2 Multilayer Capacitors — 43
    1.5.3 Capacitor Q Factor — 44

    References and Further Reading — 49
    Problems — 49

Chapter 2 Transmission Lines — 53

2.1 Introduction — 53
2.2 Plane Waves — 53
    2.2.1 Plane Waves in a Lossless Medium — 53

	2.2.2 Plane Waves in a Good Conductor	55
2.3	**Lumped Element Representation of Transmission Lines**	**56**
2.4	**Transmission Line Equations and Parameters**	**57**
	2.4.1 Definition of Attenuation and Phase Constant	59
	2.4.2 Definition of Transmission Line Characteristic Impedance	59
	2.4.3 Definition of Transmission Line Reflection Coefficient	59
	2.4.4 Definition of Voltage Standing Wave Ratio, VSWR	60
	2.4.5 Definition of Return Loss	61
	2.4.6 Lossless Transmission Line Parameters	61
	2.4.7 Lossless Transmission Line Terminations	62
	2.4.8 Simulating Reflection Coefficient and VSWR in Genesys	64
	2.4.9 Return Loss, VSWR, and Reflection Coefficient Conversion	64
2.5	**RF and Microwave Transmission Media**	**67**
	2.5.1 Free Space Characteristic Impedance and Velocity of Propagation	67
	2.5.2 Physical Transmission Lines	68
2.6	**Coaxial Transmission Line**	**70**
	2.6.1 Coaxial Transmission Lines in Genesys	73
	2.6.2 Using the RG8 Coaxial Cable Model in Genesys	74
2.7	**Microstrip Transmission Lines**	**76**
	2.7.1 Microstrip Transmission Lines in Genesys	78
2.8	**Stripline Transmission Lines**	**80**
2.9	**Waveguide Transmission Lines**	**82**
	2.9.1 Waveguide Transmission Lines in Genesys	86
2.10	**Group Delay in Transmission Lines**	**89**
	2.10.1 Comparing Group Delay of Various Transmission lines	89
2.11	**Transmission Line Components**	**91**
	2.11.1 Short-Circuited Transmission Line	91
	2.11.2 Modeling Short-Circuited Microstrip Lines	92
	2.11.3 Open-Circuited Transmission Line	94
	2.11.4 Modeling Open-Circuited Microstrip Lines	95
	2.11.5 Distributed Inductive and Capacitive Elements	96
	2.11.6 Distributed Microstrip Inductance and Capacitance	97
	2.11.7 Step Discontinuities	98

	2.11.8 Microstrip Bias Feed Networks	99
	2.11.9 Distributed Bias Feed	100
2.12	**Coupled Transmission Lines**	**102**
	2.12.1 Directional Coupler	105
	2.12.2 Microstrip Directional Coupler Design	107
	References and Further Reading	110
	Problems	110

Chapter 3 Network Parameters and the Smith Chart — 113

3.1	**Introduction**	**113**
	3.1.1 Z Parameters	113
	3.1.2 Y Parameters	114
	3.1.3 h Parameters	115
	3.1.4 ABCD Parameters	116
3.2	**Development of Network S-Parameters**	**117**
3.3	**Using S Parameter Files in Genesys**	**120**
	3.3.1 Scalar Representation of the S Parameters	123
3.4	**Development of the Smith Chart**	**124**
	3.4.1 Normalized Impedance on the Smith Chart	126
	3.4.2 Admittance on the Smith Chart	128
3.5	**Lumped Element Movements on the Smith Chart**	**130**
	3.5.1 Adding a Series Reactance to an Impedance	130
	3.5.2 Adding a Shunt Reactance to an Impedance	132
3.6	**VSWR Circles on the Smith Chart**	**134**
3.7	**Adding a Transmission Line in Series with an Impedance**	**137**
3.8	**Adding a Transmission Line in Parallel with an Impedance**	**139**
	3.8.1 Short Circuit Transmission Lines	140
	3.8.2 Open Circuit Transmission Lines	141
3.9	**Open and Short Circuit Shunt Transmission Lines**	**141**

| | References and Further Reading | 144 |
| | Problems | 144 |

Chapter 4 Resonant Circuits and Filters — 147

4.1 Introduction — 147

4.2 Resonant Circuits — 147
  4.2.1 Series Resonant Circuits — 147
  4.2.2 Parallel Resonant Circuits — 149
  4.2.3 Resonant Circuit Loss — 150
  4.2.4 Loaded Q and External Q — 151

4.3 Lumped Element Parallel Resonator Design — 152
  4.3.1 Effect of Load Resistance on Bandwidth and Q_L — 154

4.4 Lumped Element Resonator Decoupling — 155
  4.4.1 Tapped Capacitor Resonator — 156
  4.4.2 Tapped Inductor Resonator — 157

4.5 Practical Microwave Resonators — 158
  4.5.1 Transmission Line Resonators — 159
  4.5.2 Microstrip Resonator Example — 162
  4.5.3 Genesys Model of the Microstrip Resonator — 164

4.6 Resonator Series Reactance Coupling — 166
  4.6.1 One Port Microwave Resonator Analysis — 167
  4.6.2 Smith Chart Qo Measurement of the Microstrip Resonator — 171

4.7 Filter Design at RF and Microwave Frequency — 175
  4.7.1 Filter Topology — 176
  4.7.2 Filter Order — 177
  4.7.3 Filter Type — 178
  4.7.4 Filter Return Loss and Passband Ripple — 180

4.8 Lumped Element Filter Design — 183
  4.8.1 Low Pass Filter Design Example — 183
  4.8.2 Physical Model of the Low Pass Filter in Genesys — 185
  4.8.3 High Pass Filter Design Example — 187
  4.8.4 Physical Model of the High Pass Filter in Genesys — 188
  4.8.5 Tuning the High Pass Filter Response — 189
  4.8.6 S Parameter File Tuning with VBScript — 190

4.9 Distributed Filter Design — 195

	4.9.1	Microstrip Stepped Impedance Low Pass Filter Design	195
	4.9.2	Lumped Element to Distributed Element Conversion	195
	4.9.3	Electromagnetic Modeling of the Stepped Impedance Filter	200
	4.9.4	Reentrant Modes	204
	4.9.5	Microstrip Coupled Line Filter Design	205
	4.9.6	Electromagnetic Analysis of the Edge Coupled Filter	207
	4.9.7	Enclosure Effects	210

References and Further Reading — 212
Problems — 213

Chapter 5 Power Transfer and Impedance Matching — 217

5.1 Introduction — 217

5.2 Power Transfer Basics — 217
  5.2.1 Maximum Power Transfer Conditions — 218
  5.2.2 Maximum Power Transfer with Purely Resistive Source and Load Impedance — 220
  5.2.3 Maximum Power Transfer Validation in Genesys — 222
  5.2.4 Maximum Power Transfer with Complex Load Impedance — 224

5.3 Analytical Design of Impedance Matching Networks — 225
  5.3.1 Matching a Complex Load to Complex Source Impedance — 227
  5.3.2 Matching a Complex Load to a Real Source Impedance — 234
  5.3.3 Matching a Real Load to a Real Source Impedance — 242

5.4 Introduction to Broadband Matching Networks — 247
  5.4.1 Analytical Design of Broadband Matching Networks — 247
  5.4.2 Broadband Impedance Matching Using N-Cascaded L-Networks — 253
  5.4.3 Derivation of Equations for Q and the number of L-Networks — 257

5.5 Designing with Q-Curves on the Smith Chart — 259
  5.5.1 Q-Curve Matching Example — 261

5.6 Limitations of Broadband Matching — 264
  5.6.1 Example of Fano's Limit Calculation — 265

5.7	**Matching Network Synthesis**	**266**
	5.7.1 Filter Characteristics of the L-networks	266
	5.7.2 L-Network Impedance Matching Utility	267
	5.7.3 Network Matching Synthesis Utility in Genesys	270
	5.7.4 Effect of Finite Q on the Matching Networks	272
	References and Further Reading	275
	Problems	275

Chapter 6 Analysis and Design of Distributed Matching Networks   277

6.1	**Introduction**	**277**
6.2	**Quarter-Wave Matching Networks**	**277**
	6.2.1 Analysis of Quarter-Wave Matching Networks	278
	6.2.2 Analytical Design of Quarter-Wave Matching Networks	281
6.3	**Quarter-Wave Network Matching Bandwidth**	**286**
	6.3.1 Effect of Load Impedance on Matching Bandwidth	286
	6.3.2 Quarter-Wave Network Matching Bandwidth and Power Loss in Genesys	290
6.4	**Single-Stub Matching Networks**	**292**
	6.4.1 Analytical Design of Series Transmission Line	293
	6.4.2 Analytical Design of Shunt Transmission Line	295
	6.4.3 Single-Stub Matching Design Example	296
	6.4.4 Automated Calculation of Line and Stub Lengths	298
	6.4.5 Development of Single-Stub Matching Utility	299
6.5	**Graphical Design of Single-Stub Matching Networks**	**301**
	6.5.1 Smith Chart Design Using an Open Circuit Stub	301
	6.5.2 Smith Chart Design Using a Short Circuit Stub	303
6.6	**Design of Cascaded Single-Stub Matching Networks**	**304**
6.7	**Broadband Quarter-Wave Matching Network Design**	**307**
	References and Further Reading	318
	Problems	319

Chapter 7 Single Stage Amplifier Design — 321

7.1 Introduction — 321

7.2 Maximum Gain Amplifier Design — 322
  7.2.1 Transistor Stability Considerations — 323
  7.2.2 Stabilizing the Device in Genesys — 325
  7.2.3 Finding Simultaneous Match Reflection Coefficients and Impedances — 328

7.3 Analytical and Graphical Impedance Matching Techniques — 328
  7.3.1 Analytical Design of the Input Matching Network — 329
  7.3.2 Synthesis Based Input Matching Networks — 331
  7.3.3 Synthesis Based Output Matching Networks — 333
  7.3.4 Ideal Model of the Maximum Gain Amplifier — 334

7.4 Physical Model of the Amplifier — 336
  7.4.1 Transistor Artwork Replacement — 337
  7.4.2 Amplifier Physical Design and Layout — 339
  7.4.3 Optimization of the Amplifier Response — 343
  7.4.4 Optimization Setup Procedure — 344

7.5 Specific Gain Amplifier Design — 347
  7.5.1 Specific Gain Match — 347
  7.5.2 Specific Gain Design Example — 351
  7.5.3 Graphical Impedance Matching Circuit Design — 356
  7.5.4 Assembly and Simulation of the Specific Gain Amplifier — 358

7.6 Low Noise Amplifier Design — 360
  7.6.1 Noise Circles — 362
  7.6.2 LNA Design Example — 365
  7.6.3 Analytical Design of the LNA Input Matching Network — 367
  7.6.4 Analytical Design of the LNA Output Matching Network — 368
  7.6.5 Linear Simulation of the Low Noise Amplifier — 371
  7.6.6 Amplifier Noise Temperature — 373

7.7 Power Amplifier Design — 376
  7.7.1 Data Sheet Large Signal Impedance — 377
  7.7.2 Power Amplifier Matching Network Design — 379
  7.7.3 Input Matching Network Design — 379

7.7.4	Output Matching Network Design	382

References and Further Reading	386
Problems	387

Chapter 8 Multi-Stage Amplifier Design and Yield Analysis — 391

8.1 Introduction — **391**

8.2 Two-Stage Amplifier Design — **391**
    8.2.1 First Stage Matching Network Design — 392
    8.2.2 Design of the Amplifier Input Matching Network — 393
    8.2.3 Second Stage Matching Network Design — 394
    8.2.4 Inter-Stage Matching Network Design — 395
    8.2.5 Second Stage Output Matching Network — 396

8.3 Two-Stage Amplifier Simulation — **396**

8.4 Parameter Sweeps — **398**

8.5 Monte Carlo and Sensitivity Analysis — **400**

8.6 Yield Analysis — **405**
    8.6.1 Design Centering — 407

8.7 Low Noise Amplifier Cascade — **408**
    8.7.1 Cascaded Gain and Noise Figure — 408
    8.7.2 Impedance Match and the Friis Formula — 410
    8.7.3 Reducing the Effect of Source Impedance Variation — 412

8.8 Summary — 413

References and Further Reading	414
Problems	414

Appendix — 417

Appendix A	Straight Wire Parameters for Solid Copper Wire	417
Appendix B.1	Γ_i Line Generation	418
Appendix B.2	Q_L Lines on the Smith Chart	420

Appendix B.3	Ideal Q Circle on the Smith Chart	422
Appendix B.4	Q_0 Measurement on the Smith Chart	424
Appendix C	VBScript file listing for the Matching Utility of Chapter 5	425
Appendix D	VBScript file listing for the Line and Stub Matching Utility of Chapter 6	434

Index 439

About the Authors 445

Foreword

Unlike many traditional books on RF and microwave engineering written mainly for the classroom, this book adopts a practical, hands-on approach to quickly introduce students and engineers unfamiliar with this topic to this subject matter. The authors make extensive use of Electronic Design Automation (EDA) tools to illustrate the foundational principles of RF and microwave engineering. The use of EDA methodologies in the book closely parallels the latest tools and techniques commonly used in industry to accelerate the design of RF/microwave systems and components to meet demanding specifications and ensure high yields.

This book provides readers a solid understanding of RF and microwave concepts such as Smith chart, S-parameters, transmission lines, impedance matching, resonators, filters and amplifiers. More importantly, it details how to use EDA tools to synthesize, simulate, tune, and optimize these essential components into a design flow as practiced in the industry. For explanatory purposes, the authors made the judicious choice of an easy-to-use and fully featured EDA tool that is also very affordable. This ensures that the skills learned in this book can be easily and immediately put into practice without the barriers of having to acquire costly and complex EDA tools.

Genesys from Keysight Technologies is that tool and it was chosen not only for its low cost, but because it provides the ideal combination of capabilities; in circuit synthesis, simulation and optimization; MATLAB scripting, RF system design, and electromagnetic and statistical analysis. The tool is a mature, well trusted solution that has successfully proven itself in the design of state-of-the-art RF and microwave test instrumentation and been time-tested by a large following of users worldwide for over 20 years.

The investment in learning the foundational RF/microwave skills and EDA techniques taught in this book provides engineers and students with valuable knowledge that will remain relevant and sought-after for a long time to come.

I wish such a book had been available when I first started my career as a microwave component designer. Without a doubt it would have made gaining RF and microwave insights much quicker than the countless hours I invested using the "cut-and-try" method on the bench.

How-Siang Yap
Keysight EEsof EDA Genesys Planning and Marketing
1400 Fountaingrove Parkway
Santa Rosa, CA 95403, USA

Preface

Microwave Engineering can be a fascinating and fulfilling career path. It is also an extremely vast subject with topics ranging from semiconductor physics to electromagnetic theory. Unlike many texts on the subject this book does not attempt to cover every aspect of Microwave Engineering in a single volume. This text book is the first volume of a two-part series that covers the subject from a computer aided design standpoint. The first volume covers introductory topics which are appropriate to be addressed by linear simulation methods. This includes topics such as lumped element components, transmission lines, impedance matching, and basic linear amplifier design. The second volume focuses on subject matter that is better learned through non-linear computer simulation. This includes topics such as oscillators, mixers, and power amplifier design.

Almost all subject matter covered in the text is accompanied by examples that are solved using the Genesys linear simulation software by Agilent. University students will find this a potent learning tool. Practicing engineers will find the book very useful as a reference guide to quickly setup designs using the Genesys software. The authors thoroughly cover the basics as well as introducing CAD techniques that may not be familiar to some engineers. This includes subjects such as the frequent use of the Genesys equation editor and Visual Basic scripting capability. There are also topics that are not usually covered such as techniques to evaluate the Q factor of one port resonators and yield analysis of microwave circuits.

The organization of the book is as follows: Chapter 1 presents a general explanation of RF and microwave concepts and components. Engineering students will be surprised to find out that resistors, inductors, and capacitors at high frequencies are no longer ideal elements but rather a network of circuit elements. For example, a capacitor at one frequency may in fact behave as an inductor at another frequency. In chapter 2 the transmission line theory is developed and several important parameters are defined. It is shown how to simulate and measure these parameters using Genesys software. Popular types of transmission lines are introduced and their parameters are examined. In Chapter 3 network parameters and the application of Smith chart as a graphical tool in dealing with impedance behavior and reflection coefficient are discussed.

Description of RF and microwave networks in terms of their scattering parameters, known as S- Parameters, is introduced. The subject of lumped and distributed resonant circuits and filters are discussed in Chapter 4. Using the Genesys software a robust technique is developed for the evaluation of Q factor form the S- Parameters of a resonant circuit. An introduction to the vast subject of filter synthesis and the electromagnetic simulation of distributed filters are also treated in this chapter. In Chapter 5 the condition for maximum power transfer and the lumped element impedance matching are considered. The analytical equations for matching two complex impedances with lossless two-element networks are derived. Both analytical and graphical techniques are used to design narrowband and broadband matching networks. The Genesys impedance matching synthesis program is used to solve impedance matching problems. The VBScript programming techniques developed in this chapter can be used by students to generate their own synthesis applications within the Genesys software. In Chapter 6 both narrowband and broadband distributed matching networks are analytically and graphically analyzed. In Chapter 7 single-stage amplifiers are designed by utilizing four different impedance matching objectives. The first amplifier is designed for maxim gain where the input and the output are conjugately matched to the source and load impedance; the second amplifier is designed for specific gain where the input or the output is mismatched to achieve a specific gain less than its maximum; the third amplifier is a low noise amplifier where the transistor is selectively mismatched to achieve a specific Noise Figure; and the fourth amplifier is a power amplifier where the transistor is selectively mismatched to achieve a specific amount of output power. In Chapter 8 a two-stage amplifier is designed by utilizing a direct interstage matching network. Monte Carlo and Yield analysis techniques are also introduced in this chapter. Finally a brief introduction to cascade analysis is presented.

Ali A. Behagi
Stephen D. Turner
March 2015

Chapter 1

RF and Microwave Concepts and Components

1.1 Introduction

An electromagnetic wave is a propagating wave that consists of electric and magnetic fields. The electric field is produced by stationary electric charges while the magnetic field is produced by moving electric charges. A time-varying magnetic field produces an electric field and a time-varying electric field produces a magnetic field. The characteristics of electromagnetic waves are frequency, wavelength, phase, impedance, and power density. In free space, the relationship between the wavelength and frequency is given by Equation (1-1).

$$\lambda = \frac{c}{f} \qquad (1\text{-}1)$$

In the MKS system, λ is the wavelength of the signal in meters, c is the velocity of light approximately equal to 300,000 kilometers per second, and f is the frequency in cycles per second, or Hz.

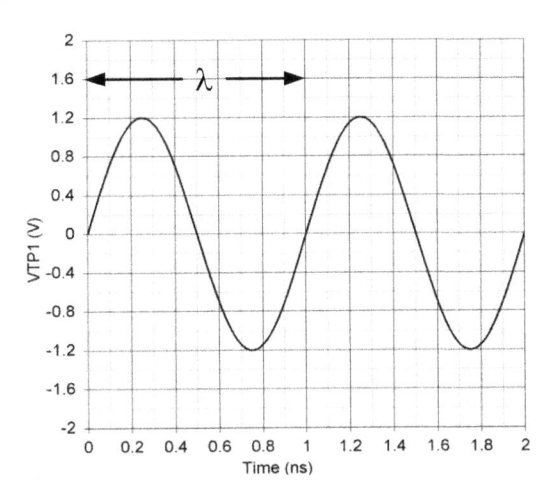

Figure 1-1 A time varying voltage waveform

The electromagnetic spectrum is the range of all possible frequencies of electromagnetic radiation. They include radio waves, microwaves, infrared radiation, visible light, ultraviolet radiation, X-rays and gamma rays. In the field of RF and microwave engineering the term RF generally refers to

Radio Frequency signals with frequencies in the 3 KHz to 300 MHz range. The term Microwave refers to signals with frequencies from 300 MHz to 300 GHz having wavelengths from 1 meter to 1 millimeter. The RF and microwave frequencies form the spectrum of all radio, television, data, and satellite communications. Figure 1-2 shows a spectrum chart highlighting the RF and microwave frequencies up through the extremely high frequency, EHF, range or 300 GHz. This text will focus on the RF and microwave frequencies as the foundation for component design techniques. The application of Keysight Technologies's Genesys software will enhance the student's understanding of the underlying principles presented throughout the text. The practicing engineer will find the text an invaluable reference to the RF and microwave theory and techniques by using Genesys software. The numerous Genesys examples enable the setup and design of many RF and microwave circuit design problems.

Wavelength	100 km	10 km	1 km	100 m	10 m	1 m	10 cm	1 cm	1 mm
Frequency	3 KHz	30 KHz	300 KHz	3 MHz	30 MHz	300 MHz	3 GHz	30 GHz	300 GHz
	VLF Very Low Frequency	LF Low Frequency	MF Medium Frequency	HF High Frequency	VHF Very High Frequency	UHF Ultra High Frequency	SHF Super High Frequency	EHF Extremely High Frequency	

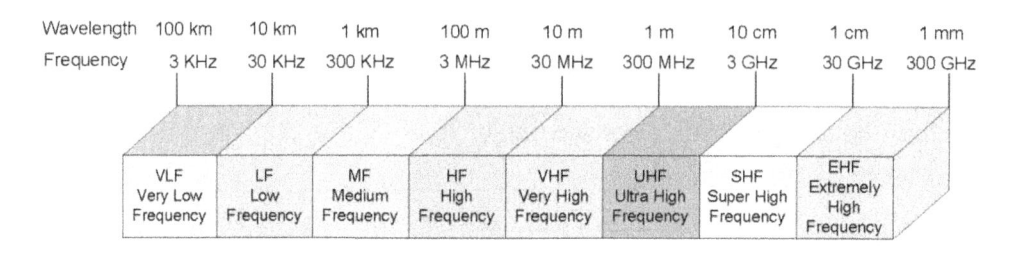

Figure 1-2 Electromagnetic spectrums from VLF to EHF

The spectrum chart of Figure 1-2 is intended as a general guideline to the commercial nomenclature for various sub bands. There is typically overlap across each of the boundaries as there is no strict dividing line between the categories. The RF frequencies typically begin in the very low frequency, VLF, range through the very high frequency, VHF, range. Microwaves are typically the ultra high frequency, UHF, super high frequency, SHF and extremely high frequency, EHF, frequency ranges. During World War II microwave engineers developed a detailed classification of the microwave frequencies into a band-letter designation. In 1984 the Institute of Electrical and Electronics Engineers, IEEE, agreed to standardize the letter designation of the microwave frequencies. These designators and their frequency ranges are shown in Table 1-1.

Band Designator	L Band	S Band	C Band	X Band	Ku Band	K Band	Ka Band
Frequency Range GHz	1 to 2	2 to 4	4 to 8	8 to 12	12 to 18	18 to 27	27 to 40

Table 1-1 Microwave band letter designators

Engineering students spend much of their formal education learning the basics of inductors, capacitors, and resistors. Many are surprised to find that as we enter the high frequency, HF, part of the electromagnetic spectrum these components are no longer a singular (ideal) element but rather a network of circuit elements. Components at RF and microwave frequencies become a network of resistors, capacitors, and inductors. This leads to the complication that the component's characteristics become quite frequency dependent. For example, we will see in this chapter that a capacitor at one frequency may in fact be an inductor at another frequency.

1.2 Straight Wire, Flat Ribbon, and Skin Depth

In this section we will begin with a basic examination of the straight wire inductance and move into more complete characterization of inductors. Similarly we will look at resistor and capacitor design and their implementation at RF and microwave frequencies. Discrete resistors, capacitors, and inductors are often referred to as lumped elements. RF and microwave engineers use the terminology to differentiate these elements from those designed in distributed printed circuit traces. Distributed component design is introduced in Chapter 2.

1.2.1 Calculation of Straight Wire Inductance

A conducting wire carrying an AC current produces a changing magnetic field around the wire. According to Faraday's law the changing magnetic field induces a voltage in the wire that opposes any change in the current flow. This opposition to change is called self inductance. At high frequencies even a short piece of straight wire possesses frequency dependent resistance and inductance behaving as a circuit element.

Example 1.2-1: Calculate the inductance of a three inch length of AWG #28 copper wire in free space.

Solution: The straight wire inductance can be calculated from the empirical Equation (1-2) [1].

$$L = K\ell \left(\ln \frac{4\ell}{D} - 0.75 \right) nH \qquad (1\text{-}2)$$

where:
  ℓ = Length of the wire
  D = Diameter of the wire (from Appendix A).
  $K = 2$ for dimensions in cm and $K = 5.08$ for dimensions in inches

Using Appendix A the diameter of the AWG#28 wire is found to be 0.0126 inches. Solving Equation (1-2) the inductance is calculated.

$$L = 5.08\,(3) \left(\ln \frac{4\,(3)}{0.0126} - 0.75 \right) = 93.1\ nH$$

It is interesting to examine the reactance of the wire. We know that the reactance is a function of the frequency and is related to the inductance by the following equation.

$$X_L = 2\pi f L \quad \Omega \qquad (1\text{-}3)$$

where: f is the frequency in Hz and L is the inductance in Henries

Calculating the reactance at 60Hz, 1MHz, and 1GHz we can see how the reactive component of the wire increases dramatically with frequency. At 60Hz the reactance is well below 1Ω while at microwave frequencies the reactance increases to several hundred ohms.

60 Hz: $X_L = 2\pi\,(60)(93.1 \cdot 10^{-9}) = 35\ \mu\Omega$
1 MHz: $X_L = 2\pi\,(10^6)(93.1 \cdot 10^{-9}) = 0.58\ \Omega$
1 GHz: $X_L = 2\pi\,(10^9)(93.1 \cdot 10^{-9}) = 585\ \Omega$

1.2.2 Straight Wire Inductance in Genesys

To analyze the **Example 1.2-1** in Genesys, create a schematic and add the straight wire model from the Parts Library, as shown in Figure 1-3.

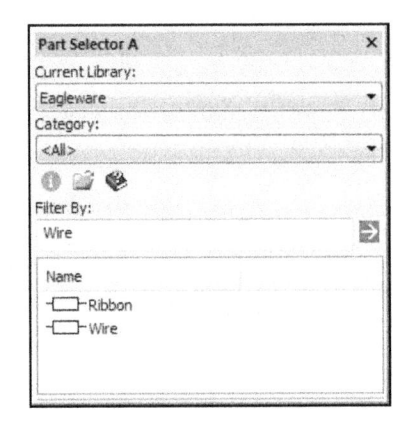

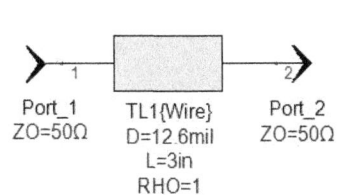

Figure 1-3 Part Selector and schematic of the straight wire

Attach 50 Ohm input and output ports, set the wire diameter to 12.6 mils, wire length to 3 inches and Rho=1, as shown in Figure 1-3. Rho is not the actual resistivity of the wire but rather the resistivity of the wire relative to copper. Because we are modeling a copper wire the value should be set to one. Table 1-2 provides a reference of common materials in terms of their actual resistivity and the relative resistivity to copper.

Material	Resistivity Relative to Copper	Actual Resistivity Ω-meters	Actual Resistivity Ω-inches
Copper, annealed	1.00	$1.68 \cdot 10^{-8}$	$6.61 \cdot 10^{-7}$
Silver	0.95	$1.59 \cdot 10^{-8}$	$6.26 \cdot 10^{-7}$
Gold	1.42	$2.35 \cdot 10^{-8}$	$9.25 \cdot 10^{-7}$
Aluminum	1.64	$2.65 \cdot 10^{-8}$	$1.04 \cdot 10^{-6}$
Tungsten	3.25	$5.60 \cdot 10^{-8}$	$2.20 \cdot 10^{-6}$
Zinc	3.40	$5.90 \cdot 10^{-8}$	$2.32 \cdot 10^{-6}$
Nickel	5.05	$6.84 \cdot 10^{-8}$	$2.69 \cdot 10^{-6}$
Iron	5.45	$1.00 \cdot 10^{-7}$	$3.94 \cdot 10^{-6}$
Platinum	6.16	$1.06 \cdot 10^{-7}$	$4.17 \cdot 10^{-6}$
Tin	52.8	$1.09 \cdot 10^{-7}$	$4.29 \cdot 10^{-6}$
Nichrome	65.5	$1.10 \cdot 10^{-6}$	$4.33 \cdot 10^{-5}$

Table 1-2 Resistivities of common materials relative to copper

Create a Linear analysis to analyze the circuit's impedance versus frequency. The Linear Analysis Properties window is shown in Figure 1-4.

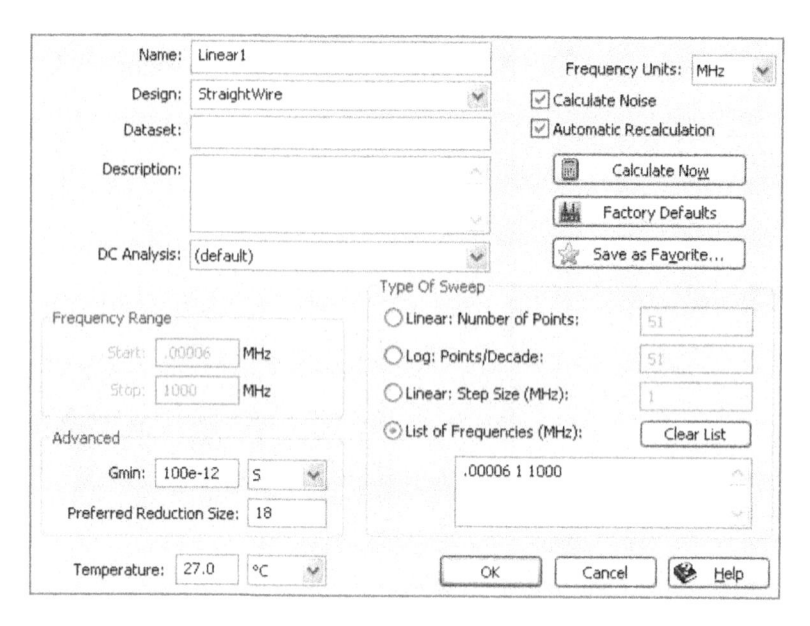

Figure 1-4 Linear analysis properties for simulation of wire impedance

Create a list of frequencies under the *Type of Sweep* setting. Enter the frequencies of 60 Hz, 1 MHz, 1 GHz. When an analysis is run the results are written to a Dataset. The results of a Dataset may then be sent to a graph or tabular output for visualization. In this example an Equation Editor is used to post process the solutions in the Dataset. Create an Equation Editor and type equations for the calculation of the wire's reactance and inductance, as shown in Figure 1-5.

```
1    using("Linear1_Data")
2    reactance=im(Linear1_Data.ZIN1)
3    inductance=reactance./(F*2*PI)
```

Figure 1-5 Equation Editor displaying the calculation of Inductance

More complex workspaces may contain multiple Datasets. It is a good practice to specify which Dataset is used to collect data for post processing. This is accomplished with the {using ("Linear1_Data")} statement of line 1 in the Equation Editor. Genesys has a built-in function ZIN1 to calculate the impedance of the circuit at each analysis frequency. Line 2 defines the reactance as the imaginary part of the impedance. Line 3 calculates the inductance from the reactance using Equation (1-3). The frequency, F, is the independent variable created by the Linear Analysis. Note the use of the dot (.) notation in the Equation Editor. There is a dot after Linear1_Data and reactance of lines 2 and 3. The dot means that these variables are not singular quantities but are arrays of values. There is a calculated array value for each independent variable, F. The Equation Editor is an extremely powerful feature of the Genesys software and is used frequently throughout this text. It is an interactive mathematical processor similar to MATLAB by MathWorks. There are two different syntaxes that may be used to define equations and perform post processing operations using the Equation Editor: the Engineering Language and the MATLAB Script. The Engineering Language is a simple structured format as shown in Figure 1-5. The MATLAB Script is compatible with the m-file syntax that is used in MATLAB. It is very convenient for students and engineers that are proficient in MATLAB. Both types will be used throughout this book to demonstrate the use of both languages. Under Equation command either Engineering Language or MATLAB Script can be selected. The simulated value of variables can be displayed in a Workspace Variables window or sent to a tabular output. Figure 1-6 shows the Workspace Variables at 3 different frequencies and the same variables in tabular output.

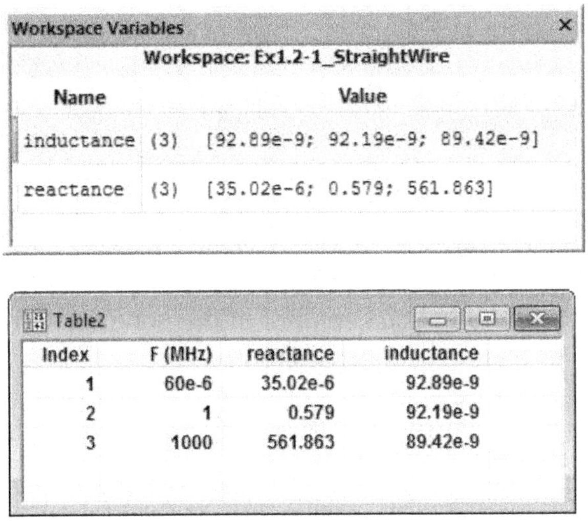

Workspace Variables
Workspace: Ex1.2-1_StraightWire

Name	Value
inductance (3)	[92.89e-9; 92.19e-9; 89.42e-9]
reactance (3)	[35.02e-6; 0.579; 561.863]

Table2

Index	F (MHz)	reactance	inductance
1	60e-6	35.02e-6	92.89e-9
2	1	0.579	92.19e-9
3	1000	561.863	89.42e-9

Figure 1-6 Workspace Variables window and output Table

Compare the results of the simulation of Figure 1-6 with the calculated values on page 4. We can see that at 60 Hz and 1 MHz the reactance and resulting inductance are very close to the calculated values. At 1 GHz however, the values begin to diverge. The simulated reactance is 24 Ω less than the calculated reactance. Equation (1-2) is useful for calculating the basic inductance at low frequency but as the frequency enters the microwave region, the value begins to change. This is normal and due to the skin effect of the conductor. The skin effect is a property of conductors where, as the frequency increases, the current density concentrates on the outer surface of the conductor.

1.2.3 Skin Depth in Conductors

At RF and microwave frequencies, due to the larger inductive reactance caused by the increase in flux linkage toward the center of the conductor, the current in the conductor is forced to flow near the conductor surface. As a result the amplitude of the current density decays exponentially with the depth of penetration from the surface. Figure 1-7 shows the cross section of a cylindrical wire with the current density area shaded.

Cross Section Area
of Wire at DC

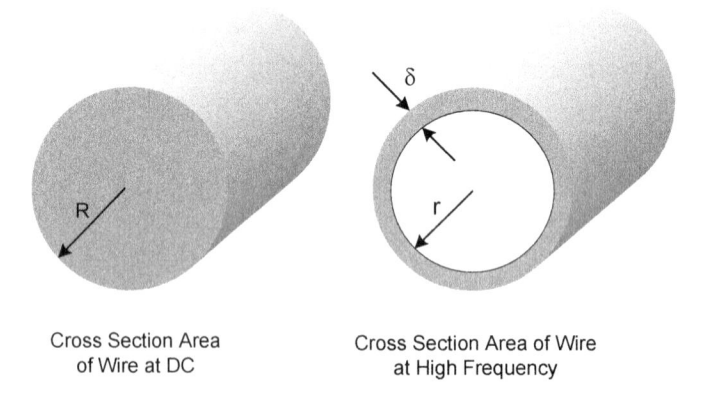

Cross Section Area of Wire
at High Frequency

Figure 1-7 Cross section of current flow in the conductor showing effect of skin depth

At low frequencies the entire cross sectional area is carrying the current. As the frequency increases to the RF and microwave region, the current flows much closer to the outside of the conductor. At the higher end of microwave frequency range, the current is essentially carried near the surface with

RF and Microwave Concepts and Components

almost no current at the central region of the conductor. The skin depth, δ, is the distance from the surface where the charge carrier density falls to 37% of its value at the surface. Therefore 63% of the RF current flows within the skin depth region. The skin depth is a function of the frequency and the properties of the conductor as defined by Equation (1-4). As the cross sectional area of the conductor effectively decreases the resistance of the conductor will increase [7].

$$\delta = \sqrt{\frac{\rho}{\mu \pi f}} \qquad (1\text{-}4)$$

where:
δ = skin depth
ρ = resistivity of the conductor
f = frequency
μ = permeability of the conductor

Use caution when solving Equation (1-4) to keep the units of ρ and μ consistent. Table 1-2 contains values of resistivity in units of Ω-meters and Ω-inches. The permeability μ is the permeability of the conductor. It is a property of a material to support a magnetic flux. Some reference tables will show relative permeability. In this case the relative permeability is normalized to the permeability of free space which is: $4\pi \cdot 10^{-7}$ Henries per meter. The relationship between relative permeability to the actual permeability is given in Equation (1-5). Most conductors have a relative permeability μ_r very close to one. Therefore conductor permeability μ is often given the same value as μ_o.

$$\mu = \mu_r \, \mu_o \qquad (1\text{–}5)$$

where:
μ = actual permeability of the material
μ_r = relative permeability of material
μ_o = permeability of free space

Example 1.2-2: Calculate the skin depth of copper wire at a frequency of 25 MHz.

Solution: Using Equation (1-4), and converting the permeability from H/m to H/inch, the skin depth for copper wire ($\rho = 6.61 \cdot 10^{-7}$ Ω-inch) is:

$$\delta = \sqrt{\frac{6.61 \cdot 10^{-7}}{(3.19 \cdot 10^{-8})\pi (25 \cdot 10^6)}} = 5.14 \cdot 10^{-4} \text{ inches}$$

Although we have been considering the skin depth in a circular wire, skin depth is present in all shapes of conductors. A thick conductor is affected more by skin effect at lower frequencies than a thinner conductor. One of the reasons that engineers are concerned about skin effect in conductors is the fact that as the resistance of the conductor increases, so does the thermal heating in the wire. Heat can be destructive in high power RF circuits causing burn out of conductors and potentially hazardous to personnel. Also the frequency dependence of the skin depth may make it difficult to maintain the impedance of transmission line structures.

1.2.4 Wire Resistance at Different Frequencies

As the frequency increases, the current is primarily flowing in the region of the skin depth. It can be visualized from Figure 1-7 that a wire would have greater resistance at higher frequencies due to the skin effect. The resistance of a length of wire is determined by the resistivity and the geometry of the wire as defined by Equation (1-6).

$$R = \frac{\rho \ell}{A} \ \Omega \qquad (1\text{-}6)$$

where:
    ρ = Resistivity of the wire
    ℓ = Length of the wire
    A = Cross sectional area

Example 1.2-3: Calculate the resistance of a 12 inch length of AWG #24 copper wire at DC and at 25 MHz.

Solution: The radius of the wire can be found in Appendix A. The DC resistance is then calculated using Equation (1-6).

$$R_{DC} = \frac{(6.61 \cdot 10^{-7}) \cdot 12}{\pi \left(\frac{0.0201}{2}\right)^2} = 0.025 \ \Omega$$

To calculate the resistance at 25 MHz the cross sectional area of the conduction region must be redefined by the skin depth of Figure 1-7. We can refer to this as the effective area, A_{eff}.

$$A_{eff} = \pi (R^2 - r^2) \qquad r = R - \delta. \qquad (1\text{-}7)$$

For the AWG#24 wire at 25 MHz the A_{eff} is calculated as:

$$A_{eff} = \pi \left(\frac{0.0201}{2}\right)^2 - \pi \left(\left(\frac{0.0201}{2}\right) - 5.14 \cdot 10^{-4}\right)^2 = 3.14 \cdot 10^{-5} \ in^2$$

Then apply Equation (1-6) to calculate the resistance of the 12 inch wire at 25 MHz.

$$R_{25MHz} = \frac{12 (6.61 \cdot 10^{-7})}{3.14 \cdot 10^{-5}} = 0.253 \ \Omega$$

Notice the resistance at 25 MHz is much greater than the resistance at DC.

1.2.5 Calculation of Flat Ribbon Inductance

Flat ribbon style conductors are very common in RF and microwave engineering. Flat ribbon conductors are encountered in RF systems in the form of low inductance ground straps. Flat ribbon conductors are encountered in microwave integrated circuits (MIC) as gold bonding straps. When a very low inductance is required the flat ribbon or copper strap is a good choice. The flat ribbon inductance can be calculated from the empirical Equation (1-8).

$$L = K \ell \left[\ln\left(\frac{2\ell}{W+T}\right) + 0.223 \left(\frac{W+T}{\ell}\right) + 0.5 \right] \ nH \qquad (1\text{-}8)$$

where:
ℓ = The length of the wire
K = 2 for dimensions in cm and K = 5.08 for dimensions in inches
W = the width of the conductor
T = the thickness of the conductor

Example 1.2-4: Calculate the inductance of the 3 inch Ribbon at 60 Hz, 1 MHz, and 1 GHz. Make the ribbon 100 mils wide and 2 mils thick.

Solution: The schematic of the Ribbon in Genesys is shown in Figure 1-8 along with the table with the reactance and inductance. Note that the same length of ribbon as the AWG#28 wire has almost half of the inductance value. Also note that the inductance value, 49.69 nH, does not change over the 60 Hz to 1 GHz frequency range. It is why the ribbon is desirable as a low inductance conductor for use in RF and microwave applications.

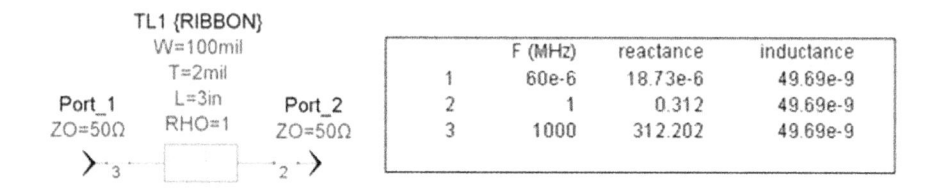

	F (MHz)	reactance	inductance
1	60e-6	18.73e-6	49.69e-9
2	1	0.312	49.69e-9
3	1000	312.202	49.69e-9

Figure 1-8 Flat ribbon schematic and tabular output

1.3 Physical Resistors

The resistance of a material determines the rate at which electrical energy is converted to heat. In Table 1-2 we have seen that the resistivity of material is specified in Ω-meters rather than Ω/meter. This facilitates the calculation of resistance using Equation (1-6). At low frequency or logic circuits we often treat resistors as ideal components.

Example 1.3-1: Plot the impedance of a 50 Ω ideal resistor and a leaded resistor in Genesys over a frequency range of 0 to 2 GHz.

Solution: For the 50 Ω ideal resistor the plot is shown in Figure 1-9. This plot shows a constant resistance at all frequencies.

RF and Microwave Concepts and Components   13

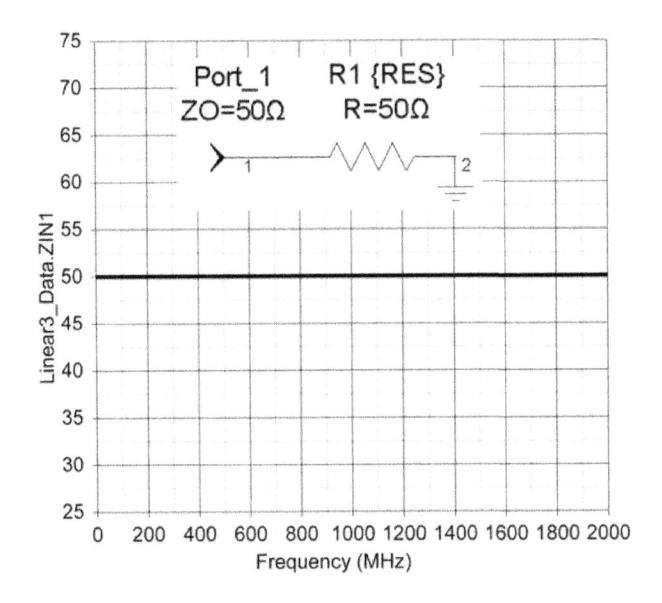

Figure 1-9 Ideal 50Ω resistor impedance versus frequency

At RF and microwave frequencies however, leaded resistors also possess inductive and capacitive elements. The stray inductance and capacitance associated with leaded resistors are often called parasitic elements. Consider the leaded resistor as shown in Figure 1-10. For such a 1/8 watt leaded resistor it is not uncommon for each lead to have about 10 nH of inductance. The body of the resistor may exhibit 0.5 pF capacitance between the leads. Designing the network in Genesys reveals an interesting result of the impedance versus frequency response. The impedance plotted in Figure 1-10 shows the impedance for a 10 Ω, 50 Ω, 500 Ω, and 1 kΩ leaded resistors swept from 0 to 2.0 GHz. By tuning the parasitic elements in Genesys we find that the low value resistors are influenced more by the lead inductance. The high value resistors are influenced more by the parasitic capacitance.

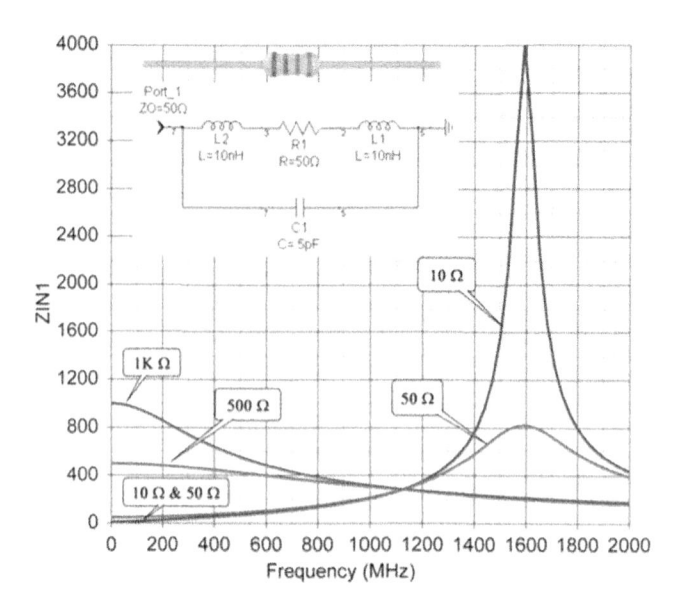

Figure 1-10 Leaded resistor impedance versus frequency

1.3.1 Chip Resistors

Thick film resistors are used in most contemporary electronic equipment. The thick film resistor, often called chip resistor, comes close to eliminating much of the inductance that plagues the leaded resistor. The chip resistor works well with popular surface mount assembly techniques preferred in modern electronic manufacturing. Figure 1-11 shows a typical thick film chip resistor along with a cross section of its design.

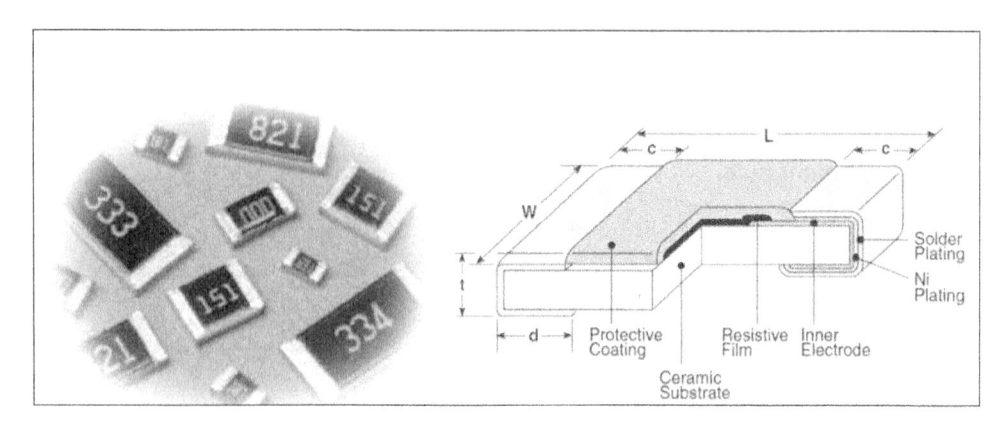

Figure 1-11 Thick film chip resistors and dimensions (*courtesy of KOA Speer Electronics*)

RF and Microwave Concepts and Components

There are many types of chip resistors designed for specific applications. Common sizes and power ratings are shown in Table 1-3.

Size	Length x Width	Power Rating
0201	20mils x 10mils	50mW
0402	40mils x 20mils	62mW
0603	60mils x 30mils	100mW
0805	80mils x 50mils	125mW
1206	120mils x 60mils	250mW
2010	200mils x 100mils	500mW
2512	250mils x 120mils	1W

Table 1-3 Standard thick film resistor size and approximate power rating

The thick film resistor is comprised of a carbon based film that is deposited onto the substrate. Contrasted with a thin film resistor that is typically etched onto a substrate or printed circuit board, the thick film resistor usually handles higher power dissipation. The ends of the chip have metalized wraps that are used to attach the resistor to a circuit board. There are companies that specialize in developing CAD models of components such as Modelithics, Inc.

Example 1.3-2: Plot the impedance of 1 kΩ, 0603 size chip resistor, manufactured by KOA, from 0 to 3 GHz.

Solution: The KOA resistor model is shown in the Modelithics Library.

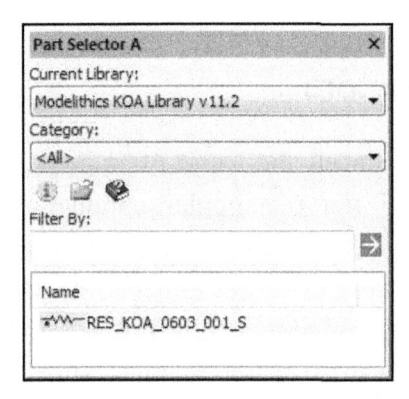

Figure 1-12 Modelithics SELECT library in Genesys

Solution: Create a schematic with the resistor and sweep the impedance from 0 to 3 GHz. The resulting schematic and response are shown in Figure 1-13.

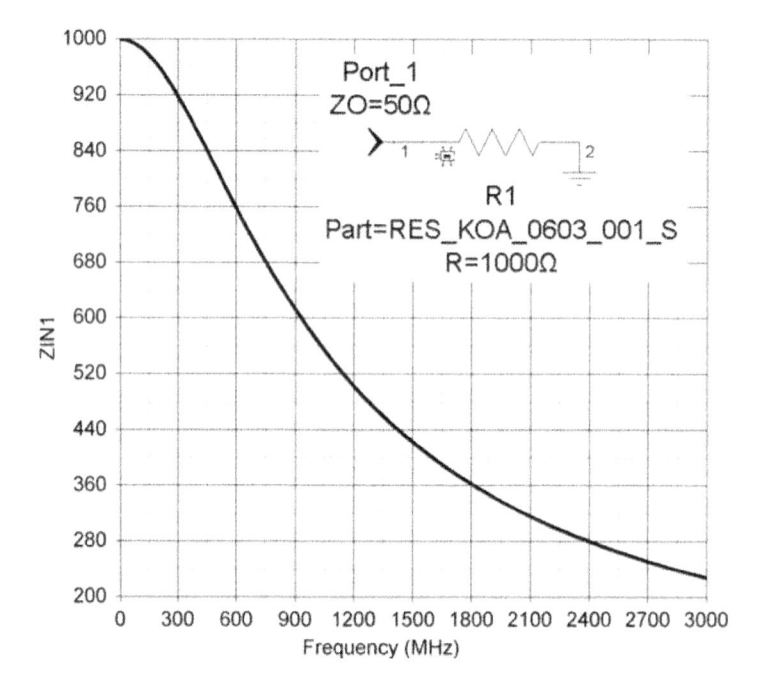

Figure 1-13 Modelithics chip resistor model schematic and impedance versus frequency

Note the roll off of the impedance with increasing frequency. This suggests that the chip resistor does have a parasitic capacitance that is in parallel with the resistor similar to the discrete model of Figure 1-10.

1.4 Physical Inductors

In section 1.2 we introduced the topic of inductance. The inductance of straight cylindrical and flat rectangular conductors was examined. The primary method of increasing inductance is not to simply keep increasing the length of a straight conductor but rather form a coil of wire. Forming a coil of wire increases the magnetic flux linkage and greatly increases the overall inductance. Because of the greater surrounding magnetic flux, inductors store energy in the magnetic field. Lumped element inductors are used in bias circuits, impedance matching networks, filters, and resonators.

RF and Microwave Concepts and Components 17

As we will see throughout this section inductors are realized in many forms including: air-core, toroidal and very small chip inductors. The concept of Q factor is introduced and will come up frequently in RF and microwave circuit design. It is a unit-less figure of merit that is used in circuits in which both reactive and resistive elements coexist. Because we know that individual air wound inductors, shown in Figure 1-14, are actually networks that are made up of resistors, inductors, and capacitors, each individual component is also characterized by a Q factor.

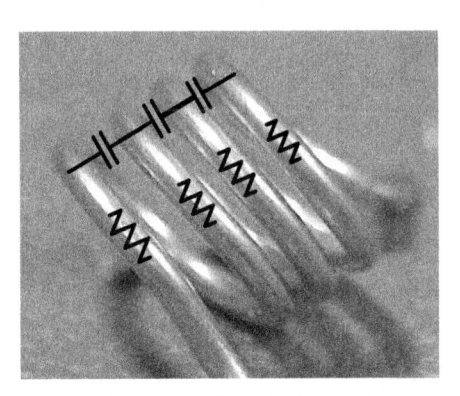

Figure 1-14 Air wound inductor showing the wire resistance and inter-winding capacitance

Basically, the higher the Q factor, the less loss or resistance exists in the energy storage property. The inductor quality factor Q is defined as:

$$Q = \frac{X}{R_S} \qquad (1\text{-}9)$$

where:
    X is the reactance of the inductor
    R_s is the resistance in the inductor

At low RF frequencies the resistance comes primarily from the resistivity of the wire and as such is quite low. At higher frequencies the skin effect and inter-winding capacitance begin to influence the resistance and reactance thus causing the Q factor to decrease. In most applications we want as high a component Q factor as possible. We can increase the Q factor of inductors by using larger diameter wire or by silver plating the wire. In a multi-turn coil, the windings can be separated to reduce the inter-winding capacitance which in turn will increase the Q factor. Winding the coil on a magnetic core can increase the Q factor.

1.4.1 Air Core Inductors

Forming a wire on a removable cylinder is the basic realization of the air core inductor. When designing an air-core inductor, use the largest wire size and close spaced windings to result in the lowest series resistance and high Q. The basic empirical equation to calculate the inductance of an air core inductor is given by Equation (1-10) [2].

$$L = \frac{(17)N^{1.3}(D+D1)^{1.7}}{(D1+S)^{0.7}} \qquad (1\text{-}10)$$

where:
- N = Number of turns of wire
- D = Core form diameter in inches
- $D1$ = Wire diameter in inches
- L = Coil inductance in nH
- S = Spacing between turns in inches

Example 1.4-1: As an interesting comparison with Example 1.2.1 calculate the amount of inductance that we can realize in that same three inches of wire if we wind it around a core to form an inductor. Choose a core form of 0.095 inches diameter as a convenient form to wrap the wire around.

Solution: First we need to calculate the approximate number of turns that we can expect to have with the three inch length of wire. We know that the circumference of a circle is related to the diameter by the following equation.

$$Circumference = \pi \, (Diameter) = \pi \, (0.095) = 0.2985 \ \ inches$$

With a circumference of 0.2985 inches we can calculate the approximate number of turns that we can wrap around the 0.095 inch core with three inches of wire.

$$N = \frac{3}{0.2985} = approximately \ 10 \ turns$$

RF and Microwave Concepts and Components 19

From Equation (1-10) we can see that the spacing between the turns has a strong effect on the value of the inductance that we can expect from the coil. When hand winding the coil, it may be difficult to maintain an exact spacing of zero inches between the turns. Therefore it is useful to solve Equation (1-10) in terms of a variety of coil spacing so that we can see the effect on the inductance. The Equation Editor in Genesys can be used to solve Equation (1-10) for a variety of coil spacing. Add an Equation Editor to the Workspace as shown in Figure 1-15.

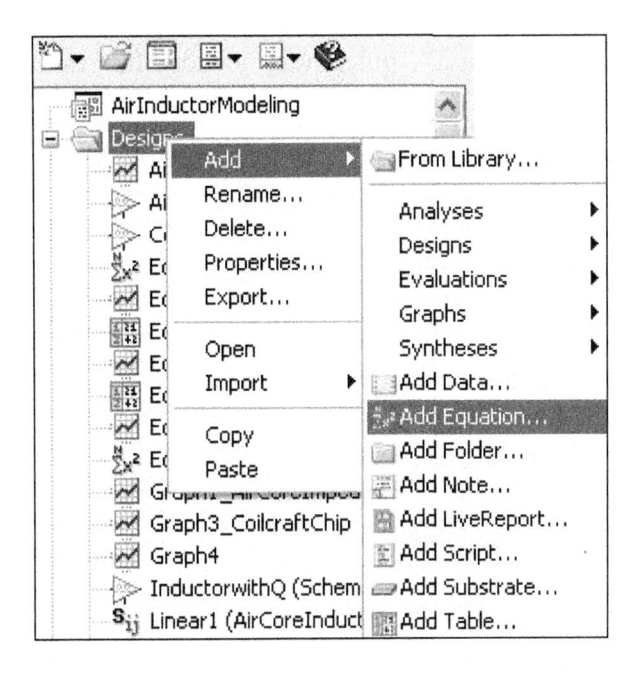

Figure 1-15 Adding an Equation Editor to the Genesys workspace

Create an Equation set similar to that shown in Figure 1-16.

```
1  'Calculate Inductance Given Number of Turns
2  D=.095      'Core Diameter in inches
3  D1=.0126    'Wire Diameter in inches
4  N=10        'Number of turns
5  Spacing=[0;.002;.004;.006;.008;.010]
6  Inductance_nH=(17*(N^1.3)*((D+D1)^1.7))/((D1+Spacing)^.7)
7  Coil_Length=(D1*N)+(Spacing.*(N-1))
```

Figure 1-16 Equation Editor to calculate the air core coil inductance

Note that the coil spacing variable, Spacing, has been defined as an array variable. Placing a semicolon between the values makes the array organized in a column format. This is handy for viewing the results in tabular format. Placing a comma between the values would organize the array in a row format. The array variables are displayed under the variable column as Real [6], indicating that these values are a six element array containing scalar values. As an aid in forming the coil, the overall coil length is also calculated with the Equation Editor. The coil length is simply the summation of the overall wire thickness times the number of turns and the spacing between the turns. A quick way to add the results to a table is shown in Figure 1-17.

```
Workspace Variables
            Workspace: Ex1.4-1A_AirInductorModel1
    Name                        Value
Coil_Length   (6)  [0.126; 0.144; 0.162; 0.18; 0.198; 0.216]
D                  0.095
D1                 0.013
Inductance_nH (6)  [163.784; 147.735; 135.037; 124.701; 116.097; 108.806]
N                  10
Spacing       (6)  [0; 2e-3; 4e-3; 6e-3; 8e-3; 0.01]
```

Figure 1-17 Send the variable to a Table directly from the Equation Editor

Right click on any variable and add it to an existing or new table. Add the inductance, spacing, and coil length to the table as shown in Figure 1-18.

Index	Inductance_nH	InductanceCalculator.Spacing	Coil_Length
1	163.784	0	0.126
2	147.735	2e-3	0.144
3	135.037	4e-3	0.162
4	124.701	6e-3	0.18
5	116.097	8e-3	0.198
6	108.806	0.01	0.216

Figure 1-18 Coil Inductance versus coil spacing and length

The table shows that the coil inductance with no spacing between the turns is 163.78 nH and it is 0.126 inches long. Contrast this to the 93.1 nH inductance with the same three inches of wire in a straight length. We can clearly see the dramatic impact of the magnetic flux linkage in increasing the inductance by forming the wire into a coil. The Figure also shows the strong influence of the inter-winding capacitance in influencing the inductance of the coil. Just 10 mils spacing between the turns reduces the coil's inductance from 163.78 nH to 108.8 nH. It is clearly important to consider the turn spacing when analyzing the inductor's performance. In practice this is an effective means to tune the inductor's value in circuit. When designing and building the inductor it is necessary to solve Equation (1-10) for the number of turns given a desired value of inductance. Figure 1-19 shows an Equation Editor setup to solve for the number of turns, N.

```
1  'Calculate Number of Turns Given the Inductance
2  D=.095                            'Core Diameter in inches
3  D1=.0126                          'Wire Diameter in inches
4  Inductance_nH=163.784             'Wire inductance in nH
5  Spacing=[0;.002;.004;.006;.008;.010]
6  N=((((D1+Spacing)^.7)*Inductance_nH)/(17*((D+D1)^1.7)))^.7692
7  Coil_Length=(D1*N)+(Spacing.*(N-1))
```

Figure 1-19 Equation Editor to calculate the air core number of turns

Note that there is one subtle difference in line 12, the calculation of the coil length. In this case both variables, Spacing and N, are array variables. When multiplying array variables use a period in front of the multiplication sign to signify that this is an operation on arrays. This makes sure that the correct array index is maintained between the variables. In the previous Equation Editor of Figure 1-16 this notation was not necessary because N was a constant.

Number_of_Turns.N	Number_of_Turns.Spacing	Number_of_Turns.Coil_Length
9.999	0	0.126
10.825	0.002	0.156
11.599	0.004	0.189
12.332	0.006	0.223
13.029	0.008	0.26
13.696	0.01	0.3

Figure 1-20 Number of turns versus coil spacing and length

1.4.2 Modeling the Air Core Inductor in Genesys

The air core inductor is modeled in Genesys using the AIRIND1 model as shown in Figure 1-21.

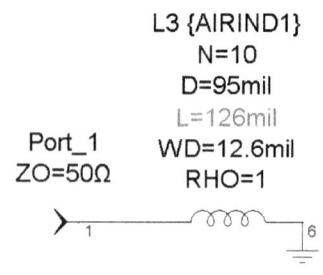

```
             L3 {AIRIND1}
                N=10
                D=95mil
                L=126mil
   Port_1    WD=12.6mil
   ZO=50Ω      RHO=1
```

Figure 1-21 Genesys model of the air core inductor

Note the required parameters: number of turns, wire diameter, core diameter, and coil length. The coil spacing cannot be entered directly but is accounted for by the overall coil length for a given number of turns. Make the coil length a variable so that we can analyze the inductance as a function of the coil length. Simulate the value of inductance at a fixed frequency of 1 MHz. Set up a fixed frequency Linear Analysis as shown in Figure 1-23. Simulate the value of inductance at a fixed frequency of 1 MHz.

Figure 1-22 Fixed frequency linear analysis at 1MHz

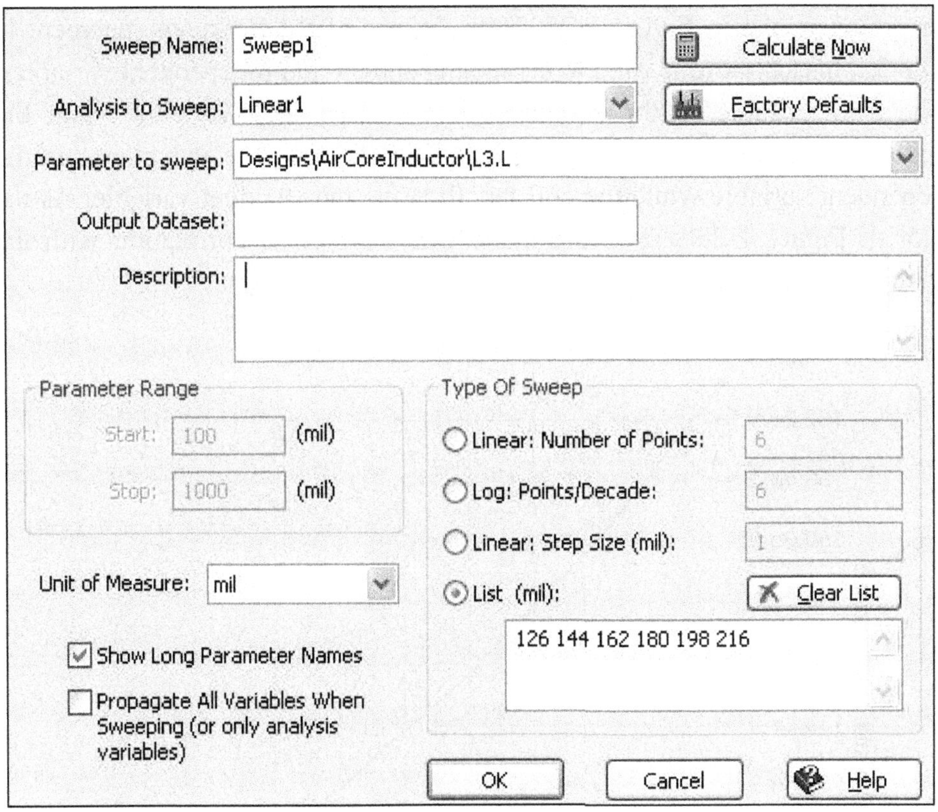

RF and Microwave Concepts and Components 23

Figure 1-23 Adding the Parameter Sweep to the workspace

To vary the length of the inductor we will use the Parameter Sweep capability in Genesys. Add a Parameter Sweep to the Workspace as shown in Figure 1-23.

```
1    using("Sweep1_Data")
2    reactance=im(Sweep1_Data.ZIN1)
3    inductance=reactance./((Sweep1_Data.F)*2*PI)
4    setindep(*inductance","Sweep1_Data.L3_L_Swp_F")
```

Figure 1-24 Equation Editor to work with parameter sweep dataset

On the Parameter-to-Sweep entry make sure that the coil length is selected from the drop down box. Under the Type-of-Sweep, select list and enter the six coil lengths that were calculated in the table of Figure 1-18. Finally create an Equation Editor as shown in Figure 1-24 to calculate the coil's inductance from the simulated reactance and plot it as a function of the coil length as shown in Figure 1-25. Note the use of the Setindep statement in line 5. This keyword is used to set the dependent and independent variables. The first variable in the statement is the dependent variable while the second is the independent variable. In this example the inductance is the dependent variable while the coil length is the independent variable. As the plot of Figure 1-25 shows the model has very close correlation with the inductance calculated with Equation (1-10).

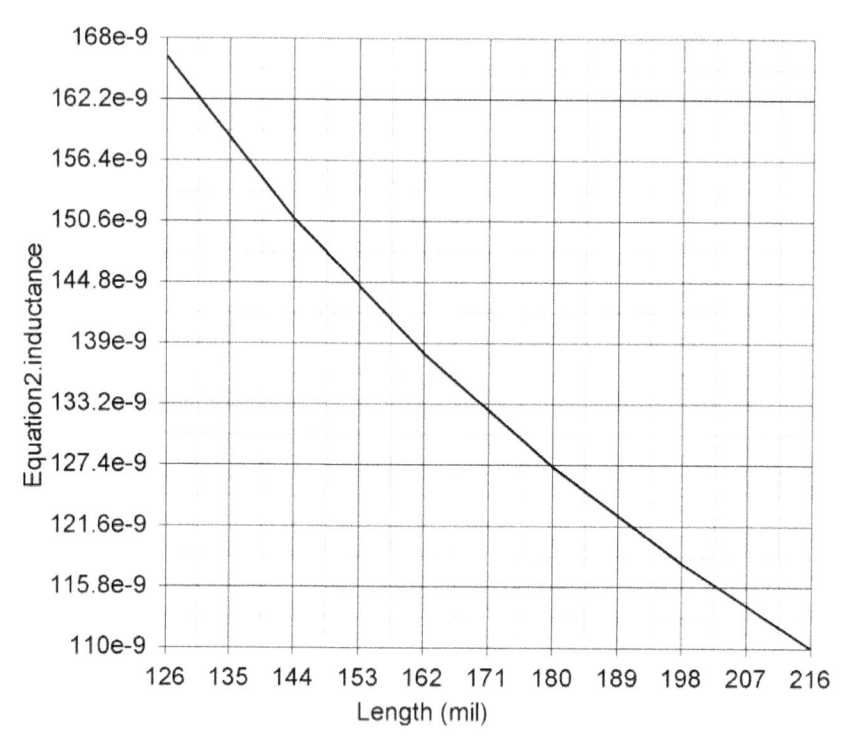

Figure 1-25 Plot of inductance versus coil length for the air inductor model

Change the Linear Analysis to a linear frequency sweep with 401 points over a range of 1 MHz to 1300 MHz. Plot the impedance of the inductor across the frequency range as shown in Figure 1-26. Note the interesting spike, or increase in impedance that occurs around 1098.6 MHz. This is the parallel, self resonant, frequency of the inductor.

RF and Microwave Concepts and Components

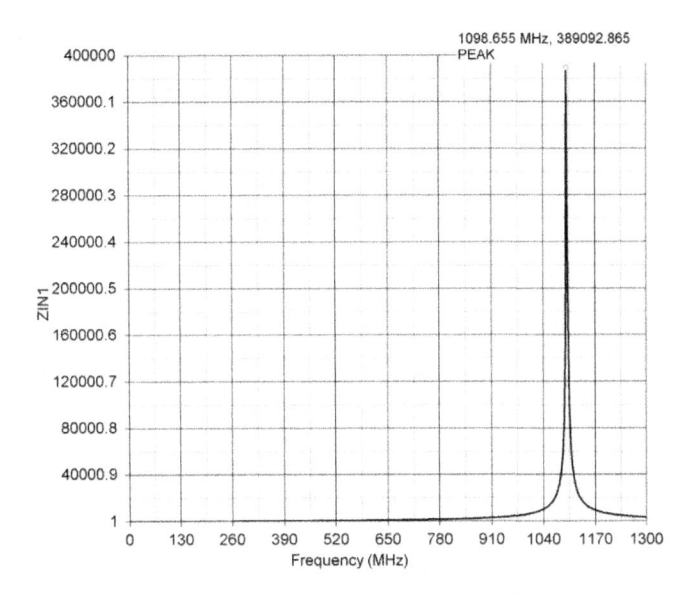

Figure 1-26 Impedance of the air core inductor as a function of frequency

The inductor is not an ideal component or a pure inductance but rather a network that includes parasitic capacitance and resistance. As an example, create a simple RLC network that gives an equivalent impedance response to Figure 1-26. One such circuit is shown in Figure 1-27.

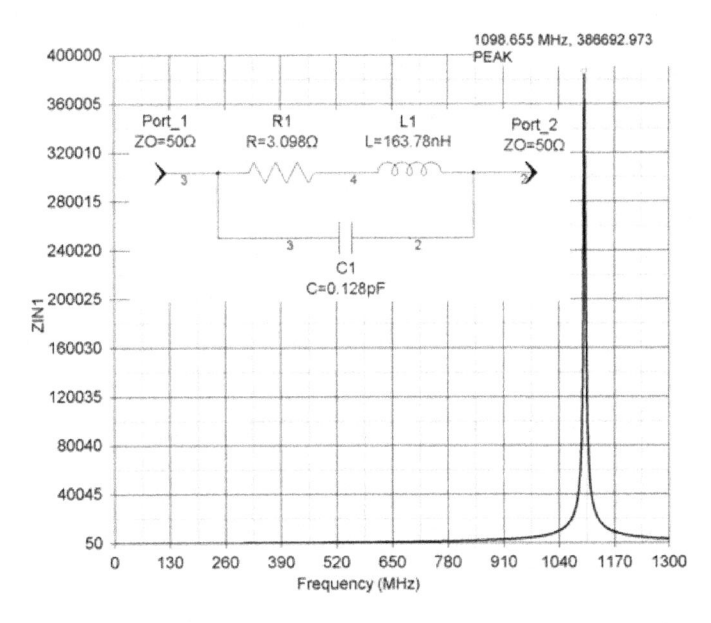

Figure 1-27 Equivalent ideal element network of the air core inductor

Resonant circuits are covered in detail in chapter 4 but it is important to understand that each individual component such as the air core inductor has its own resonant frequency. The resonant frequency is the frequency at which the inductive reactance and capacitive reactance are equal and cancel one another. When this condition occurs in the inductor it is a parallel resonant circuit which results in a very high real impedance. If we plot the reactance along with the impedance a very interesting response is obtained. This response is shown in Figure 1-28.

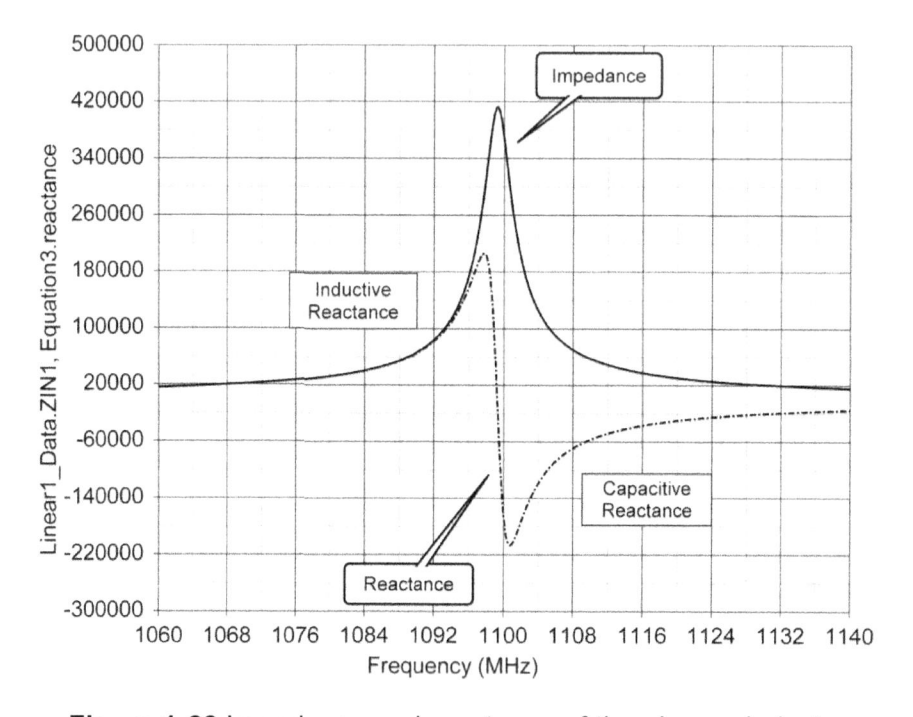

Figure 1-28 Impedance and reactance of the air core inductor

The reactance, up to the resonant frequency, is positive but beyond resonance the reactance becomes negative. From basic circuit theory we know that a negative reactance is associated with a capacitor. Therefore above the Self Resonant Frequency, SRF, the inductor actually becomes a capacitor. In practice we want to make sure that our inductor really behaves like an inductor. A good design practice is to keep this self resonant frequency about four times higher than the frequency of operation. However using the inductor near its resonant frequency might make a good choke. A choke is a high reactance inductor often used to feed voltage to a circuit in

RF and Microwave Concepts and Components 27

which all RF energy is blocked from the DC side of the circuit. The inductor manufacturer will typically specify SRF of the inductor. It is important to remember that the inductor's SRF is the parallel resonant frequency; not the series resonant frequency.

1.4.3 Inductor Q Factor

Example 1.4-2: Setup an Equation Editor to calculate the Q factor of the inductor based on Equation (1-9).

Solution: The Equation Editor is shown in Figure 1-29.

```
1    using("Linear3_Data")
2    reactance=im(Linear3_Data.ZIN1)
3    inductance=abs(reactance./(F*2*PI))
4    resistance=re(Linear3_Data.ZIN1)
5    Qfactor=abs(reactance)/resistance
```

Figure 1-29 Equation Editor to calculate the Q factor of the inductor

The plot of inductor Q factor versus frequency is shown in Figure 1-30.

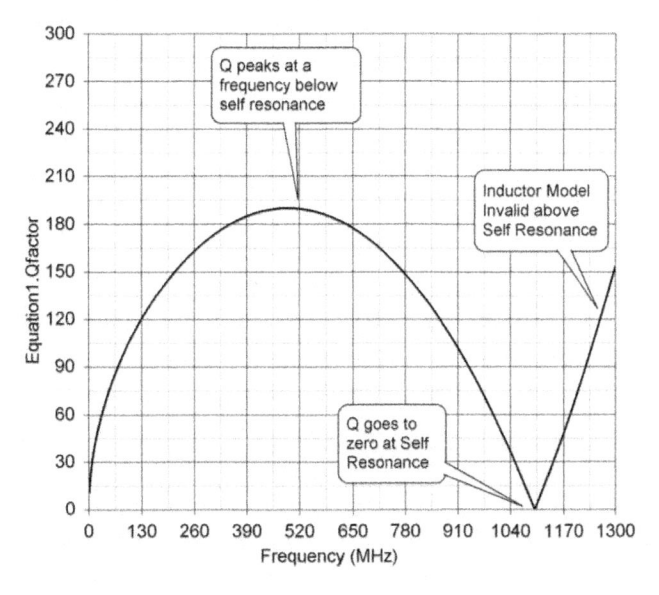

Figure 1-30 Air core inductor model Q factor versus frequency

It is interesting to note that the Q factor peaks at a frequency well below the self resonant frequency of the inductor. The actual frequency at which the Q factor peaks will vary among inductor designs but is usually ranges from 2 to 5 times less than the SRF. Close winding spacing results in inter-winding capacitance, which lowers the self-resonant frequency of the inductor. Thus there is a tradeoff between maximum Q factor and high self resonant frequency. It is also noteworthy that the Q factor goes to zero at the self resonant frequency. Figure 1-30 shows the Q factor monotonously increasing beyond the self resonant frequency. This is erroneous and is due to the fact that the model used to simulate the inductor's performance is invalid beyond the self resonance. The Air Core inductor model uses a simplified network similar to the one shown in Figure 1-27. Beyond self resonance the complexity and number of ideal elements need to increase in order to accurately model the inductor. A resistor needs to be added in series with the capacitor to begin to model the Q factor because the inductor is becoming a capacitor above self resonance. For most practical work however the native model will work fine because we should be using the inductor well below the self resonant frequency. A technique commonly used by microwave engineers to increase the Q factor of an inductor is to silver plate the wire. This can be modeled by setting Rho = 0.95 in the inductor model.

1.4.4 Chip Inductors

The inductor core does not have to be air. Other materials may be used as the core of an inductor. Similar in size to the chip resistor there is a large assortment of chip inductors that are popular in surface mount designs. The chip inductor is a form of dielectric core inductor. There are a variety of modeling techniques used for chip inductors. One of the more popular modeling techniques is with the use of S parameter files. The subject of S parameters is covered in chapter 3. At this point consider the S parameter file as an external data file that contains an extremely accurate network model of the component. Most component manufacturers provide S parameter data files for their products. It is a good practice to always check the manufacturer's website for current S parameter data files. Coilcraft, Inc is one manufacturer of chip inductors. A typical chip inductor is built with extremely small wire formed on a ceramic form as shown in Figure 1-31.

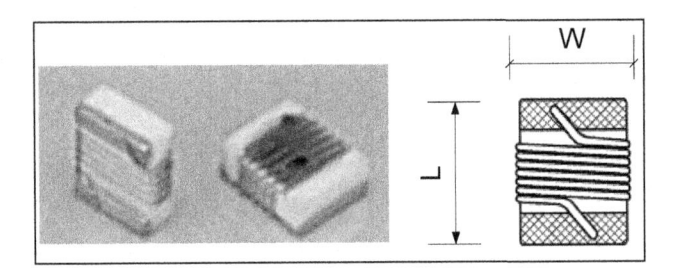

Figure 1-31 chip inductors (*courtesy of Coilcraft, Inc*)

Chip inductors are manufactured in many standard chip sizes as shown in Table 1-4.

Size	Length x Width
0201	20mils x 18mils
0302	34mils x 15mils
0402	44mils x 20mils
0603	69mils x 30mils
0805	90mils x 50mils
1008	105mils x 80mils
1206	140mils x 56mils
1812	195mils x 100mils

Table 1-4 Standard chip Inductor size and approximate power rating

The characteristic differences among the various sizes are more difficult to quantify than the chip resistors. A careful study of the data sheets is required for the proper selection of a chip inductor. In general the larger chip inductors will have higher inductance values. Often the smaller chip inductors will have higher Q factor. The impedance and self resonant frequency can vary significantly across the sizes as well as the current handling capability.

1.4.5 Chip Inductor Simulation in Genesys

The Genesys library has a collection of S parameter files for the Coilcraft chip inductors. Use the Part Selector to navigate to the Coilcraft chip inductors and select the 180 nH 1008 series inductor. The Part Selector and inductor schematic are shown in Figure 1-32.

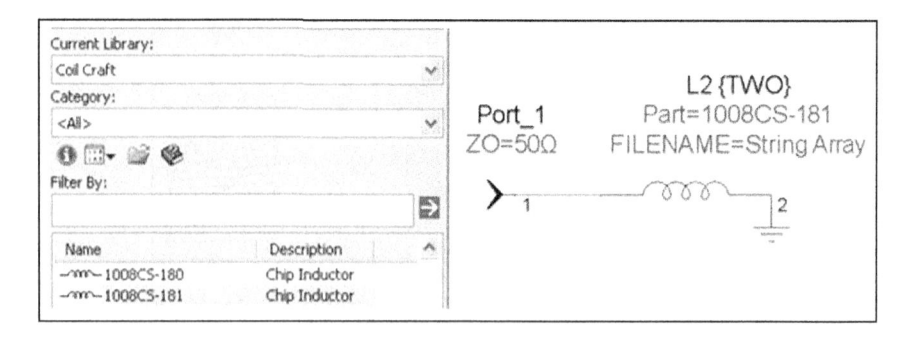

Figure 1-32 Part selector to locate chip inductor and the circuit schematic

Example 1.4-3: Setup a Linear Analysis and plot the impedance of the inductor from 1 MHz to 1040 MHz.

Solution: Create an Equation Editor to calculate and plot the inductance from the impedance as shown in Figure 1-33. For reference the manufacturer's specifications for the inductance is overlaid on the plot.

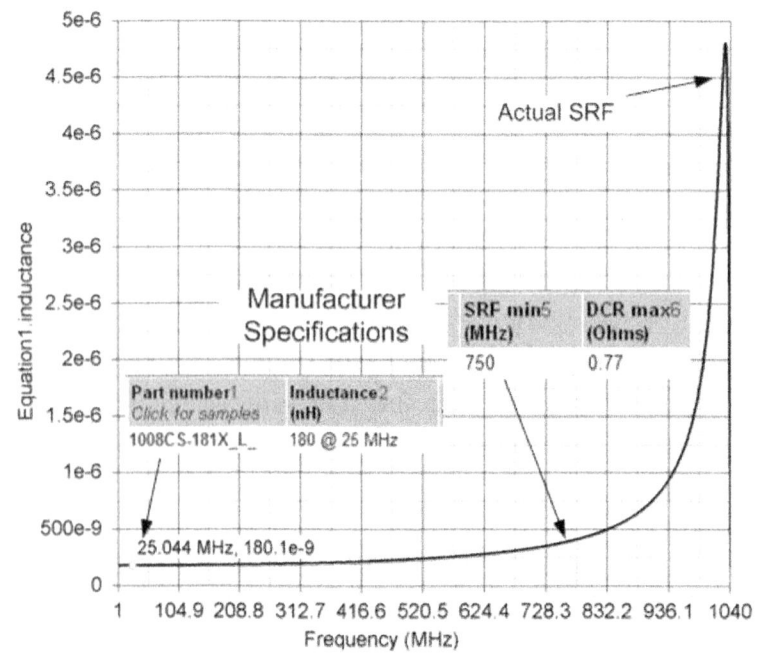

Figure 1-33 Plot of the 180 nH chip inductor inductance versus frequency

RF and Microwave Concepts and Components 31

The manufacturer specifies the inductance as 180 nH at a frequency of 25 MHz. From the marker on the plot we see that the simulation of the S parameter file has very close correlation measuring 180.1 nH. From the plot we can see that the nominal inductance value remains close to specification up to about 500 MHz. The self resonant frequency (parallel resonant) is specified as a minimum of 750 MHz. The actual SRF could be higher but it is guaranteed not be be less than 750 MHz. We can see from the plot in Figure 1-33 that the actural SRF is closer to 1 GHz. From Figure 1-34 we see that the manufacturer specifies a minimum Q factor of 45 at a frequency of 100 MHz.

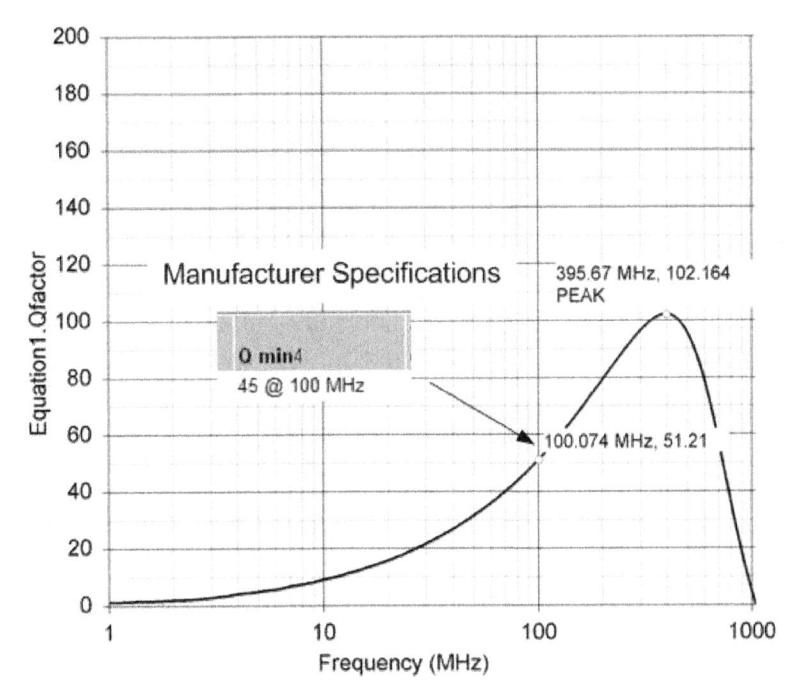

Figure 1-34 180 nH chip inductor Q factor versus frequency

The plot of the Q factor derived from the S parameter file shows that the Q factor is 51.21 near 100 MHz. Manufacturers will often plot the Q vs frequency on a logarithmic scale. It is very easy to change the x-axis to a logarithmic scale on the rectangular graph properties window in Genesys. This allows us to have a visual comparison to the manufacturer's catalog plot.

1.4.6 Magnetic Core Inductors

We have seen that the inductance of a length of wire can be increased by forming the wire into a coil. We can make an even greater increase in the inductance by replacing the air core with a magnetic material such as ferrite or powdered iron. Two popular types of magnetic core inductors are the rod core and toroidal core inductors shown in Figure 1-35.

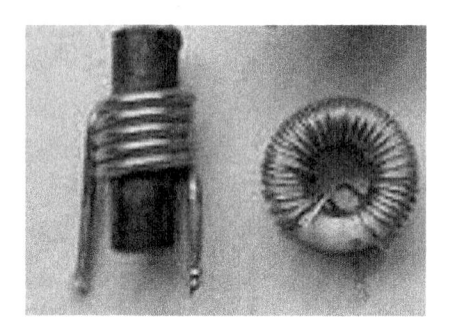

Figure 1-35 Rod and toroidal magnetic core inductors

The magnetic field around an inductor is characterized by the magnetic force H, and the magnetic flux B. They are related by the level of the applied signal and the permeability, μ, of the core material. This relationship is given by Equation (1-11).

$$B = \mu H \qquad (1\text{-}11)$$

where:

B = Flux density in Gauss
H = Magnetization intensity in Oersteds
μ = Permeability in Webers/Ampere-turn

This relationship is nonlinear in that as H increases, the amount of flux density will eventually level off or saturate. We will consider the linear region of this relationship throughout the discussion of this text. In an iron core inductor the permeability of the magnetic core is much higher than an air core and produces a high flux density. This magnetic flux density for each type of inductor is shown in Figure 1-36.

RF and Microwave Concepts and Components 33

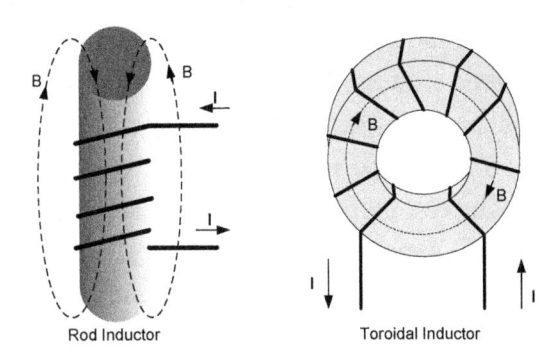

Figure 1-36 Magnetic flux densities for rod and toroidal magnetic core inductors

The rod inductor has magnetic flux outside of the core as well as inside the core. Rod inductors that are used in tuned circuits generally require a metal shield around the inductor to contain this magnetic flux so that it does not interfere or couple to adjacent circuits and other inductors. The toroidal inductor flux remains primarily inside the core material. This suggests that the toroidal inductor experiences less loss and should have higher Q factor. This also gives the toroidal inductor a self shielding characteristic and does not require a metallic enclosure. Because of its self-shielding properties and high Q factor the toroidal inductor is one of the most popular of all magnetic core inductors. However one advantage of the rod inductor is that it is much easier to tune. The coil can be wound on a hollow plastic cylindrical form in which the magnetic rod can be placed inside. The rod is then free to move longitudinally which can tune the inductance value.

Core materials are characterized by their permeability. Core permeability can vary quite a bit with frequency and temperature and can be confusing to specify for a given application. The stability of the permeability can change with the magnetic field due to DC current or RF drive through the inductor. As the frequency increases the permeability eventually reduces to the same value as air. Therefore iron core inductors are used only up to about 200 MHz. In general powdered iron can handle higher RF power without saturation and permanent damage. Ferrite cores have much higher permeability. The higher permeability of ferrite results in higher inductance values but lower Q factors. This characteristic can be advantageous in the design of RF chokes and broad band transformers. For inductors used in tuned circuits and filters, however, the higher Q factor of powdered iron is preferred. The powdered iron cores are manufactured in a variety of mixes to achieve different characteristics. The iron powders are made of hydrogen reduced iron and have greater permeability and lower Q factor. These cores are often used in RF chokes, electromagnetic interference (EMI) filters, and switched mode power supplies. Carbonyl iron tends to have better temperature stability and more constant permeability over a wide range of power. At the same time the Carbonyl iron maintains very good Q factor making them very popular in RF circuits. These characteristics lead to the popularity of toroidal inductors of Carbonyl iron for the manufacture of RF inductors. There is a wide variety of sizes and mixtures of Carbonyl iron that are used in the design of toroidal inductors. A few of the popular sizes that are manufactured by Micrometals Inc. are shown in Table 1-5.

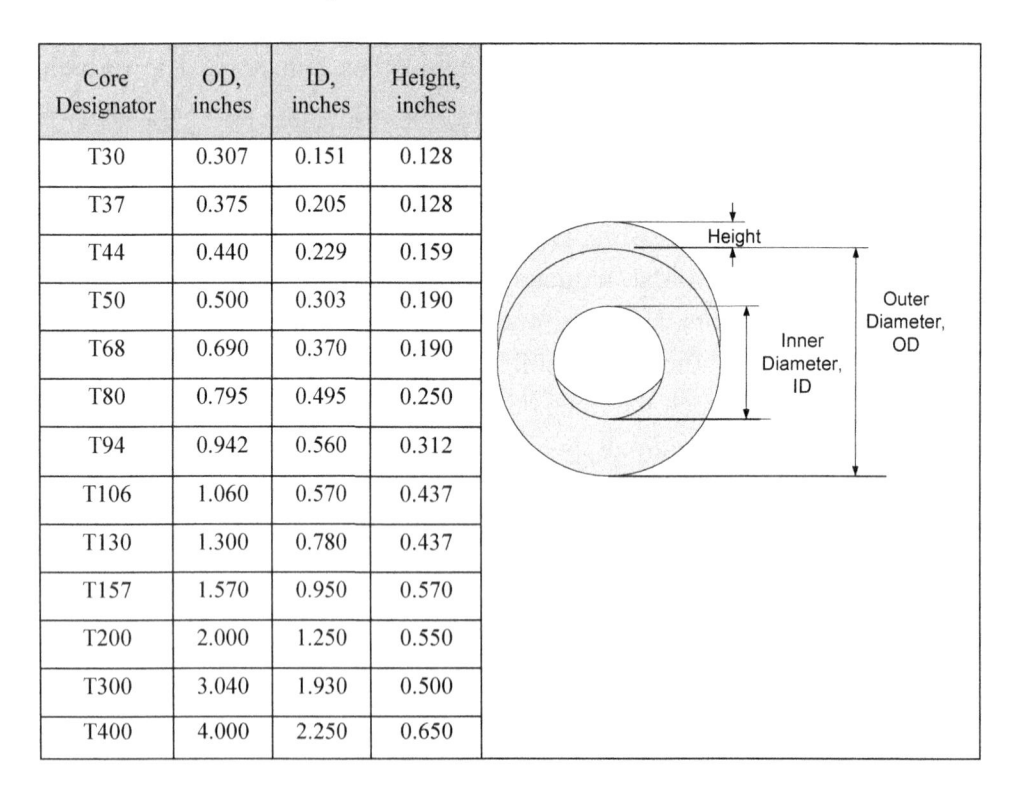

Core Designator	OD, inches	ID, inches	Height, inches
T30	0.307	0.151	0.128
T37	0.375	0.205	0.128
T44	0.440	0.229	0.159
T50	0.500	0.303	0.190
T68	0.690	0.370	0.190
T80	0.795	0.495	0.250
T94	0.942	0.560	0.312
T106	1.060	0.570	0.437
T130	1.300	0.780	0.437
T157	1.570	0.950	0.570
T200	2.000	1.250	0.550
T300	3.040	1.930	0.500
T400	4.000	2.250	0.650

Table 1-5 Partial listing of popular toroidal cores with designators

The inductance per turn of a toroidal inductor is directly related to its permeability and the ratio of its cross section to flux path length as given by Equation (1-12) [4].

$$L = \frac{4\pi N^2 \mu A}{length} \quad nH \qquad (1\text{-}12)$$

where:

    L_{nH} = inductance
    μ = permeability
    A = cross sectional area
    length = flux path length
    N = number of turns

As Equation (1-12) shows the inductance is proportional to the square of the turns. A standard specification used by toroid manufacturers for the calculation of inductance is the inductive index, A_L. The inductive index is typically given in units of nH/turn. The inductance can then be defined by Equation (1-13).

$$L = N^2 A_L \quad nH \qquad (1\text{-}13)$$

Some manufacturers specify the A_L in terms of uH or mH. To convert among the three quantities use the following guideline.

$$\frac{1\ nH}{turn} = \frac{10\ uH}{100\ turns} = \frac{1\ mH}{1000\ turns} \qquad (1\text{-}14)$$

The various powdered iron mixes are optimized for good Q factor and temperature stability over certain frequency bands. A partial summary of some popular mixtures is shown in Table 1-6. Powdered Iron cores have a standard color code and material sub-type designator. The toroid is painted with the appropriate color so that the mixture can be identified. A given A_L is dependent on both the size of the toroid and the material mix.

Material Mix Designator	Material Permeability	Magnetic Material	Color Code	Frequency Range	Temperature Stability (ppm/°C)
-17	4.0	Carbonyl	Blue/Yellow	20 – 200 MHz	50
-10	6.0	Carbonyl W	Black	10 – 100 MHz	150
-6	8.5	Carbonyl SF	Yellow	2.0 – 30 MHz	35
-7	9.0	Carbonyl TH	White	1.0 – 20 MHz	30
-2	10.0	Carbonyl E	Red	0.25 – 10 MHz	95
-1	20.0	Carbonyl C	Blue	0.15 – 2.0 MHz	280
-3	35.0	Carbonyl HP	Grey	0.02 – 1.0 MHz	370

Table 1-6 Partial listing of powdered iron core mixes and suggested frequency range

Because of the complex properties of the core material, the determination of the Q factor of a toroidal inductor can be difficult. It is not simply the magnetic material properties alone, but also the wire winding loss as well that determines the overall Q factor. These losses can then vary greatly with frequency, flux density, and the toroid and wire size. The optimal Q factor occurs when the winding losses are equal to the core losses [4]. In general for a given inductance value and core mix a larger toroid will produce larger Q factors. Conversely for a given toroid size higher Q factor is achieved at higher frequency as the permeability decreases. The skin effect of the wire in the windings can have a significant impact on the achievable Q factor. Just as we have seen with the air core inductor, the inter-turn capacitance will also have a limiting effect on the resulting Q factor as well as the self resonant frequency. Manufacturers often provide a set of Q curves that the designer can use as a design guide for determining the toroidal inductor Q factor. Figure 1-37 shows a typical set of optimal Q curves for Carbonyl W core material at various toroid sizes.

RF and Microwave Concepts and Components   37

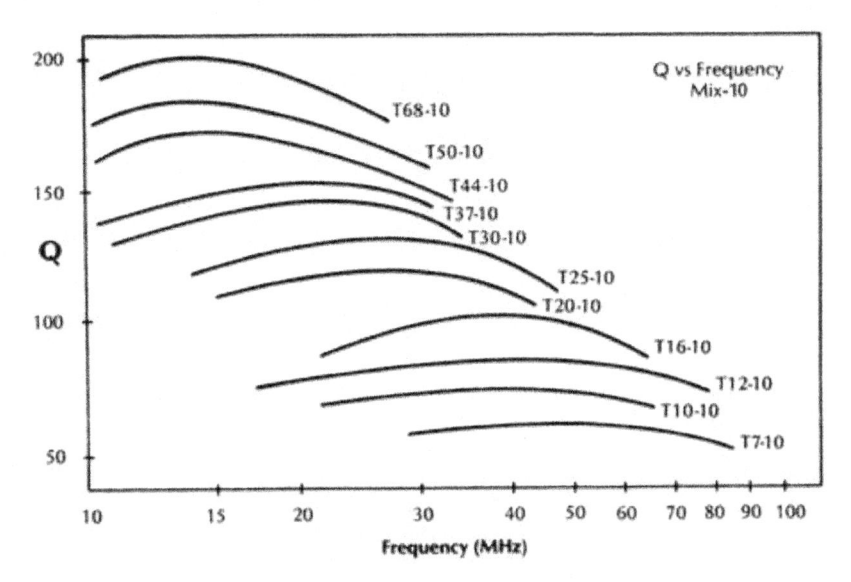

Figure 1-37 Optimal Q factor versus toroid size for core mix -10 (*courtesy Micrometals Inc.*)

Example 1.4-4: Design a 550 nH inductor using the Carbonyl W core of size T30. Determine the number of turns and model the inductor in Genesys.

Solution: From the manufacturer's data sheet the A_L value is 2.5 for a T30-10 toroidal core. Rearranging Equation (1-13) to solve for the number of turns we find that 14.8 turns are required.

$$N = \sqrt{\frac{L}{A_L}} = \sqrt{\frac{550\ nH}{2.5}} = 14.8$$

To reduce the winding loss we want to use the largest diameter of wire that will result in a single layer winding around the toroid. Equation (1-15) will give us the wire diameter.

$$d = \frac{\pi\ ID}{N + \pi} \tag{1-15}$$

where:
    d = Diameter of the wire in inches
    ID = Inner diameter of the core in inches (from Table 1-5)
    N = Number of turns

therefore,

$$d = \frac{\pi(0.151)}{14.8 + \pi} = \frac{0.4744}{19.942} = 0.0238 \; inches$$

From Appendix A, AWG#23 wire is the largest diameter wire that can be used to wind a single layer around the T30 toroid. Normally AWG#24 is chosen because this is a more readily available standard wire size. The toroidal inductor model in Genesys requires a few more pieces of information. The Genesys model requires that we enter the total winding resistance, core Q factor, and the frequency for the Q factor, F_q. As an approximation, set F_q to about six times the frequency of operation. In this case set F_q to (6)·25 MHz = 150 MHz. Then tune the value of Q to get the best curve fit to the manufacturer's Q curve. We know that we have 14.8 turns on the toroid but we need to calculate the length of wire that these turns represent. The approximate wire length around one turn of the toroid is calculated from the following equation.

$$Length = [(2)\,Height + (OD - ID)](\#turns) \quad (1\text{-}16)$$

Using the dimensions for the T30 toroid from Table 1-6 we can calculate the total length of the wire as 6.10 inches.

$$[(2)(0.128) + (0.307 - 0.151)] \cdot 14.8 = 6.10 \; inches$$

Then use the techniques covered in section 1.2.3 to calculate the resistance of the 6.10 inches of wire taking into account the skin effect. To get a better estimate of the actual inductor Q, use the F_q frequency rather than the operating frequency for the skin effect calculation. The resistance at 150 MHz for the AWG#24 wire is calculated as 0.30 Ω. Use an Equation Editor to calculate the Q and inductance. Figure 1-38 shows the schematic of the toroidal inductor and the simulated Q.

RF and Microwave Concepts and Components 39

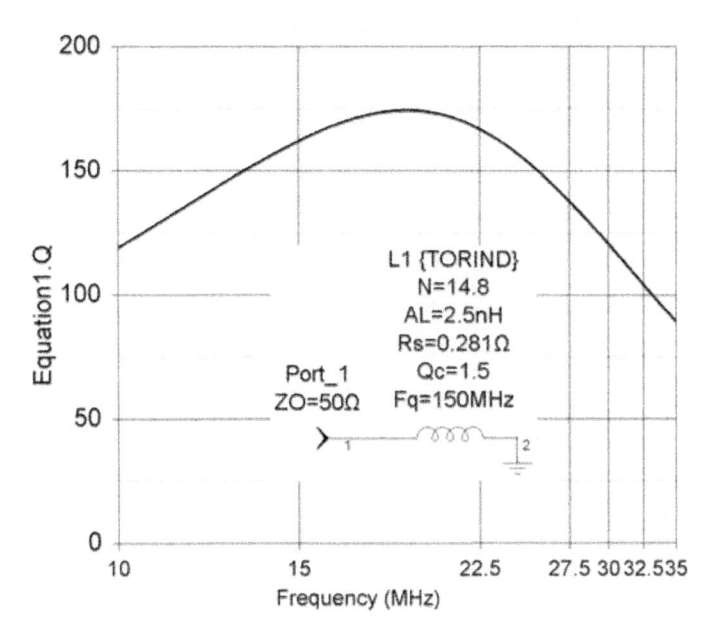

Figure 1-38 Genesys toroidal inductor model and simulated Q factor

The Qc parameter of the model has been tuned to give a reasonably good fit with the manufacturer's (T30) curve of Figure 1-37. Figure 1-39 shows that the simulated self resonant frequency of the inductor model is near 150 MHz.

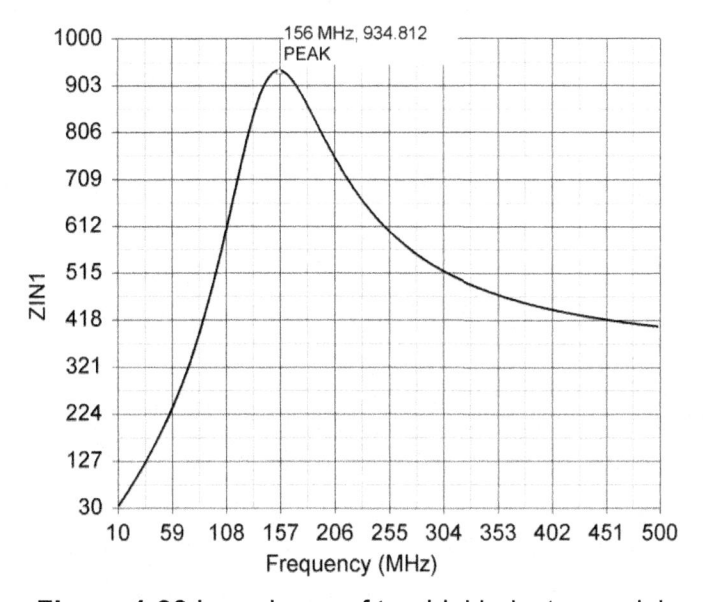

Figure 1-39 Impedance of toroidal inductor model

Figure 1-40 shows the inductance that was calculated from the impedance. The inductance is exactly 550 nH at 10 MHz and begins to increase slightly to 563 nH at the design frequency of 25 MHz. Given the myriad of variables associated with the toroidal inductor, this simple model gives a good first order model of the actual inductor and will enable accurate simulation of filter or resonator circuits.

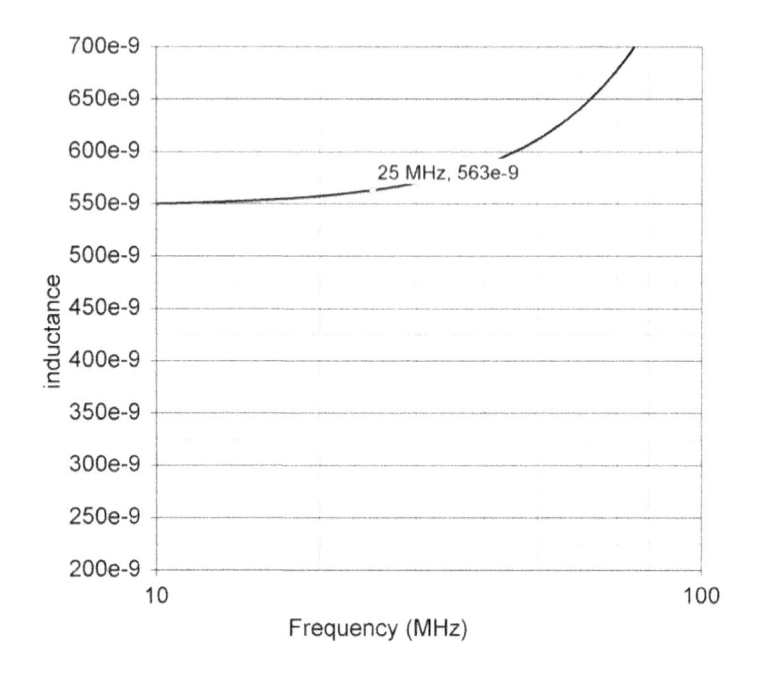

Figure 1-40 Inductance value of the toroidal inductor model

1.5 Physical Capacitors

The capacitor is an electrical energy storage component. The amount of energy that can be stored is dependent on the type and thickness of the dielectric material and the area of the electrodes or plates. Capacitors take on many physical forms throughout electrical circuit designs. These range from leaded bypass capacitors in low frequency applications to monolithic forms in millimeter wave applications. Table 1-7 summarizes many of the applications in which capacitors are found. The table also shows some of the types of materials in which the capacitors are manufactured.

RF and Microwave Concepts and Components   41

Application	Dielectric Type	Notes
Audio Frequency Coupling	Aluminum Electrolytic	Very High Capacitance
	Tantalum	High Capacitance for given size
	Polyester/ Polycarbonate	Medium capacitance, low cost
Power Supply Filtering	Aluminum Electrolytic	High Capacitance, high ripple current
RF Coupling	Ceramic NPO (COG)	Small, low loss, low cost
	Ceramic X7R	Small, low cost, higher loss than COG
	Polystyrene	Very low loss in RF range, larger than ceramic
Tuned Circuits, Resonators	Silver Mica	Low loss, low tolerance for RF applications
	Ceramic NPO (COG)	Low loss, low tolerance, not as good as silver mica.

Table 1-7 Applications for various types of capacitors

As Table 1-7 shows, the ceramic capacitors dominate the higher RF and microwave frequency applications. Two of the most popular of the ceramic capacitors are the single layer and multilayer ceramic capacitors, as shown in Figure 1-41. These capacitors are available with metalized terminations so that they are compatible with a variety of surface mount assembly techniques from hand soldering to wire bonding and epoxy attachment.

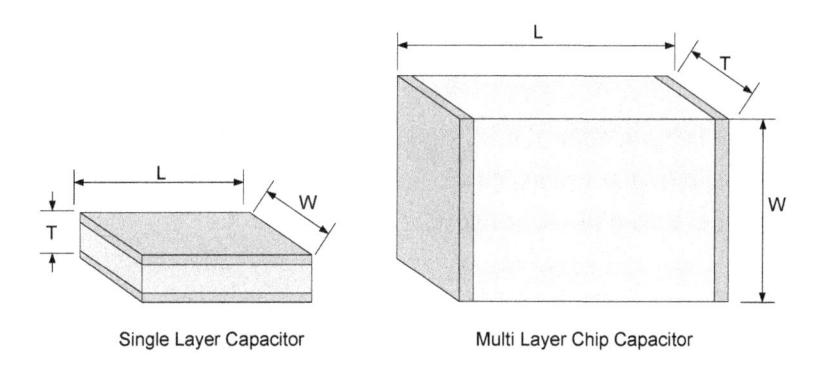

Figure 1-41 Single layer and multi layer chip capacitor dimensions

1.5.1 Single Layer Capacitor

The single layer capacitor is one of the simplest and most versatile of the surface mount capacitors. It is formed with two plates that are separated by a single dielectric layer as shown in Figure 1-42. Most of the electric field (E) is contained within the dielectric however there is a fraction of the E field that exists outside of the plates. This is known as the fringing field.

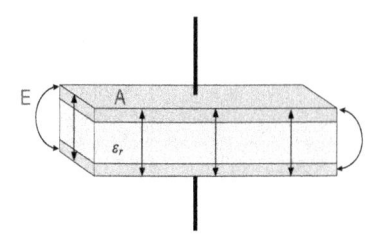

Figure 1-42 Single layer parallel plate capacitor

The capacitance formed by a dielectric material between two parallel plate conductors is given by Equation 1-17 [1].

$$C = (N-1)\left(\frac{KA\varepsilon_r}{t}\right)(FF) \quad pF \qquad (1\text{-}17)$$

where, A = plate area
ε_r = relative dielectric constant
t = separation
K = unit conversion factor; 0.885 for cm and 0.225 for inches
FF = fringing factor; 1.2 when mounted on microstrip
N = number of parallel plates.

Example 1.5-1: Consider the design of a single layer capacitor from a dielectric that is 0.010 inches thick and has a dielectric constant of three. Each plate is cut to 0.040 inches square.

Solution: When the capacitor is mounted with at least one plate on a large printed circuit board track, a value of 1.2 is typically used in calculation. The Equation Editor in Genesys can be used to solve Equations (1-17) as shown in Figure 1-43.

```
1    'Single Layer Capacitor Solution
2    er=3        'Dielectric Constant
3    t=.01       'Dielectric Thickness
4    A=.0016     'Plate Area
5    N=2         'Number of Plates
6    FF=1.2      'Fringe Factor
7    K=.225      'constant for inches
8    C=(N-1)*FF*((K*A*er)/t)
```

Figure 1-43 Single layer capacitance calculation

The single layer capacitor can be modeled in Genesys using the Thin Film Capacitor.

$$C = (2-1)\left(\frac{(.225)(0.04 \cdot 0.04)(3)}{0.010}\right)(1.2) = 0.13\ pF$$

The ceramic dielectrics used in capacitors are divided into two major classifications. Class 1 dielectrics have the most stable characteristics in terms of temperature stability. Class 2 dielectrics use higher dielectric constants which result in higher capacitance values but have greater variation over temperature. The temperature coefficient is specified in either percentage of nominal value or parts per million per degree Celsius (ppm/°C). Ceramic materials with a high dielectric constant tend to dominate RF applications with a few exceptions. NPO (negative-positive-zero) is a popular ceramic that has extremely good stability of the nominal capacitance versus temperature.

Dielectric Material	Dielectric Constant
Vacuum	1.0
Air	1.004
Mylar	3
Paper	4 - 6
Mica	4 - 8
Glass	3.7 - 19
Alumina	9.9
Ceramic (low ε_r)	10
Ceramic (high ε_r)	100 – 10,000

Table 1-8 Dielectric constants of materials

1.5.2 Multilayer Capacitors

Multilayer capacitors, shown in Figure 1-44, are very popular in surface mount designs. They are physically larger than single layer capacitors and can be attached by hand or by automatic pick-and-place machines. As Figure 1-44 shows, the multilayer chip capacitor is a parallel array of capacitor plates in a single package. Due to this type of construction the chip capacitor can handle higher voltages than the single layer capacitor. The insulation resistance of the capacitor is its ability to oppose the flow of electricity and is a function of the dielectric material and voltage. The insulation resistance is typically specified as a minimum resistance value in MΩ at a specified working voltage. The working voltage rating, WVDC, is the maximum DC voltage at which the capacitor can operate over the lifetime of the capacitor. The AC voltage rating is approximately one half of the WVDC value. The dielectric withstand voltage (DWV) is the electrical strength of the dielectric at 2.5 times the rated voltage. This is a maximum short term over-voltage rating and is usually specified as a length of time that the dielectric can withstand the 2.5 times the WVDC value without arcing through [6].

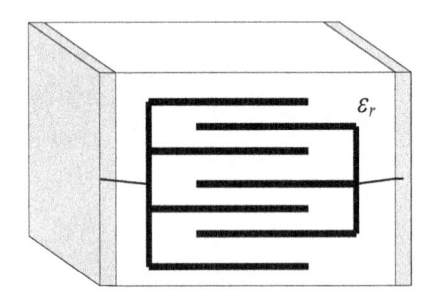

Figure 1-44 Multilayer chip capacitor construction

There is no physical model for the multilayer chip capacitor in Genesys. The designer must rely on S parameter files or Modelithic models as we have used for the chip resistor in section 1.3.1. Two of the major manufacturers of multilayer chip capacitors are American Technical Ceramics, ATC, and Dielectric Laboratories Inc., DLI. These manufacturers provide S parameter files for their capacitors that are readily available on the company websites. They also provide helpful software applications that can aid the designer in making decisions for the proper selection of chip capacitors in specific applications.

1.5.3 Capacitor Q Factor

RF losses in the dielectric material of a capacitor are characterized by the dissipation factor. The dissipation factor is also referred to as the loss tangent and is the ratio of energy dissipated to the energy stored over a period of time. It is essentially the capacitor's efficiency rating. The dissipation factor and other ohmic losses lead to a parameter known as the Equivalent Series Resistance, ESR. The dissipation factor is the reciprocal of the Q factor. Just as we have seen with resistors and inductors, the physical model of a capacitor is a network of R, L, and C components.

Example 1.5-2: Calculate the Q factor versus frequency for the physical model of an 8.2 pF multilayer chip capacitor shown in Fig. 1-45.

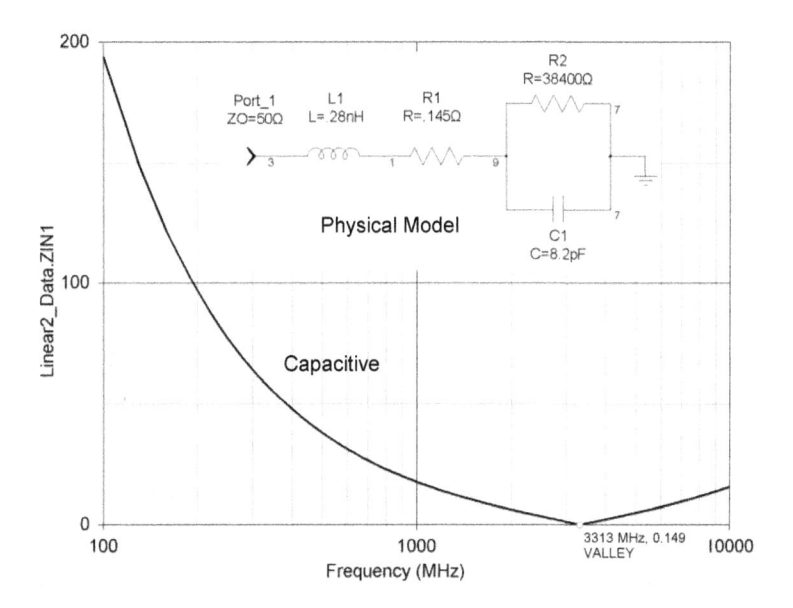

Figure 1-45 Physical model of the 8.2 pF chip capacitor and impedance

The capacitor has a series inductance and resistance component along with a resistance in parallel with the capacitance. The parallel resistor sets the losses in the dielectric material. The series resistance and inductance represent any residual lead inductance and ohmic resistances.

Solution: The values entered for the physical model and the Q factor can be obtained from the capacitor manufacturer. The plot of Figure 1-45 shows the impedance of the capacitor versus frequency. Note that the impedance decreases as would be expected until the self resonant frequency is reached. Above the self resonant frequency the impedance begins to increase suggesting that the capacitor is behaving as an inductor. The self resonant frequency is due to the series inductance resonating with the capacitance. At resonance the reactance cancels leaving only the resistances *R1* and *R2*. The parallel resistance, *R2*, can be converted to an equivalent series resistance by Equation (1-18).

$$R2' = \frac{1}{1+Q^2} R2 \qquad (1\text{-}18)$$

These two series resistances can then be added to find the equivalent series resistance, ESR, as defined by Equation (1-19).

$$ESR = R1 + R2' \qquad (1\text{-}19)$$

The capacitor Q factor is then calculated by Equation (1-20). X_T is the total series reactance of the inductive and capacitive reactance.

$$Q = \frac{X_T}{ESR} \qquad (1\text{-}20)$$

As Figure 1-45 shows, the 8.2 pF multilayer chip capacitor has a series resonant frequency of 3313 MHz. A marker is placed on the trace indicating the impedance at the SRF as 0.149 Ω. Because the reactance is cancelled at the SRF, this impedance essentially becomes the ESR of the capacitor. R2 of Figure 1-45 is extremely frequency dependent. This means that the capacitor's Q factor is also extremely frequency dependent. An improved model for analyzing the capacitor's characteristics over a wide frequency range is to use the Genesys model for capacitor with Q. Using this model the Q factor of the capacitor can be made proportional to the square root of the applied frequency. The capacitor Q factor can be calculated from the impedance using the Equation Editor as shown in Figure 1-46. The resistive and inductive components are used to define the Q factor.

```
1    using("Linear1_Data")
2    reactance1=im(Linear1_Data.ZIN1)
3    resistance1=re(Linear1_Data.ZIN1)
4    Qphysical=abs(reactance1)./resistance1
5    capacitance_model=1/(abs(reactance1).*F*2*3.14)
```

Figure 1-46 Equation Editor used to calculate Q factor and capacitance

RF and Microwave Concepts and Components 47

Figure 1-47 shows the large dependence on the capacitor Q with frequency. At 1006 MHz the Q factor is 120 while at 100 MHz the Q factor is greater than 1000. The Q factor goes to zero at the self resonant frequency. Above the self resonant frequency the Q is undefined.

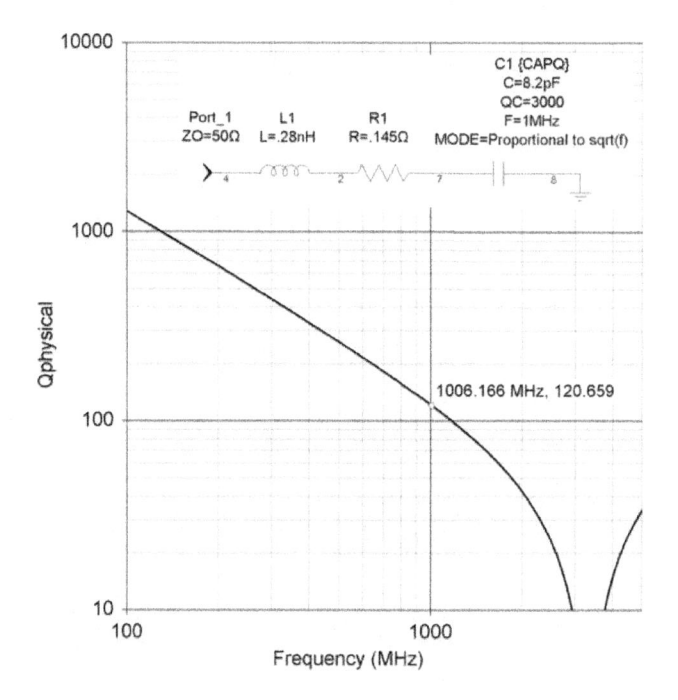

Figure 1-47 Calculated Q factor of 8.2 pF chip capacitor and equivalent model

The Equation Editor of Figure 1-46 also calculates the effective capacitance from the total reactance of the model using Equation (1-21).

$$C = \frac{1}{2\pi F X_T} \tag{1-21}$$

The plot of Figure 1-48 reveals some interesting characteristics about the chip capacitor. From 100 MHz to 300 MHz the capacitance value is fairly constant at 8.217 pF. As the frequency increases above 300 MHz we see that the capacitance actually increases. The parasitic inductive reactance of the capacitor package actually makes the effective capacitance greater than its nominal value.

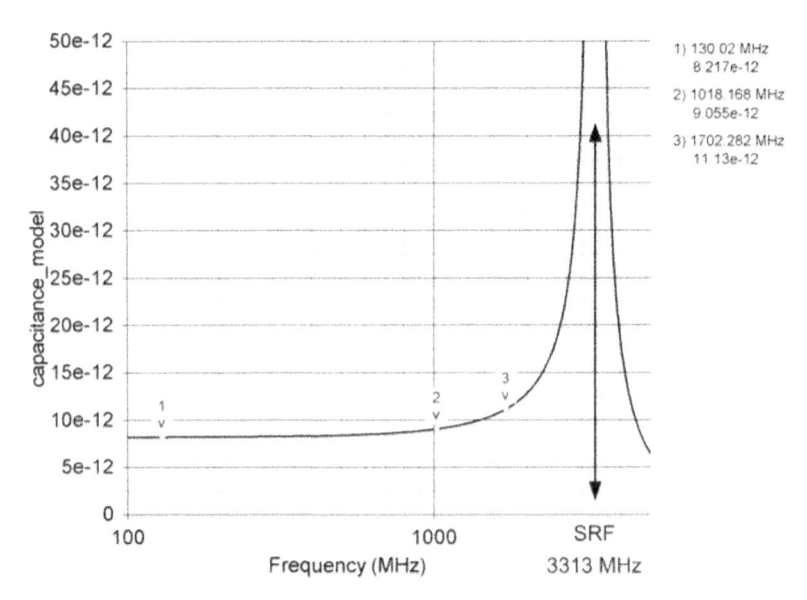

Figure 1-48 Effective capacitance of the 8.2 pF chip capacitor

This is a property of the capacitor that is not always intuitive. As the frequency approaches the self resonant frequency the capacitance rapidly approaches infinity. The capacitor actually becomes nearly a short circuit to RF at the self resonant frequency. This is an important property of the capacitor that is used frequently in RF and microwave design. In RF coupling or bypass capacitor applications, capacitors are very often used at or near their self resonant frequency. The capacitor's self resonant frequency is due to the series resonant circuit. In applications requiring bypassing over a wide range of frequencies it is often necessary to use several capacitors each selected to have a uniquely spaced self resonant frequency. In filter and other tuned circuit applications where we want the chip to appear as an 8.2 pF capacitor, we clearly must stay well below the self resonant frequency of the capacitor. A typical rule-of-thumb is to use the capacitor over a frequency range up to 35% of the self resonant frequency. Therefore the 8.2 pF chip capacitor with a self resonant frequency of 3313 MHz would be used as a capacitor in tuned circuits up to a frequency of about 1160 MHz. Figure 1-48 shows that at frequencies above 1160 MHz the capacitance is extremely nonlinear. However the capacitor could be used as a DC blocking, RF coupling capacitor that would efficiently pass microwave energy near its SRF of 3300 MHz.

References and Further Reading

[1] Ali A. Behagi and Stephen D. Turner, *Microwave and RF Engineering,* BT Microwave LLC, State College, PA, 2011

[2] Paul Lorrain, Dale P. Corson, and Francois Lorrain, *Electromagnetic Fields and Waves*, W.H. Freeman and Company, New York, 1988

[3] *Design Guide, Microwave Components* Inc., P.O. Box 4132, South Chelmsford, MA 01824

[4] *Iron Powder Cores for High Q Inductors*, Micrometals, Inc.

[5] Keysight Technologies, *Genesys 2014.03, Users Guide*, www.keysight.com

[6] *Capacitors for RF Applications,* Dielectric Laboratories, Inc.,2777 Rt.20 East, Cazenovia, NY. 13035

[7] R. Ludwig and P. Bretchko, *RF Circuit Design -Theory and Applications*, Prentice Hall, New Jersey, 2000

[8] William Sinnema and Robert McPherson, *Electronic Communication,* Prentice Hall Canada, 1991

[9] M.F. "Doug" DeMaw, *Ferromagnetic Core Design & Application Handbook*, MFJ Publishing Co., Inc. Starkville, MS. 39759, 1996

[10] *The RF Capacitor Handbook,* American Technical Ceramics, One Norden Lane, Huntington, New York 11746

Problems

1-1. Calculate the wavelength of an electromagnetic wave operating at a frequency of 428MHz.

1-2. Calculate the inductance of a 5 inch length of AWG #30 straight copper wire.

1-3. Using the Equation Editor, calculate the reactance of the wire from Problem 2 at 10 Hz, 10 MHz, and 10 GHz. Create a Linear analysis in Genesys and display the wire impedance vs. frequency.

1-4. Calculate the resistance of a 12 inch length of AWG #24 copper wire at DC and at 25MHz

1-5. Find the skin depth and the resistance of a 2 meter length of copper coaxial line at 2 GHz. The inner conductor radius is 1 mm and the outer conductor is 4 mm.

1-6. Calculate the inductance of a 5 inch length of copper flat ribbon conductor. The dimensions of the ribbon are 0.100 inches in width and 0.002 inches thick.

1-7. Model a chip resistor (size 0603) with a resistance of 50Ω in Genesys. Consider an application in which 50Ω impedance must be maintained with +10%. Create a Linear Analysis and determine the maximum usable frequency of the chip resistor.

1-8. Design an air core inductor with an inductance value of 84nH. Use a copper wire of 0.050 inch diameter wound on a core diameter or 0.100 inch. Determine the number of turns required assuming a tight spaced winding.

1-9. Using the inductor from Problem 8 and the techniques developed in Section 1.4.1 examine the change in the coil inductance as the turn spacing is increased from zero to 0.10 inch in 0.002 inch increments.

1-10. Using the inductor from Problem 8, determine the self resonant frequency of the inductor and comment on the maximum frequency in which the inductor may be used in a tuned circuit application.

1-11. Using the inductor from Problem 8, determine the maximum Q factor of the inductor and the frequency at which the maximum Q factor is obtained.

1-12. In Genesys, select a chip inductor from the CoilCraft library with an inductance value of 80nH. Determine the maximum Q factor of the inductor and comment on the maximum usable frequency of the inductor in a filter application.

1-13. Design a 1mH toroidal inductor on a Carbonyl W core size T30. Determine the maximum wire size that could be used to realize a single layer winding.

1-14. Using the inductor from Problem 13, model the inductor in Genesys and determine the approximate self resonant frequency. Comment on the maximum usable frequency of the inductor in the front end of a radio receiver.

1-15. A 0.05pF capacitor is required to couple a transistor to the resonator of a microwave oscillator. Design a single layer capacitor using a 0.020 inch thick dielectric with e_r=2.2. Determine the dimensions of the capacitor assuming square footprint is desired.

1-16. For the single layer capacitor of Problem 15, determine the dimensions of the capacitor with a dielectric constant e_r=10.2.

1-17. In Genesys, model a 47pF chip capacitor using the ATC 0603 model from the Modelithics library. Determine the Q factor of the capacitor at a frequency of 1000MHz.

1-18. Determine the self resonant frequency of the capacitor in Problem 17 and comment on the maximum frequency that this capacitor could be used in a tuned circuit. At what frequency does the capacitor have the lowest amount of energy loss?

# Chapter 2

Transmission Lines

2.1 Introduction

Transmission lines play an important role in designing RF and microwave networks. In chapter 1 we have seen that, at high frequencies where the wavelength of the signal is smaller than the dimension of the components, even a small piece of wire acts as an inductor and affects the performance of the network. In this chapter we present the lumped-element equivalent model of a transmission line and analytically define several important transmission line parameters such as transmission and reflection coefficients, characteristic impedance, propagation constant, attenuation constant, phase constant, voltage standing wave ratio, return loss, velocity factor and group delay. It is demonstrated how to simulate and measure these parameters using the Genesys software. Popular types of transmission lines, such as: coaxial lines, microstrip lines, strip lines, and waveguides are discussed. Several methods of characterizing reflection coefficients and the characteristic impedance of these transmission lines are examined. Field coupling between adjacent (coupled) transmission lines is introduced. The chapter concludes with the design of a microstrip directional coupler.

2.2 Plane Waves

Plane waves are the simplest form of electromagnetic waves in which the electric field intensity, **E**, is perpendicular to magnetic field intensity, **H**, and both are perpendicular to the direction of propagation (**E** and **H** are represented by vectors). Such a wave is also called transverse electromagnetic or TEM wave.

2.2.1 Plane Waves in a Lossless Medium

In rectangular coordinates, assuming that electric field **E** is a vector in the x direction and varies as it moves along the z direction, the Wave Equation for the electric field in a lossless medium can be written as [1]:

$$\frac{\partial^2 E_x}{\partial z^2} + k^2 E_x = 0 \qquad (2\text{-}1)$$

where,

$k = \omega\sqrt{\mu\varepsilon}$ is the wave number
ω is the angular frequency in radians per second
μ is permeability of the medium in Henries per meter
ε is permittivity of the medium in Farads per meter

In a lossless medium μ and ε are positive real numbers, therefore k is also positive and real. The solution to Equation (2-1) is of the following form.

$$E_x(z) = E^+ e^{-jkz} + E^- e^{+jkz} \tag{2-2}$$

Where E^+ and E^- are arbitrary constants determined by the boundary conditions. For the sinusoidal waveforms at frequency ω, Equation (2-2) can be written as:

$$\mathcal{E}_x(z,t) = E^+ \cos(\omega t - kz) + E^- \cos(\omega t + kz) \tag{2-3}$$

where:

$E^+ \cos(\omega t - kz)$ is the wave traveling in the forward direction
$E^- \cos(\omega t + kz)$ is the wave traveling in the reverse direction

Some of the wave characteristics are as follows:

1. Phase velocity is the velocity of a fixed point on the wave that is obtained by setting the derivative of the phase, with respect to t, equal to zero:

$$\omega - k\frac{\partial z}{\partial t} = 0$$

$$v_p = \frac{\partial z}{\partial t} = \frac{\omega}{k} = \frac{1}{\sqrt{\mu\varepsilon}} \tag{2-4}$$

In free space: $\mu = \mu_0 = 4\pi(10^{-7})$ H/m and $\varepsilon = \varepsilon_0 = 8.854(10^{-12})$ F/m, therefore,

$$v_p = \frac{1}{\sqrt{\mu_0\varepsilon_0}} = 2.998(10^8) = c \quad meters/second$$

where c is the velocity of light in free space.

2. The wavelength, λ, is defined as the distance between two successive maximum, or minimum, points on the wave at a fixed instant of time. Therefore $k\lambda = 2\pi$ which leads to Equation (2-5).

$$\lambda = \frac{2\pi}{k} = \frac{v_p}{f} \qquad (2\text{-}5)$$

3. The plane wave impedance, η, is defined as the ratio between electric and magnetic field components travelling in the same direction. Therefore, if

$$E_x^+ = E^+ \cos(\omega t - kz)$$
$$H_y^+ = H^+ \cos(\omega t - kz)$$

then,

$$\eta = \frac{E_x^+}{H_y^+} = \frac{E^+}{H^+} \qquad (2\text{-}6)$$

Here E^+ and H^+ represent the electric and magnetic field amplitude travelling in the positive z direction in units of Volt/meter and Ampere/meter. Based on Maxwell's curl equations the plane wave impedance is given by:

$$\eta = \sqrt{\frac{\mu}{\varepsilon}} \qquad (2\text{-}7)$$

In free space the plane wave impedance is,

$$\eta_0 = \sqrt{\frac{\mu_0}{\varepsilon_0}} = 377 \ \Omega$$

2.2.2 Plane Waves in a Good Conductor

Metallic conductors used in microwave networks are not perfect but they are considered to be very good conductors. In a material with conductivity σ the current density due to conduction is given by:

$$J = \sigma E \quad (2\text{-}8)$$

where σ is the conductivity of the conductor in S/m.

In a good conductor the conductive current is much greater than the displacement current, therefore, by ignoring the displacement current, the propagation constant can be written as [1]:

$$\gamma = \alpha + j\beta = \sqrt{\frac{\omega\mu\sigma}{2}} + j\sqrt{\frac{\omega\mu\sigma}{2}} \quad (2\text{-}9)$$

The skin depth for a good conductor is defined as the inverse of the attenuation constant:

$$\delta = \frac{1}{\alpha} = \frac{1}{\sqrt{\pi f \mu \sigma}} \quad (2\text{-}10)$$

where:
$\quad\delta$ is the skin depth in meters
$\quad$f is the frequency in Hertz
$\quad\mu$ is the permeability in H/m
$\quad\sigma$ is the conductivity in S/m

In a good conductor the positive traveling wave is of the form:

$$e^{-\alpha z}\cos(\omega t - \beta z) \quad (2\text{-}11)$$

Notice that when the wave travels a distance equal to the skin depth, $z = \delta = 1/\alpha$, the signal amplitude drops to $e^{-1}\cos(\omega t - \beta z) = 0.368\cos(\omega t - \beta z)$ which is 36.8% of the original signal's amplitude at $z = 0$.

2.3 Lumped Element Representation of Transmission Lines

A lumped model of a small section of parallel wire transmission line, of physical length dz, is shown in Figure 2-1. Any transmission line of a given length can be considered a cascade of many sections of the length dz.

Transmission Lines

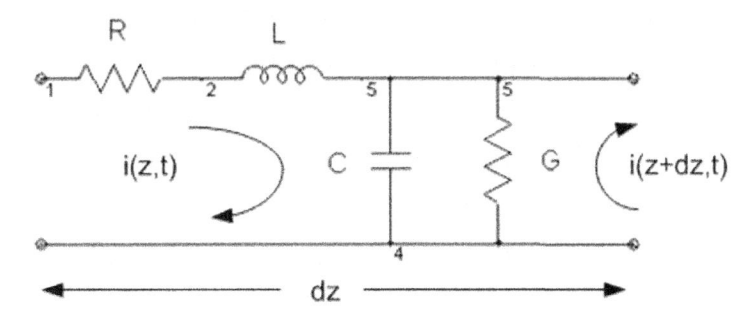

Figure 2-1 Lumped model representation of transmission line section

In Figure 2-1, R is the per unit length series resistance of both lines in Ω/m, L is the per unit length series inductance of both lines in Henries/meter, G is the per unit length shunt conductance of the line in Siemens/meter, and C is the per unit length capacitance of the line in Farads/meter. At higher frequencies, where the wavelength of the signal is smaller than the physical dimension of the network, the voltages and currents along a uniform transmission line are functions of position and time. For a TEM wave traveling in the z direction, the voltage and current are given by:

$$V(z,t) = re\left[V(z)e^{j\omega t}\right] \qquad (2\text{-}12)$$

$$I(z,t) = re\left[I(z)e^{j\omega t}\right] \qquad (2\text{-}13)$$

The quantities *V(z)* and *I(z)* are complex functions of z along the transmission line and ω is the frequency of the source in radians per second.

2.4 Transmission Line Equations and Parameters

For sinusoidal steady-state excitations, Kirchhoff's voltage and current laws along the line yield the following equations:

$$\frac{dV(z)}{dz} + (R + j\omega L)\,I(z) = 0 \qquad (2\text{-}14)$$

$$\frac{dI(z)}{dz} + (G + j\omega C)\,V(z) = 0 \qquad (2\text{-}15)$$

By taking the derivative of both sides of Equation (2-14) with respect to z, and substituting for $\frac{dI(z)}{dz}$ from Equation (2-15), we have:

$$\frac{d^2V(z)}{dz^2} - (R + j\omega L)(G + j\omega C) V(z) = 0 \qquad (2\text{-}16)$$

Similarly, by taking the derivative of both sides of Equation (2-15) with respect to z, and substituting for $\frac{dV(z)}{dz}$ from Equation (2-14), we have:

$$\frac{d^2I(z)}{dz^2} - (R + j\omega L)(G + j\omega C) I(z) = 0 \qquad (2\text{-}17)$$

By defining the complex propagation constant:

$$\gamma = \sqrt{(R + j\omega L)(G + j\omega C)} \qquad (2\text{-}18)$$

Equations (2-16) and (2-17) can be redefined as:

$$\frac{d^2V(z)}{dz^2} - \gamma^2 V(z) = 0 \qquad (2\text{-}19)$$

$$\frac{d^2I(z)}{dz^2} - \gamma^2 I(z) = 0 \qquad (2\text{-}20)$$

Equations (2-19) and (2-20) are known as voltage and current wave equations respectively. It is easy to show that the solution to Equation (2-19) is of the form:

$$V(z) = Ae^{-\gamma z} + Be^{\gamma z} \qquad (2\text{-}21)$$

where:
    A and B are constants
    $Ae^{-\gamma z}$ is the incident voltage traveling in the +z direction
    $Be^{\gamma z}$ is the reflected voltage traveling in the -z direction

Similarly it can be shown that the solution to Equation (2-20) is of the form:

$$I(z) = \frac{\gamma}{(R + j\omega L)} \left[Ae^{-\gamma z} - Be^{\gamma z} \right] \qquad (2\text{-}22)$$

where:

$\dfrac{\gamma}{R+j\omega L} Ae^{-\gamma z}$ is the incident current traveling in the $+z$ direction

$\dfrac{\gamma}{R+j\omega L} Be^{\gamma z}$ is the reflected current traveling in the $-z$ direction

2.4.1 Definition of Attenuation and Phase Constant

The complex propagation constant defined in Equation (2-18), is written as:

$$\gamma = \alpha + j\beta \qquad (2\text{-}23)$$

where the real part α is defined as the attenuation constant in Nepers per meter (1 Neper = 8.686 dB) and the imaginary part β is defined as the phase constant in radians per meter.

2.4.2 Definition of Transmission Line Characteristic Impedance

The characteristic impedance of a transmission line, Z_0, is defined as the ratio of the incident voltage to incident current. Therefore by dividing the incident voltage by the incident current, in Equations (2-21) and (2-22), and replacing γ from Equation (2-18), the transmission line characteristic impedance is written as:

$$Z_o = \dfrac{(R+j\omega L)}{\gamma} = \sqrt{\dfrac{R+j\omega L}{G+j\omega C}} \qquad (2\text{-}24)$$

2.4.3 Definition of Transmission Line Reflection Coefficient

It is important to understand the concepts of incident (forward) and reflected (backward) wave propagation in transmission lines. In high power RF systems there can be potentially dangerous high voltage peaks that can occur at points along the transmission line when the incident and reflected waves are in phase and add together. The voltage reflection coefficient,

$\Gamma(z)$, of a transmission line along the z axis is defined as the ratio of reflected to incident voltage as shown in Equation (2-25).

$$\Gamma(z) = \frac{Be^{\gamma z}}{Ae^{-\gamma z}} = \frac{B}{A}e^{2\gamma z} = \Gamma_o e^{2\gamma z} \qquad (2\text{-}25)$$

where Γ_o is the load reflection coefficient at z = 0, namely:

$$\Gamma_o = \Gamma(0) = \frac{B}{A} \qquad (2\text{-}26)$$

Notice that when there is no reflection the load reflection coefficient $\Gamma_o = 0$.

2.4.4 Definition of Voltage Standing Wave Ratio, VSWR

VSWR is the ratio of the maximum to minimum value of the standing wave. The VSWR is a very common quantity for describing the percentage of power reflected by a given load impedance. The forward and reflected waves travel in opposite directions to form a standing wave pattern. The maximum value of the standing wave is given by Equation (2-27) while the minimum value is given by Equation (2-28).

$$|V(z)|_{max} = |A|(1+|\Gamma_o|) \qquad (2\text{-}27)$$

$$|V(z)|_{min} = |A|(1-|\Gamma_o|) \qquad (2\text{-}28)$$

The ratio of the maximum to minimum voltage is related to the reflection coefficient as shown in Equation (2-29).

$$VSWR = \frac{|V(z)|_{max}}{|V(z)|_{min}} = \frac{1+|\Gamma_o|}{1-|\Gamma_o|} \qquad (2\text{-}29)$$

Solving the equation for the magnitude of the reflection coefficient, Γo results in Equation (2-30).

$$|\Gamma_o| = \frac{VSWR-1}{VSWR+1} \qquad (2\text{-}30)$$

Transmission Lines

Notice that when the reflection coefficient is zero, VSWR = 1. This VSWR is commonly presented as 1:1 ratio.

2.4.5 Definition of Return Loss

When the transmission line is mismatched to the load, a portion of the incident power is reflected back to the source. This can be considered a loss of power absorbed by the load. Therefore the return loss, RL, in dB, is defined as:

$$RL\ (dB) = -20 \log |\Gamma| \tag{2-31}$$

For a matched line $|\Gamma| = 0$ and RL = ∞ dB whereas for $|\Gamma| = 1$, RL = 0 dB.

2.4.6 Lossless Transmission Line Parameters

From Figure 2-1 we can see that a transmission line is considered lossless when R = G = 0. For a lossless transmission line the propagation constant reduces to:

$$\gamma = j\beta = j\omega \sqrt{LC} \tag{2-32}$$

therefore,

$$\beta = \omega \sqrt{LC} \tag{2-33}$$

and

$$\alpha = 0 \tag{2-34}$$

The characteristic impedance of a lossless transmission line i.e. R = G = 0 is obtained from Equation (2-24) as:

$$Z_o = \sqrt{\frac{L}{C}} \tag{2-35}$$

Similarly the voltage reflection coefficient in Equation (2-25) reduces to:

$$\Gamma = \Gamma_o e^{2j\beta z} \tag{2-36}$$

where: $\Gamma_0 = B/A$, the load reflection coefficient.

Notice from Equations (2-32) through Equation (2-34) for lossless transmission lines the attenuation constant is zero and the phase constant is linearly proportional to frequency. The propagation constant is purely imaginary, and the characteristic impedance is a positive real number.

2.4.7 Lossless Transmission Line Terminations

When a lossless transmission line of length d and characteristic impedance Z_0 is terminated in an arbitrary load Z_L, the input impedance of the line is given by [1]:

$$Z_{in} = Z_o \frac{Z_L + j Z_o \tan \beta d}{Z_o + j Z_L \tan \beta d} \qquad (2\text{-}37)$$

where β is the phase constant of the line

In the following sections we discuss the input impedance of terminated transmission lines.

1. Transmission Line Terminated in Z_0

When a lossless transmission line of characteristic impedance Z_0 is terminated in a load equal to Z_0, Equation (2-37) shows that the input impedance becomes equal to Z_0. Such a line behaves like an infinitely long transmission line with no reflection. In this case all of the incident power is absorbed by the load.

2. Transmission Line Terminated in a Short Circuit

When a lossless transmission line of characteristic impedance Z_0 is terminated in a short circuit, $Z_L = 0$, Equation (2-37) shows that the input impedance becomes a purely imaginary number equal to:

$$Z_{in} = j Z_o \tan \beta d \qquad (2\text{-}38)$$

Depending on the length of the line the input impedance takes any possible reactive value from minus to plus infinity.

3. Transmission Line Terminated in an Open Circuit

When a lossless transmission line of characteristic impedance Z_0 is terminated in an open circuit, $Z_L = \infty$, Equation (2-37) shows that the input impedance becomes equal to:

$$Z_{in} = -jZ_o \cot \beta d \qquad (2\text{-}39)$$

In this case Z_{in} is also purely imaginary. Depending on the length of the line the input impedance takes any possible reactive value from minus to plus infinity.

4. Half Wavelength Transmission Lines

For a lossless transmission line of characteristic impedance Z_0 with the length $d = \lambda/2$, the input impedance from Equation (2-37) becomes equal to:

$$Z_{in} = Z_L \qquad (2\text{-}40)$$

This means that the input impedance of a transmission line of one half wavelength is equal to the load impedance regardless of the line characteristic impedance.

5. Quarter Wavelength Transmission Lines

For a lossless transmission line with the length $d = \lambda/4$, the input impedance from Equation (2-37) becomes equal to:

$$Z_{in} = \frac{Z_o^2}{Z_L} \qquad (2\text{-}41)$$

In this case the transmission line transforms the load impedance to a different impedance as defined by Equation (2-41). Such a line is called a quarter-wave transformer. Equation (2-41) can be rewritten to solve for the characteristic impedance of a quarter-wave matching section as:

$$Z_o = \sqrt{Z_{in} Z_L} \qquad (2\text{-}42)$$

2.4.8 Simulating Reflection Coefficient and VSWR in Genesys

The Genesys software has built-in functions to display VSWR, return loss, and reflection coefficient. The following example explores the simulation of input reflection coefficient, S[1,1], and VSWR.

Example 2.4-1: For the series RLC elements in Figure 2-2 measure the reflection coefficients and VSWR from 100 to 1000 MHz.

Solution: Create a Linear Analysis in Genesys and sweep the frequency from 100 to 1000 MHz to measure the VSWR at port 1 and the input reflection coefficients, S[1,1], in dB and (Mag abs+Angle) formats.

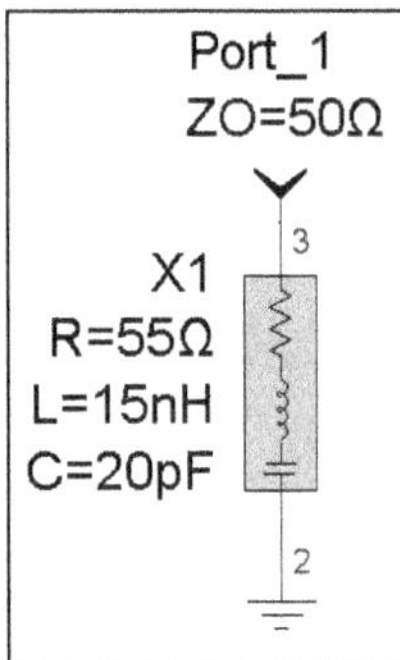

F (MHz)	VSWR1	S[1,1] (dB)	mag(S[1,1])	ang(S[1,1]) (deg)
100	3.514	-5.084	0.557	-52.175
200	1.503	-13.933	0.201	-65.292
300	1.106	-25.944	0.05	18.321
400	1.42	-15.207	0.174	64.69
500	1.811	-10.796	0.289	64.345
600	2.245	-8.321	0.384	61.007
700	2.727	-6.682	0.463	57.291
800	3.26	-5.506	0.531	53.694
900	3.849	-4.62	0.588	50.344
1000	4.494	-3.931	0.636	47.27

Port_1, ZO=50Ω
X1, R=55Ω, L=15nH, C=20pF

Figure 2-2 Schematic and tabular output of VSWR, return loss, and reflection coefficient

2.4.9 Return Loss, VSWR, and Reflection Coefficient Conversion

Return Loss, VSWR, and Reflection Coefficient are all different ways of characterizing the wave reflection. These definitions are often used interchangeably in practice. Therefore it is important to be able to convert from one form of reflection to another. Return Loss is often used to characterize components such as filters, amplifiers, and networks. VSWR is normally used in systems such as radio and TV transmitters. Reflection coefficient is normally used in device characterization such as transistors, capacitors, inductors, etc. Equations (2-29), (2-30), and (2-31) give the conversions among these three parameters. Note, however, that the reflection coefficient is a vector quantity whose magnitude is all that is required to determine the VSWR or return loss. Therefore when calculating Γ from the VSWR or return loss we can only find the magnitude and not the angle of the reflection coefficient. Another useful parameter is the mismatch loss. The mismatch loss is a measure of power that is reflected by the load.

$$Mismatch\ Loss\ (dB) = -10\log\left(1-|\Gamma|^2\right) \qquad (2\text{-}43)$$

Example 2.4-2: Generate a table showing the return loss, the reflection coefficient, and the percentage of reflected power as a function of VSWR.

Solution: In Genesys create a schematic with a resistor. Make the resistance value a tunable variable. Set the Linear Analysis at a fixed frequency of 100 MHz. Then add a Parameter Sweep to sweep the value of the resistor.

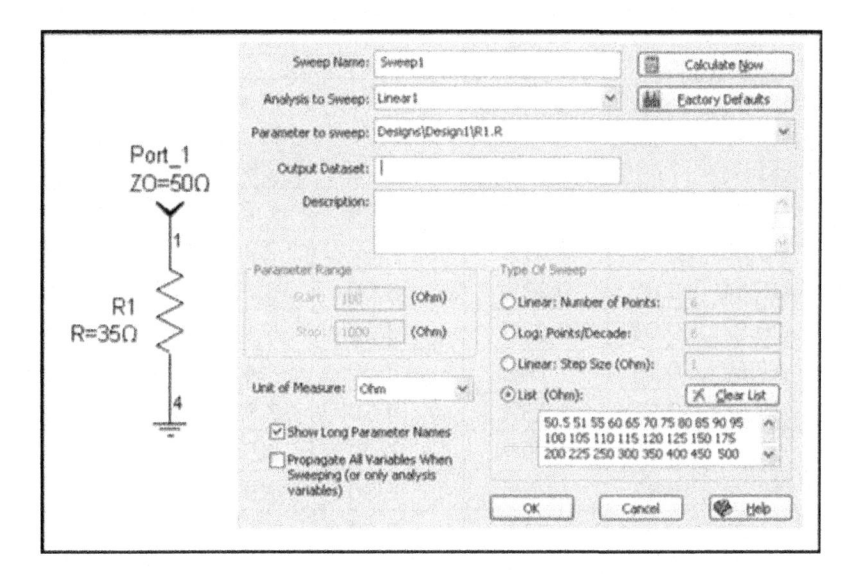

Figure 2-3 Schematic and parameter sweep for VSWR table

Edit the Parameter Sweep and select the current Linear Analysis. Then select the resistance of the Resistor element on the Parameter to Sweep drop down list. Under the Type of Sweep choose the List option and enter the discrete resistance values as shown in Figure 2-3. When the circuit is swept the 30 resistance values will create a unique VSWR, Return Loss, and

Reflection Coefficient. Then create an equation to calculate this parameter. Figure 2-4 shows the equation block for the calculation of mismatch loss.

```
1    'Enter Design Equations
2    vswr=Sweep1_Data.VSWR1
3    ReflCoef=(vswr-1)/(vswr+1)
4    mismatch=1-((ReflCoef)^2)
5    powerloss=(1-mismatch)*100
6
```

Figure 2-4 Equation block for calculation of mismatch loss

We will present the mismatch loss as a percentage of the available power that is reflected by the load. Add the equation block variable, power loss, to the output table. The results of the Genesys Table can be saved to a comma delimited text file. This text file can then be read into Excel for a more attractive formatted table as shown in Table 2-1. As we can see from the table if we can keep the VSWR less than 1.25:1 we will have less than 1% power loss due to reflective impedance mismatch.

VSWR	Return Loss (dB)	$|\Gamma|$	% Reflected Power	VSWR	Return Loss (dB)	$|\Gamma|$	% Reflected Power
1.01	-46.06	0.00	0.00%	2.40	-7.71	0.41	16.96%
1.02	-40.09	0.01	0.01%	2.50	-7.36	0.43	18.37%
1.10	-26.44	0.05	0.23%	3.00	-6.02	0.50	25.00%
1.20	-20.83	0.09	0.83%	3.50	-5.11	0.56	30.86%
1.30	-17.69	0.13	1.70%	4.00	-4.44	0.60	36.00%
1.40	-15.56	0.17	2.78%	4.50	-3.93	0.64	40.50%
1.50	-13.98	0.20	4.00%	5.00	-3.52	0.67	44.44%
1.60	-12.74	0.23	5.33%	6.00	-2.92	0.71	51.02%
1.70	-11.73	0.26	6.72%	7.00	-2.50	0.75	56.25%
1.80	-10.88	0.29	8.16%	8.00	-2.18	0.78	60.49%
1.90	-10.16	0.31	9.63%	9.00	-1.94	0.80	64.00%
2.00	-9.54	0.33	11.11%	10.00	-1.74	0.82	66.94%
2.10	-9.00	0.36	12.59%	20.00	-0.87	0.91	81.86%
2.20	-8.52	0.38	14.06%	200.00	-0.09	0.99	98.02%
2.30	-8.09	0.39	15.52%	2000.00	-0.01	1.00	99.80%

Table 2-1 Relationship among return loss, VSWR, and reflection coefficient

2.5 RF and Microwave Transmission Media

2.5.1 Free Space Characteristic Impedance and Velocity of Propagation

There are many physical transmission lines that are encountered in RF and microwave circuits and systems. In addition it should not be overlooked that free space is also a transmission medium. Therefore Equation (2-35) could be used to describe the characteristic impedance, Z_o, of free space. In order to define L and C we need to consider the inductive and capacitive properties of free space. Permeability is the ability of a transmission media to support a magnetic field. Considered as a density in free space the permeability is related to inductance and defined by Equation (2-44).

$$\mu_o = 4\pi \cdot 10^{-7} \quad Henries/meter \quad (2\text{-}44)$$

Permittivity is the ability of a transmission media to support an electric field. Considered as a density in free space the permittivity is closely related to capacitance in equation (2-45).

$$\varepsilon_o = \frac{1}{36\pi} \cdot 10^{-9} \quad Farads/meter \quad (2\text{-}45)$$

Therefore we can rewrite equation (2-35) as:

$$Z_o = \sqrt{\frac{\mu_o}{\varepsilon_o}} = \sqrt{\frac{4\pi \cdot 10^{-7}}{\frac{1}{36\pi} \cdot 10^{-9}}} = 377\ \Omega \quad (2\text{-}46)$$

Equation (2-46) shows that the characteristic impedance is related to the inductance and capacitance of the transmission media. The velocity of propagation is also related to inductance and capacitance. It can be shown that the time required for a sine wave to propagate through a unit length of lossless transmission line is related by Equation (2-47) [5].

$$t = \sqrt{LC} \quad (2\text{-}47)$$

The velocity of propagation is related to the wavelength in one period T, by:

$$v = \frac{\lambda}{T} = \frac{1}{t} = \frac{1}{\sqrt{LC}} \qquad (2\text{-}48)$$

where: T is the time period of the sine wave.

Using the free space permeability and permittivity the velocity of propagation through free space is:

$$v_o = \sqrt{\mu_o \varepsilon_o} = 2.998 \cdot 10^8 \quad meters/second \qquad (2\text{-}49)$$

From Equations (2-46) and (2-49) we can see that the characteristic impedance and velocity of propagation can be defined for any transmission media based on the inductive and capacitive properties of that media.

2.5.2 Physical Transmission Lines

Free space can be used for the propagation of radio signals across long distances but may not be as useful for the point to point connection and isolation of specific RF signals. For this purpose engineers use a variety of different transmission line media. This text is focused on a few of the most commonly used transmission media in RF and microwave circuit and system design including, microstrip, stripline, coaxial, and waveguide transmission lines. As a general differentiator, transmission media may be divided among pure TEM, quasi TEM, and non TEM propagation modes. TEM refers to the transverse electromagnetic mode of wave propagation. Signals traveling through space propagate in the TEM mode. This simply means that the magnetic field, electric field, and the direction of propagation are all orthogonal to one another. This orthogonal relationship is depicted in Figure 2-5.

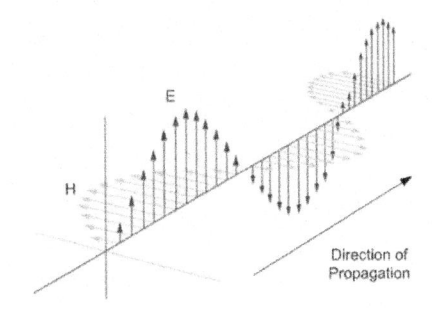

Figure 2-5 TEM wave propagation

The velocity with which a wave travels will almost always be slower in a physical transmission line than it is in free space. Engineers frequently need to trim a transmission line for a specific wavelength so it is essential that they know the velocity of propagation in the transmission line. Knowing the distributed inductive and capacitive properties of the media, Equation (2-48) could be used to determine the propagation velocity. Manufacturers of transmission lines usually specify a velocity factor that is the ratio of actual transmission line velocity v, to the velocity of free space.

$$v_f = \frac{v}{v_o} \qquad (2\text{-}50)$$

Similar to the velocity factor, manufacturers of transmission line media typically express the permittivity of the dielectric material as a relative dielectric constant. The relative dielectric constant, ε_r, is the ratio of the actual material dielectric constant to the dielectric constant of free space.

$$\varepsilon_r = \frac{\varepsilon}{\varepsilon_o} \qquad (2\text{-}51)$$

Depending on the type of transmission media the manufacturer may choose to specify either the velocity factor or the relative dielectric constant. The two quantities are related by Equation (2-52).

$$v_f = \frac{1}{\sqrt{\varepsilon_r}} \qquad (2\text{-}52)$$

There are many types of physical transmission lines ranging from twisted wire pairs to fiber optic cables. There are many good reference sources to examine the characteristics and application of transmission lines [4]. In this chapter we will focus on just a few of these transmission lines that the RF and microwave engineer will frequently encounter. They include:

- Coaxial Transmission Lines
- Microstrip Transmission Lines
- Stripline Transmission Lines
- Waveguides

Note that coaxial, microstrip, and stripline transmission lines require two conductors to transfer power from a given source to its load while waveguides require only a single hollow conductor. A brief introduction to each of the above transmission media is followed by the techniques in which these transmission lines are modeled in the Genesys software.

2.6 Coaxial Transmission Line

Coaxial transmission line, also known as coaxial cable, is one of the most common types of transmission line used in electrical equipment interconnection. A typical coaxial line consists of an inner conductor inside another cylindrical conductor and a dielectric material in between, as shown in Figure 2-6. The outer conductor, or the shield, is usually grounded to minimize RF interference and radiation loss. The coaxial transmission line supports pure TEM propagation as shown in Figure 2-6. The dielectric material is typically a form of Teflon. Teflon has very good physical strength as well as low loss and high temperature operation. For very high power and low loss applications the dielectric material may be air. In this case there must be some dielectric spacer installed at certain intervals to support the center conductor and maintain concentricity. It is also common to introduce nitrogen into the air dielectric to help prevent condensation from forming inside the cable. Condensation would greatly increase the loss and lower the voltage breakdown. Coaxial cables that are used in very high power applications generally require a larger diameter to increase the voltage breakdown.

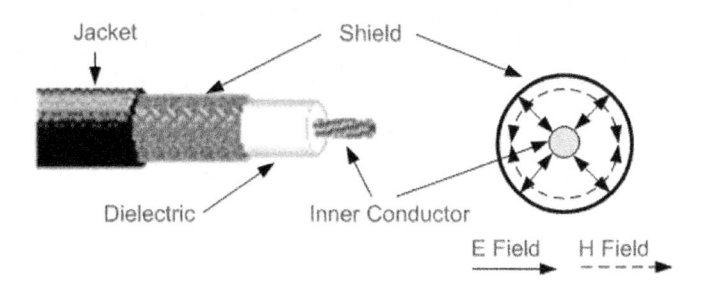

Figure 2-6 Coaxial cable construction and field orientation

Earlier in this chapter we dealt with reflective losses in transmission lines. This is mostly due to the impedance mismatch between the source and load impedance connected to each end of the coaxial cable. There also exists a dissipative loss in a coaxial cable that varies significantly with frequency. When a good impedance match exists between the source and the load it is only the cable loss that is significant. The losses in coaxial cable are primarily due to the conductor loss and the dielectric material loss. Because of its smaller cross sectional area the center conductor dominates the conductor loss contribution. The center conductor's diameter and resistivity determine the conductor loss. Skin effects which cause the RF currents to propagate near the surface of the conductor require that the center conductor have very good conductivity. Sometimes a copper conductor may have a silver plating to improve the surface conductivity. The characteristics of the dielectric material and its thickness determine the losses in the dielectric. Table 2-2 gives a brief listing of some commonly used flexible coaxial cables. Manufacturers typically specify the cable loss in dB per 100 foot lengths at various frequencies. This table gives the manufacturer's (Belden) part number along with an RG designation. The RG designation is an attempt by the U.S. government under MIL-C-17 specification to have a standard designation for the manufacture of coaxial cables [5]. The R means that the cable is intended for RF frequency usage. The G means that the cable is manufactured to general specifications. The number then identifies the unique specifications that the cable is designed to meet. Minor differences in the specification have a letter appended to the numerical designator. From Table 2-2 we can see that the cable loss is specified from 100 MHz to 900 MHz in dB per 100 ft. length. The characteristic impedance and velocity factor are also specified. It is not uncommon to have the cable's capacitance, per unit length, specified as well. Two physical characteristics included in Table 2-2 are the outer dimension (O.D.) and the shield percentage. The shield percentage is only applicable in flexible cables in which the outer conductor is actually a braid of several discrete wires that form a tight mesh. A 100% rating means that there are no air gaps in the wires that comprise the braid and therefore a perfect outer conductor is formed. The term O.D. is referred to as either the outer dimension or overall diameter. The O.D. is the overall diameter of the cable including the outer jacket; not the diameter of the dielectric itself.

Coaxial Cable	Loss in dB/100ft. @ frequency in MHz					Z_o	V_f	Capac. per ft.	O.D. inch	Shield
	100 MHz	200 MHz	400 MHz	700 MHz	900 MHz					
Belden 9913	1.30	1.80	2.70	3.60	4.20	50	.84	24.6 pF	0.405	100%
Belden 9914	1.60	2.40	3.50	5.00	5.70	50	.82	24.8 pF	0.403	100%
Belden 8214 RG-8/U	1.80	2.70	4.20	5.80	6.70	50	.78	26.0 pF	0.405	97%
Belden 8238 RG-11/U	2.00	2.90	4.20	5.80	6.70	75	.66	20.5 pF	0.405	97%
Belden 8267 RG-213/U	1.90	2.70	4.10	6.50	7.60	50	.66	30.8 pF	0.405	97%
Belden 8242 RG-9/U	2.10	3.00	4.80	6.50	7.60	51	.66	30.0 pF	0.420	98%
Belden 9258 RG-8/X	3.70	5.40	8.00	11.10	12.80	50	.80	25.3 pF	0.242	95%
Belden 84142 RG-142	3.90	5.60	8.20	11.00	12.50	50	.695	29.2 pF	0.195	98%
Belden 9273 RG-223/U	4.10	6.00	8.80	12.00	13.80	50	.66	30.8 pF	0.212	95%
Belden 8240 RG-58/U	4.50	6.80	10.00	14.00	16.00	51.5	.66	29.9 pF	0.195	95%
Belden 8259 RG-58A/U	4.90	7.30	11.50	17.00	20.00	50	.66	30.8 pF	0.193	95%
Belden 9259 RG-59/U	3.00	4.50	6.60	8.90	10.1	75	.78	17.3 pF	0.242	95%
Belden 8241 RG-59/U	3.40	4.90	7.00	9.70	11.1	75	.66	20.5 pF	0.242	95%
Belden 8216 RG-174/U	8.40	12.50	19.00	27.00	31.00	50	.66	30.8 pF	0.101	90%
Belden 9228 RG-62A/U	2.70	3.80	5.30	7.30	8.20	93	.84	13.5 pF	0.242	95%

Table 2-2 Sample table of coaxial cable specifications

2.6.1 Coaxial Transmission Lines in Genesys

The TLINE utility in Genesys is very useful for the synthesis of the generic electrical parameters of coaxial cables. Parameters such as the characteristic impedance can be determined from the basic geometrical design. There are many cases in which the engineer needs to design a custom air line coaxial feed-through in a microwave component. The TLINE utility can be used to determine the exact physical dimensions for the inner and outer conductors of a coaxial transmission line given the characteristic impedance. Figure 2-7 shows the TLINE utility program setup for the calculation of the outer conductor diameter of an air-line coaxial feed-through. The inner conductor diameter is known to be 0.050 inches and an impedance of 50 Ω is desired. A dielectric constant of 1.0 is entered for the dielectric constant. The inner conductor diameter of 50 mils is entered in the 'Parameter' section. An initial guess can be entered for the outer conductor (b) dimension. The characteristic impedance is then displayed above the cable cross section view. Iteratively select values of the (b) dimension until an acceptable value of characteristic impedance close to 50 Ω is achieved. The characteristic impedance of a coaxial line is determined by the diameter of the inner conductor (d) and the outer conductor (b) along with the relative dielectric constant, as given by Equation (2-53).

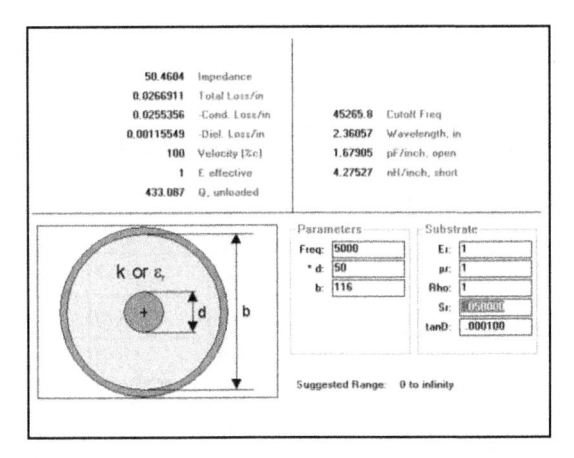

$$Z_o = \frac{60}{\sqrt{\varepsilon_r}} \cdot \ln\left(\frac{b}{d}\right) \quad (2\text{-}53)$$

Figure 2-7 TLINE utility for calculation of coaxial cable geometry Z_0

There are a variety of ways to model transmission lines in Genesys. Figure 2-8 shows the coaxial cable models that are available in the library. There are predetermined models for some of the more popular RG type coaxial cables. Alternatively you can use the generic, Coax Line, model and assign any arbitrary geometry to the model.

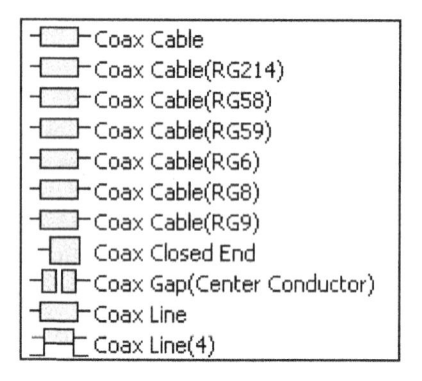

```
Coax Cable
Coax Cable(RG214)
Coax Cable(RG58)
Coax Cable(RG59)
Coax Cable(RG6)
Coax Cable(RG8)
Coax Cable(RG9)
Coax Closed End
Coax Gap(Center Conductor)
Coax Line
Coax Line(4)
```

Figure 2-8 Coaxial transmission line models available in Genesys

2.6.2 Using the RG8 Coaxial Cable Model in Genesys

To use the RG8 model in Genesys, create a simple schematic using the RG8 coaxial cable model as shown in Figure 2-9. Set the length to 1200 inches (100 ft.) to compare the cable loss in 100 ft. against the published data of Table 2-2. The simulated insertion loss in the Table of Figure 2-9 is in very close agreement with the manufacturer's data for the RG-8/U coaxial cable.

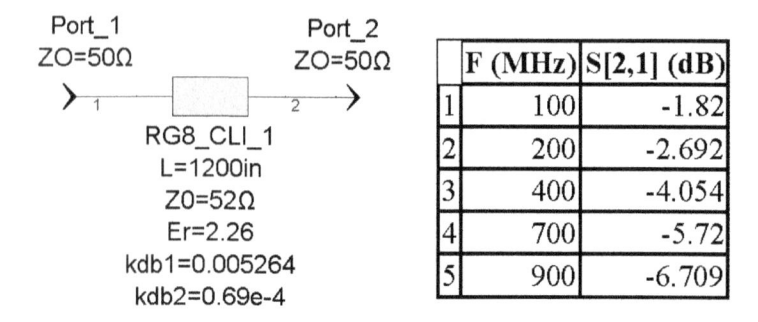

Port_1　　　　　　Port_2
ZO=50Ω　　　　　ZO=50Ω

RG8_CLI_1
L=1200in
ZO=52Ω
Er=2.26
kdb1=0.005264
kdb2=0.69e-4

	F (MHz)	S[2,1] (dB)
1	100	-1.82
2	200	-2.692
3	400	-4.054
4	700	-5.72
5	900	-6.709

Figure 2-9 RG8 coaxial cable insertion loss

Next terminate the coaxial line with an impedance of 75 Ω and examine the input return loss S[1,1] and VSWR at frequencies 100, 200, 400, 700, and 900 MHz as shown in Figure 2-10. We see that the S[1,1] angle changes significantly with frequency suggesting that we have standing waves along the transmission line.

Index	F (MHz)	mag(S[1,1])	ang(S[1,1]) (°)	db(S[1,1])	VSWR
1	100	0.102	150.544	-19.829	1.227
2	200	0.111	-41.748	-19.087	1.25
3	400	0.071	-83.147	-22.989	1.153
4	700	0.068	4.703	-23.35	1.146
5	900	0.055	-28.763	-25.249	1.116

Figure 2-10 RG8 coaxial cable terminated with 75 Ω load impedance

Notice that the VSWR is rather good because it is close to one. Remove the cable from the schematic to examine the mismatch between the source impedance of 50 Ω and the 75 Ω load impedance. This can be done by simply 'shorting' the model in its properties menu. Here we see that the actual VSWR is 1.5. We must take into account that the dissipative loss of the 100 ft. cable actually makes the VSWR look better at the source. This is an important consideration to remember when remotely looking at the VSWR of a radio antenna through a length of coaxial cable. It is important to realize that the actual power transferred to the load from the source is dependent on both reflective (mismatch loss) and dissipative (cable loss).

Index	F (MHz)	mag(S[1,1])	ang(S[1,1]) (°)	db(S[1,1])	VSWR
1	100	0.2	0	-13.979	1.5
2	200	0.2	0	-13.979	1.5
3	400	0.2	0	-13.979	1.5
4	700	0.2	0	-13.979	1.5
5	900	0.2	0	-13.979	1.5

Figure 2-11 Actual mismatch between 50 Ω source and 75 Ω load impedance

2.7 Microstrip Transmission Lines

Microstrip is a planar transmission line media in which the transmission line is etched onto the top side of a printed circuit board. The bottom side of the printed circuit board is completely metalized and grounded. The printed circuit board has a low loss dielectric material that is suitable for microwave transmission. Figure 2-12 shows a sampling of microstrip lines that represent specific circuits. The top side of the dielectric substrate is shown with the copper conductor, microstrip transmission lines.

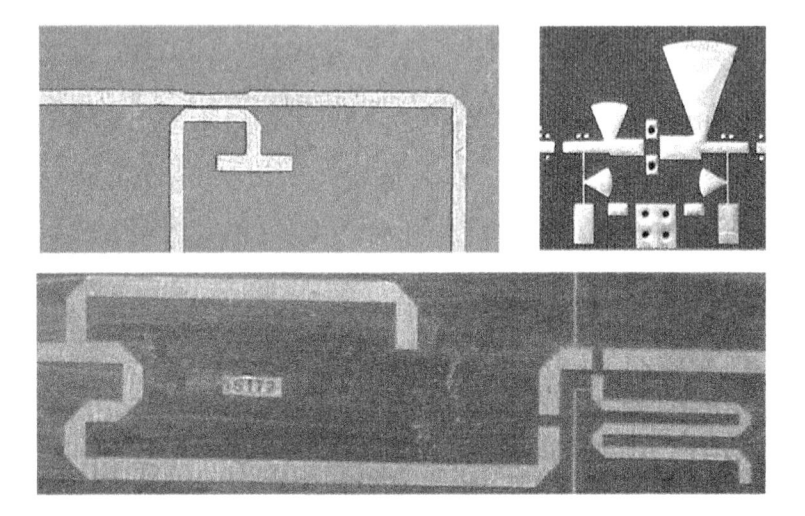

Figure 2-12 Examples of microstrip transmission lines on various dielectrics

Unlike the coaxial cable the center conductor is not shielded. This means that a portion of the electric and magnetic field is in the air space above the microstrip line as shown in Figure 2-13. Thus the propagation in microstrip is not purely TEM but rather quasi TEM. This also leads to the fact that the dielectric constant beneath the microstrip line is slightly less than the relative dielectric constant of the material. This is known as the effective dielectric constant, ε_{eff}, and is a function of the width of the microstrip line, W and the height of the substrate, h as shown in Figure 2-13. The thickness of the microstrip conductor, t, has a minor effect on ε_{eff} and is omitted from the computation. The effective dielectric constant is calculated in the empirical Equations (2-54) and (2-55) [6].

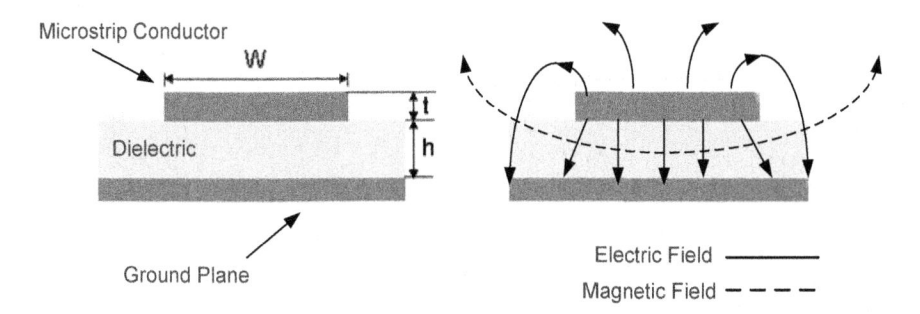

Figure 2-13 Cross sections of microstrip transmission line and electromagnetic field lines

For $\left(\dfrac{W}{h}\right) < 1$

$$\varepsilon_{eff} = \dfrac{\varepsilon_r + 1}{2} + \dfrac{\varepsilon_r - 1}{2} \cdot \left[\left(1 + 12\left(\dfrac{h}{W}\right)\right)^{-0.5} + 0.04\left(1 - \left(\dfrac{W}{h}\right)\right)^2\right] \qquad (2\text{-}54)$$

For $\left(\dfrac{W}{h}\right) \geq 1$

$$\varepsilon_{eff} = \dfrac{\varepsilon_r + 1}{2} + \dfrac{\varepsilon_r - 1}{2} \cdot \left(1 + 12\left(\dfrac{h}{W}\right)\right)^{-0.5} \qquad (2\text{-}55)$$

Similarly the characteristic impedance also has two solutions [6].

For $\left(\dfrac{W}{h}\right) < 1$

$$Z_o = \dfrac{60}{\sqrt{\varepsilon_{eff}}} \ln\left(8 \cdot \dfrac{h}{W} + 0.25 \cdot \dfrac{W}{h}\right) \qquad (2\text{-}56)$$

For $\left(\dfrac{W}{h}\right) \geq 1$

$$Z_o = \dfrac{120\pi}{\sqrt{\varepsilon_{eff}} \cdot \left[\dfrac{W}{h} + 1.393 + 0.6667 \ln\left(\dfrac{W}{h} + 1.444\right)\right]} \qquad (2\text{-}57)$$

Equations (2-56) and (2-57) do not consider the conductor thickness of the line. The effective width, W_e, of a microstrip line is the equivalent width of the line with the conductor thickness taken into account. The effective line width is defined by Equations (2-58) and (2-59).

For $\left(\dfrac{W}{h}\right) \geq \dfrac{1}{2\pi}$

$$W_e = W + \frac{t}{\pi}\left(1 + \ln\frac{2h}{t}\right) \qquad (2\text{-}58)$$

For $\left(\dfrac{W}{h}\right) < \dfrac{1}{2\pi}$

$$W_e = W + \frac{t}{\pi}\left(1 + \ln\left(\frac{4\pi \cdot W}{t}\right)\right) \qquad (2\text{-}59)$$

2.7.1 Microstrip Transmission Lines in Genesys

The TLINE utility in Genesys is also very useful for the calculation of microstrip line width for a given impedance and substrate. The physical transmission line width can be determined based on the dielectric constant and substrate thickness for the desired impedance. Note that the utility calculates and lists the effective dielectric constant for reference. Experimenting with the TLINE utility reveals that there are two additional parameters that can affect the line width calculation for a given line impedance. These are the conductor thickness, t, and the cover or box height. When using a thicker conductor for higher current handling it is recommended to model the actual conductor thickness, t, as it will have a slight impact on the line width. If the cover height becomes too close to the microstrip line it will add enough capacitance to lower the impedance of the line. The TLINE utility suggests a minimum cover height to maintain so that the cover does not add significant capacitance to the microstrip transmission line. The TLINE utility for calculation of microstrip transmission line parameters is shown in Figure 2-14.

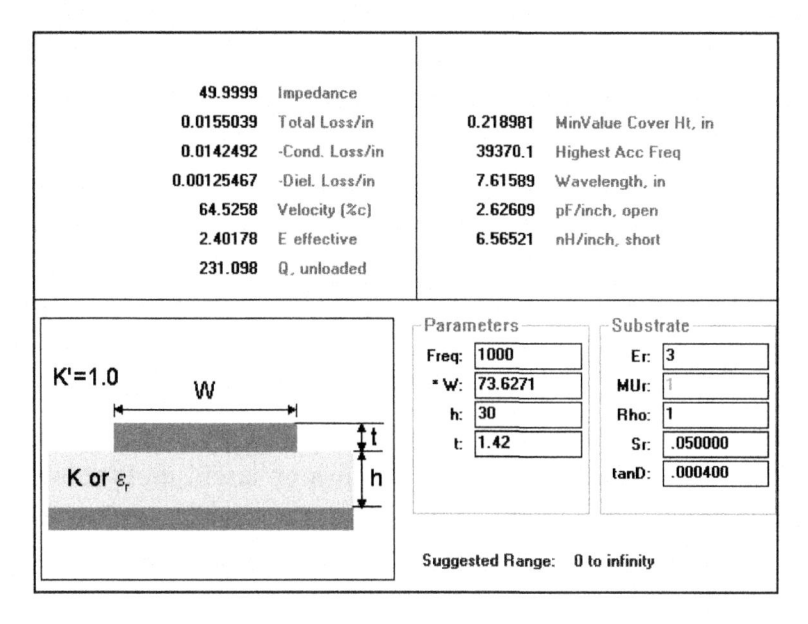

Figure 2-14 TLINE utility for calculation of microstrip line geometry Z_0

Before adding any microstrip transmission lines to a Genesys workspace schematic, a dielectric substrate must be defined. A substrate may be added to the schematic through the 'Add Substrate' menu. The substrate properties window will be displayed as shown in Figure 2-15.

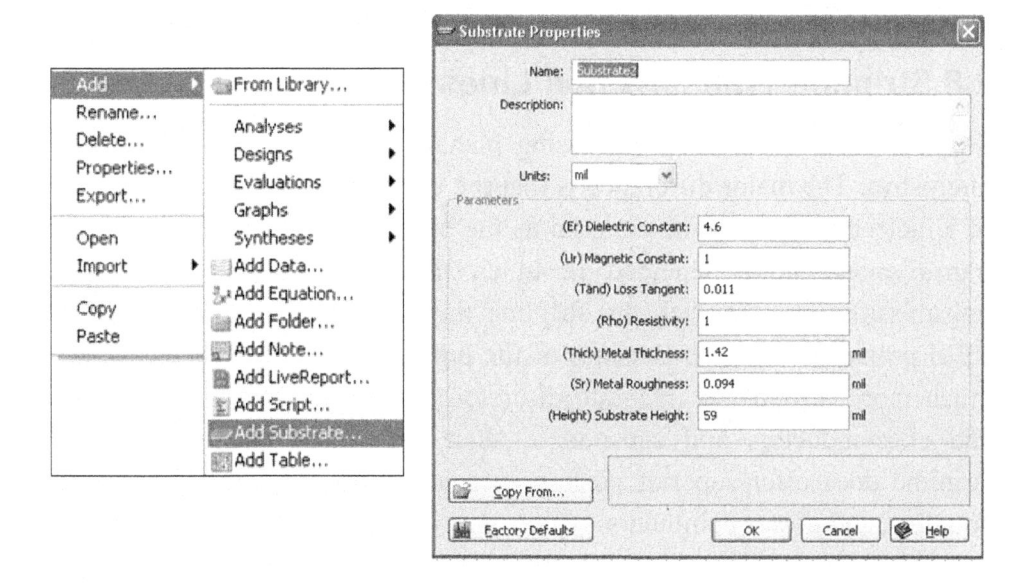

Figure 2-15 Adding a dielectric substrate to the Genesys schematic

The substrate's dielectric constant and height are the most important parameters to enter into the parameters window. The default settings can be accepted for the remaining parameters for most circuits. The dielectric loss tangent and conductor resistivity will dominate the microstrip losses of resonators, filters or other critical low loss circuits. A virtual catalog of microwave dielectrics is built into the Genesys software. Select the 'Copy From' button to select a substrate material from the Genesys library. A comprehensive collection of the major manufacturers' soft substrate material is contained in the library. Substrates can be divided into hard substrate and soft substrate types. Hard substrates are very stiff and brittle and must be cut with diamond, tipped saws or lasers. Soft substrates are very pliable and can be cut with a razor knife or scissors. All of the circuit designs and components discussed in this book are based on soft substrates. Hard substrates are normally used in semiconductor design i.e., transistors, and MMICs. Alumina is a hard substrate that is often used in microwave integrated circuit (MIC) designs. Because the top half of the microstrip is exposed it is very easy to attach discrete components. This has made microstrip the most popular form of transmission line for designing and building microwave components. There are many variants of the basic microstrip transmission line including suspended line, inverted line, and coplanar waveguide [8].

2.8 Stripline Transmission Lines

Stripline is a transmission line using planar dielectric material similar to microstrip. The major difference is that the top half of the line also consists of a dielectric of the same material as the bottom conductor. Therefore the transmission line is shielded in much the same way as the coaxial transmission line. As such the stripline transmission line supports a pure TEM propagation mode. Because of the presence of the top dielectric it is much more difficult to integrate discrete components such as transistors, and chip inductors and capacitors. As a pure TEM transmission line, stripline does offer superior performance in distributed filters, directional couplers, and power combiners.

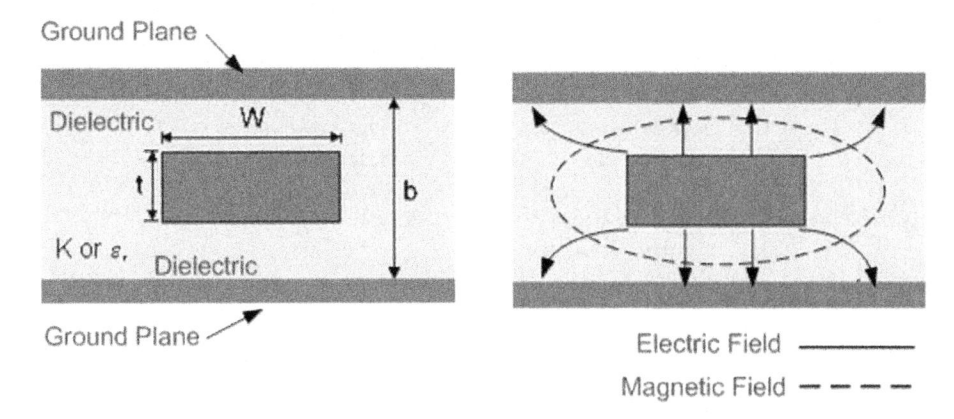

Figure 2-16 Cross section view of stripline transmission line and field lines

Because the propagation mode is pure TEM, there is no effective dielectric constant, just the relative dielectric constant of the substrate material. Soft substrates are almost entirely used for stripline transmission circuits. One popular variation is when an air dielectric is used. This is referred to as suspended substrate stripline (SSS) and is characterized by very low loss. Suspended substrate stripline is often used in filter and multiplexer circuit designs. The characteristic impedance of a stripline transmission line is given by the empirical Equation (2-59) [7].

$$Z_o = \frac{94.2}{\sqrt{\varepsilon_r}} \ln \left[\frac{1 + \frac{W}{b}}{\frac{W}{b} + \frac{t}{b}} \right] \qquad (2\text{-}60)$$

In practice a 0.060 stripline media is realized by clamping or fusing two 0.030 dielectric halves together. The circuit pattern is etched on one of the 0.030 dielectrics similar to a microstrip circuit pattern. The top half dielectric would then have all metal removed on the side which meets the bottom half circuit patterns. A special bonding film can be used to effectively glue the halves together or the dielectrics could be clamped together using sufficient pressure. The TLINE utility synthesis of a 50 Ω line in 0.060 stripline is shown in Figure 2-17.

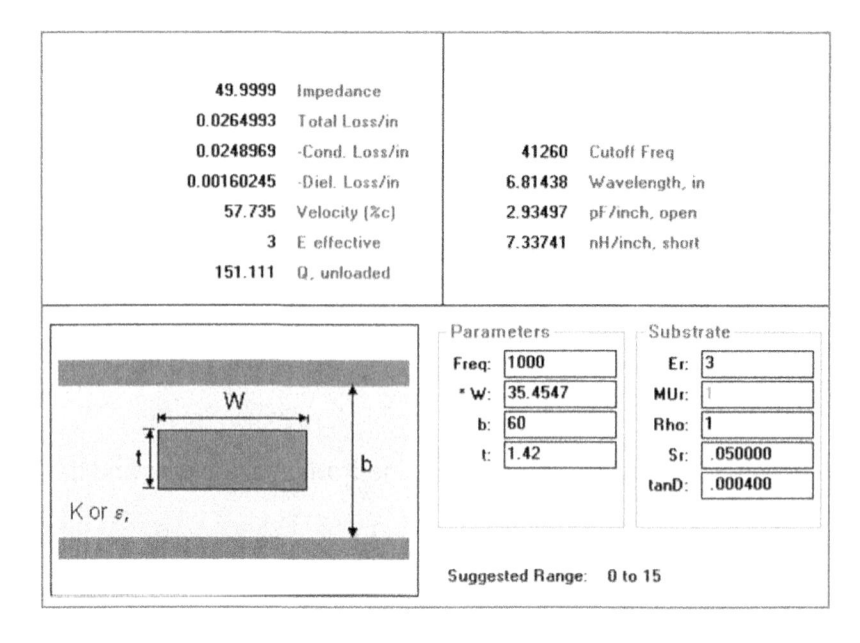

Figure 2-17 TLINE utility for calculation of stripline geometry and Z_o

2.9 Waveguide Transmission Lines

A waveguide is a special form of transmission line that is used primarily at microwave frequencies above 2 GHz. The most important difference between waveguide and other forms of transmission line is the low loss and capability to transmit very high power. Waveguides do not have a separate conductor, ground plane, or shield as the previous transmission line structures. Rather waveguide is a hollow tube in which the wave propagates through. Even though the propagation is in the air space, enclosing an RF wave in a metallic boundary causes the wave to propagate quite differently than it would in free space. Waveguide tubes can be circular or rectangular but the rectangular tube is much more popular due to its ease of manufacture. Figure 2-18 shows a typical rectangular waveguide section with flanges on each end. The flanges allow a connection to be made to a mating piece of waveguide or waveguide component. The broad dimension of the waveguide is labeled as the 'a' dimension while the narrow side is labeled the 'b' dimension. The TEM propagation mode does not exist in waveguide. The propagation must be either transverse electric (TE) or transverse magnetic (TM). The propagation modes are quite complex with many subcategories of TE and TM propagation in existence.

Figure 2-18 Rectangular waveguide section showing inside dimensions

In TE mode the electric field is transverse to the direction of propagation. This means that there is no component of the electric field in the direction of propagation. In the TM mode there is no magnetic field component in the direction of propagation. Depending on how the RF energy is launched into the waveguide there are many variations of both TE and TM propagation modes. These sub-modes are designated with the subscripts m and n ($TE_{m,n}$). The sub-mode notation describes the field patterns that exist in the waveguide. The dominant sub-mode is the $TE_{1,0}$ mode. This is the lowest frequency that the waveguide will support. Figure 2-19 shows the field pattern for the $TE_{1,0}$ mode in a rectangular waveguide. The 1,0 index signifies that there is one field variation along the broad dimension and no field variations along the narrow dimension. The remaining discussion of rectangular waveguide properties is based on the $TE_{1,0}$ mode.

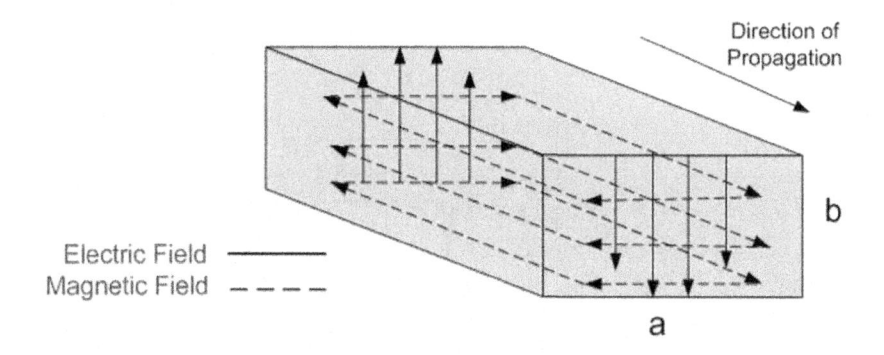

Figure 2-19 $TE_{1,0}$ dominant mode field pattern in rectangular waveguide

The wave pattern in a waveguide leads to an interesting condition where at or below a certain frequency the wave bounces from side-to-side or top-to-bottom in the waveguide tube and no longer travels in the direction of propagation. This frequency is known as the cutoff frequency. Below the cutoff frequency the waveguide transmits very little energy. As such the waveguide has a natural high pass filter characteristic. The cutoff frequency of the $TE_{1,0}$ in rectangular waveguide is given by Equation (2-61),

$$f_{c_{1,0}} = \frac{c}{2a} = \frac{3 \cdot 10^8}{2a} \qquad (2\text{-}61)$$

where: c is the speed of light and a is the wider dimension of the waveguide.

The wavelength in a rectangular waveguide it then defined as.

$$\lambda_g = \frac{\lambda}{\sqrt{\left(1-\left(\frac{\lambda}{(2a)^2}\right)\right)}} \qquad (2\text{-}62)$$

The characteristic impedance of waveguide is based on the propagation mode. The characteristic impedance must be defined separately for the transverse electric and magnetic fields as given by equations (2-63) and (2-64).

$$Z_{o\,(TE_{m,n})} = \frac{\eta}{\sqrt{1-\left(\frac{f_c}{f}\right)^2}} \qquad (2\text{-}63)$$

$$Z_{o\,(TM_{m,n})} = \eta \sqrt{1-\left(\frac{f_c}{f}\right)^2} \qquad (2\text{-}64)$$

The term, η, is the intrinsic impedance of the medium. In the case of air $\eta = 377\ \Omega$. Unlike the other transmission lines the concept of the characteristic impedance is rarely used in practice. Waveguide is typically referred and characterized by its cross section dimensions as shown in Fig. 2-18. The cross sectional dimension of commonly used rectangular waveguide is given a WR designator. Table 2-3 shows a listing of some of the more popular waveguides by WR designator.

Frequency Band, GHz	U.S. (EIA) Designator	British WG Designator	Cut Off Freq. in GHz $TE_{1,0}$	a dimension inches	b dimension inches
1.12 - 1.70	WR 650	WG 6	0.908	6.500	3.250
1.45 - 2.20	WR 510	WG 7	1.158	5.100	2.550
1.70 - 2.60	WR 430	WG 8	1.375	4.300	2.150
2.20 - 3.30	WR 340	WG 9A	1.737	3.400	1.700
2.60 - 3.95	WR 284	WG 10	2.080	2.840	1.340
3.30 - 4.90	WR 229	WG 11A	2.579	2.290	1.145
3.95 - 5.85	WR 187	WG 12	3.155	1.872	0.872
4.90 - 7.05	WR 159	WG 13	3.714	1.590	0.795
5.85 - 8.20	WR 137	WG 14	4.285	1.372	0.622
7.05 - 10.00	WR 112	WG 15	5.260	1.122	0.497
8.2 - 12.4	WR 90	WG 16	6.560	0.900	0.400
9.84 - 15.0	WR 75	WG 17	7.873	0.750	0.375
11.9 - 18.0	WR 62	WG 18	9.490	0.622	0.311
14.5 - 22.0	WR 51	WG 19	11.578	0.510	0.255
17.6 - 26.7	WR 42	WG 20	14.080	0.420	0.170
21.7 - 33.0	WR 34	WG 21	17.368	0.340	0.170
26.4 - 40.0	WR 28	WG 22	21.100	0.280	0.140
32.9 - 50.1	WR 22	WG 23	26.350	0.224	0.112
39.2 - 59.6	WR 19	WG 24	31.410	0.188	0.094
49.8 - 75.8	WR 15	WG 25	39.900	0.148	0.074
60.5 - 91.9	WR 12	WG 26	48.400	0.122	0.061
73.8 - 112.0	WR 10	WG 27	59.050	0.100	0.050

Table 2-3 Standard rectangular waveguide characteristics

To interface waveguide with coaxial, microstrip, or stripline transmission lines a special transformer, known as an adapter, must be used. Waveguide to coax adapters come in many forms. Some adapters couple energy to the E field while others couple to the H field. The adapter shown in Fig. 2-20 is an E field waveguide to coax transition. The center conductor of the coaxial connector extends into the waveguide to excite the wave propagation in the waveguide. The center conductor is approximately $\lambda_g/4$ from the back wall.

Figure 2-20 Waveguide to coax transition adapter

2.9.1 Waveguide Transmission Lines in Genesys

Because of the complex EM fields that can propagate in waveguide, modern computer aided design techniques are best handled by three dimensional EM solvers. Genesys is limited in its ability to model waveguide propagation and waveguide components but does have two models that are useful to the engineer. The first model is a straight section of waveguide in which the $TE_{1,0}$ mode is utilized. The second model is a waveguide to TEM transition which is similar to the adapter of Figure 2-20. The input and output ports used in Genesys can be thought of as coaxial ports supporting TEM propagation. Therefore we cannot attach a section of waveguide directly to a port because an impedance mismatch will exist. Consider the model of a one inch length of WR112 waveguide as used in a X Band satellite transmit system. Make the length of the waveguide tunable. From Table 2-3 we can get the width (a) and height (b) dimensions to enter into the waveguide model. Place a waveguide to TEM adapter on each side of the waveguide transmission line and sweep the insertion loss (S21) from 4 GHz to 8 GHz as shown in Figure 2-22. Note that the use of the waveguide models does require a substrate definition. In this case however the waveguide models only use the dielectric constant Er, and Rho, the resistivity of the metal walls. Typically enter values of one for both the air dielectric and the resistivity normalized to copper as shown in Figure 2-21.

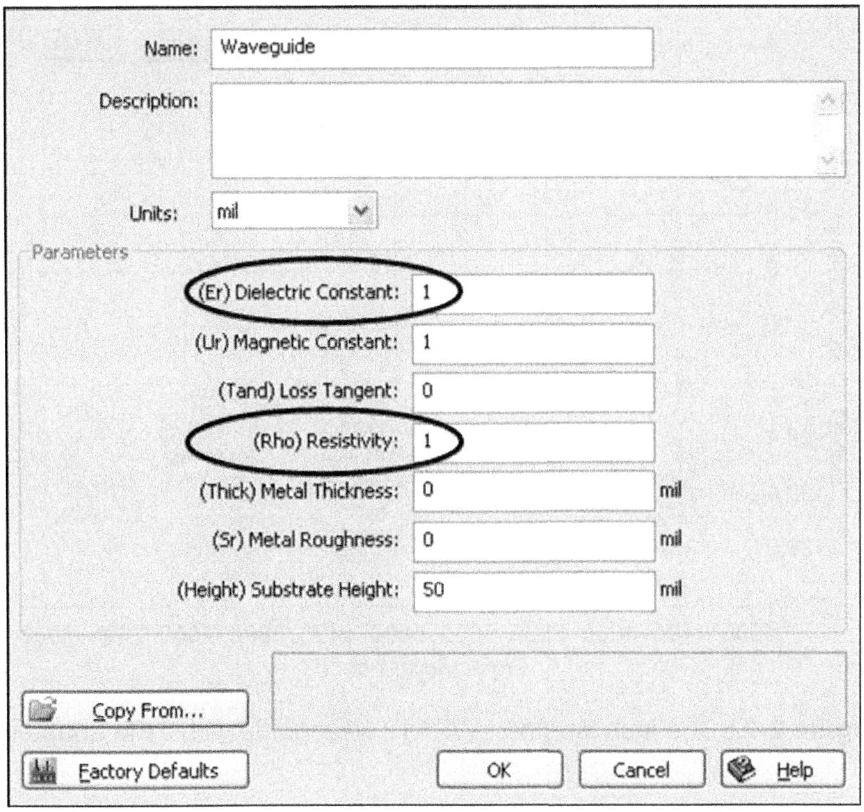

Figure 2-21 Substrate properties for Genesys waveguide models

Figure 2-22 shows the schematic and response. The insertion loss of the waveguide in its pass band at 8 GHz is extremely low. This is one of the advantages of using waveguide transmission lines as they are practically the lowest loss microwave transmission line available. Also note that the insertion loss increases as we move below the cutoff frequency. A marker is placed at the cutoff frequency of 5.26 GHz. Increase the length of the WR112 waveguide to 3 inches. Note the dramatic increase in the rejection below the cutoff frequency. The insertion loss in band is still quite low. We can see that using the waveguide below the cutoff frequency is an effective method of achieving a very good microwave high pass filter. The three inch length of WR112 waveguide with TEM adapters is shown in Figure 2-23.

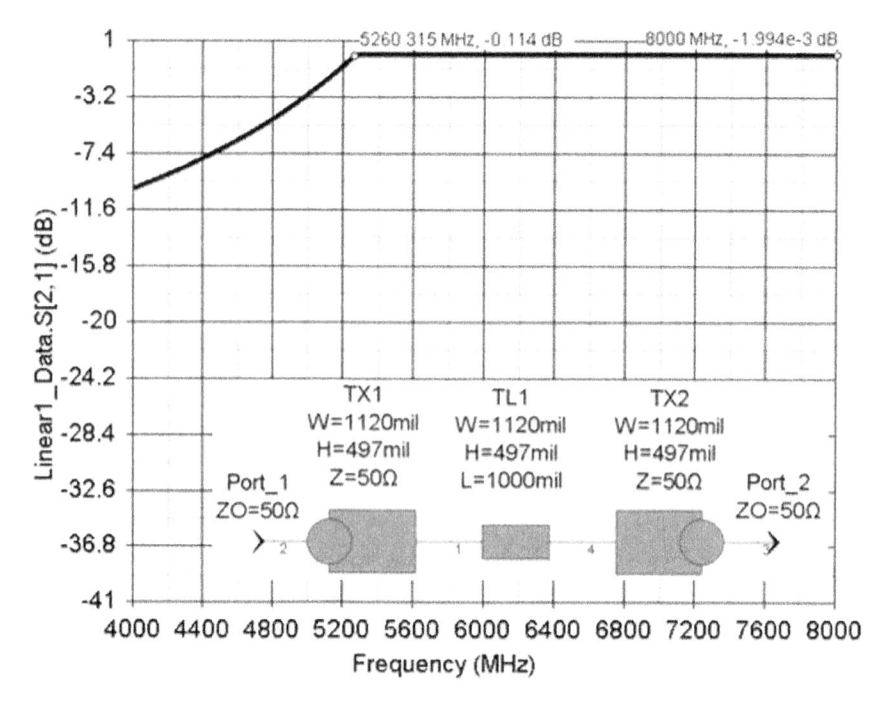

Figure 2-22 One inch length of WR112 waveguide with TEM adapters

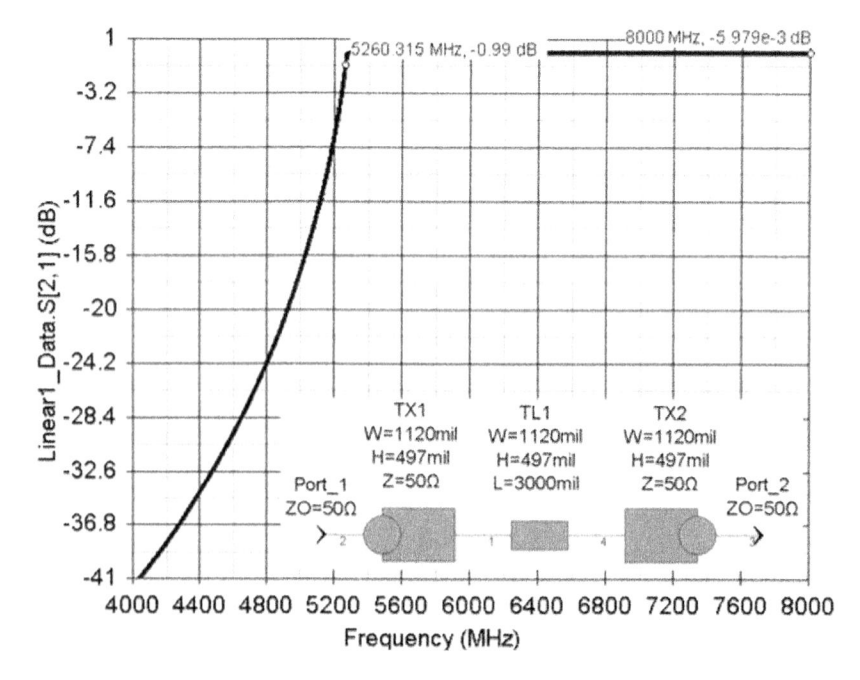

Figure 2-23 Three inch length of WR112 waveguide with TEM adapters

2.10 Group Delay in Transmission Lines

A frequently encountered concept related to the transmission line velocity factor is group delay. Group delay is a measure of the time that it takes a signal to traverse a transmission line, or its transit time. It is a strong function of the length of the line, and usually a weak function of frequency. It is expressed in units of time, picoseconds for short distances or nanoseconds for longer distances. Remember that in free space all electromagnetic signals travel at the speed of light, c, which is approximately 300,000 kilometers per second. Therefore, in free space, electromagnetic radiation travels one foot in one nanosecond, unless there is something to slow it down such as a dielectric. Mathematically the group delay is the derivative of phase versus frequency. In communication systems, the ripple in the group delay creates a form of distortion.

2.10.1 Comparing Group Delay of Various Transmission lines

Example 2.10-1: To compare group delays of various transmission lines, create a Genesys workspace with four schematics and corresponding linear analysis. In each schematic model place a 20 inch length of the previously discussed transmission lines. Use the RG8 cable for the coaxial transmission line. For the microstrip and stripline transmission lines use the Rogers RO3003 dielectric material with 1 oz. copper and 30 mil substrate thicknesses.

Solution: The TLINE utility can be used to synthesize the line widths for 50 Ω lines. Use WR430 for the waveguide transmission line. Refer to Table 2-3 for the waveguide dimensions. The schematics representing the four transmission lines are shown in Figure 2-24. Sweep the frequency from 2 GHz to 5 GHz in each of the four transmission lines. Genesys has a built-in function for the display of the group delay. The group delay (gd) function is shown in the inset of Figure 2-25.

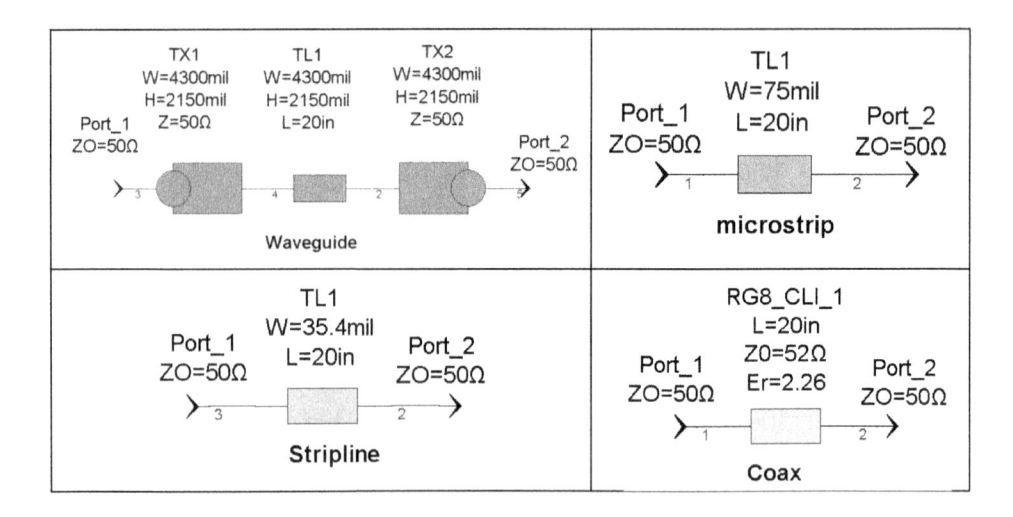

Figure 2-24 20 inch length of different transmission lines

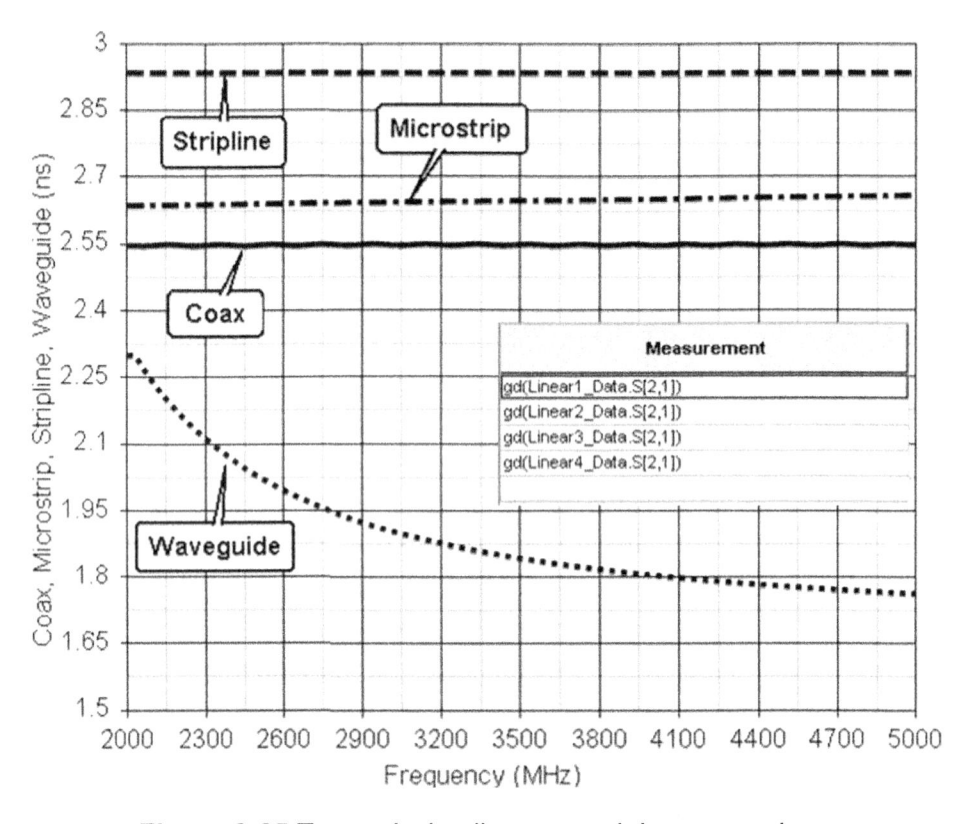

Figure 2-25 Transmission line group delay comparison

2.11 Transmission Line Components

There are many useful components that can be realized using transmission lines. These include power splitters, directional couplers, voltage and current insertion networks, as well as various filter networks. These networks can be realized with any of the physical transmission line structures. However, because of its popularity, we will explore many of these components in microstrip transmission line. These components are referred to as distributed components. It can be shown that the series inductance and shunt capacitance can be realized with distributed microstrip transmission lines. We will begin this section with an examination of the open and short-circuited microstrip transmission lines.

2.11.1 Short-Circuited Transmission Line

Equation (2-38) demonstrated that the input impedance of a lossless short-circuited transmission line is a pure imaginary function; therefore, the input reactance is given by the following equation.

$$X_{in} = Z_O \tan\theta \tag{2-65}$$

where $\theta = \beta d$ is the electrical length of the transmission line in degrees

From Equation (2-65) we can see that this reactance can change from inductive to capacitive depending on the length of the transmission line.

Example 2.11-1: In Genesys plot the reactance of a loss less short-circuited transmission line as a function of the electrical length of the line.

Solution: To plot the reactance of the short-circuited transmission line in Genesys, create a schematic with a grounded transmission line. Make the length of the transmission line a variable with any starting value in degrees. Setup a Linear Analysis with a single frequency at 1500 MHz. Then use a Parameter sweep to vary the electrical length of the transmission line from 0 to 360 degrees. Setup a graph to plot the reactance of the shorted transmission line vs. the electrical length from the Parameter sweep data set as shown in Fig. 2-26.

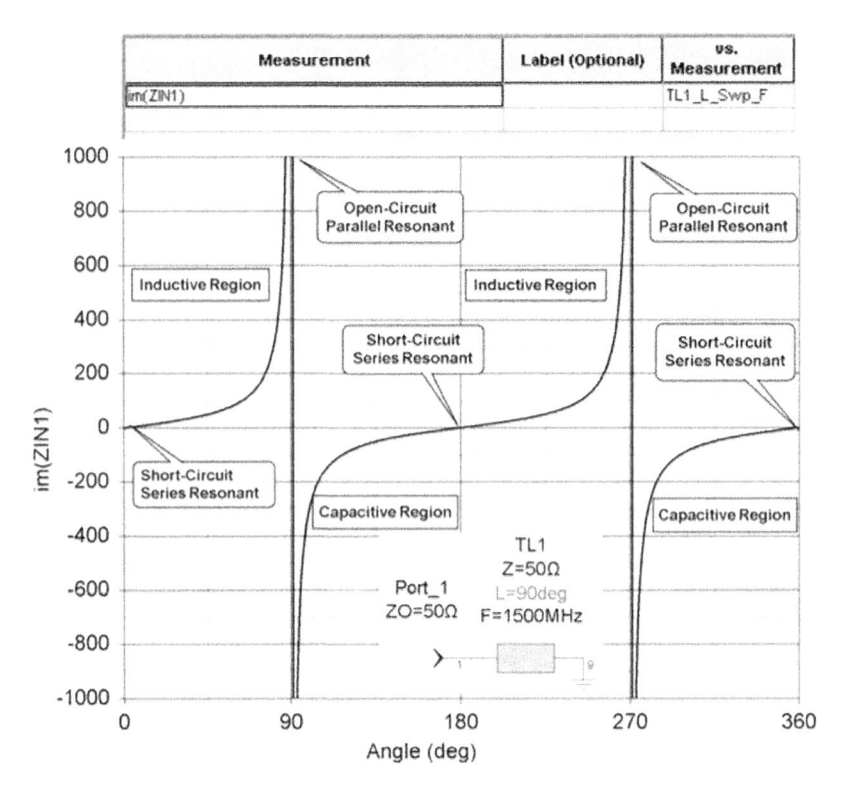

Figure 2-26 Short-circuited line reactance versus electrical length

Normally the independent variable in most linear simulations is frequency. Genesys however does allow for the easy selection of another variable to be used as the independent variable. This is shown in the Measurement Properties display of Figure 2-26. Under the 'vs. Measurement' column, the transmission line electrical length variable from the linear analysis dataset is entered. This defines the electrical length as the independent variable to result in the plot of reactance vs. electrical line length. From the reactance plot of Figure 2-26 we can see that at one quarter wavelength, $\lambda_g/4$, the short-circuited line looks like an open circuited line. From 0 to 90 degrees the line looks like an inductor. From 90° to 180° the circuit looks like a capacitor. The pattern then repeats from 180° to 360°.

2.11.2 Modeling Short-Circuited Microstrip Lines

The short-circuited microstrip line can be modeled as a microstrip transmission line connected to a grounded via hole. A via hole is made by drilling a hole in the dielectric and metalizing the inside of the hole to form a conductive path to the ground side of the dielectric. To create a 90 degree line in the microstrip substrate we must know the effective dielectric constant so that the wavelength in the dielectric, λ_g, can be calculated. The relationships between the line length and electrical degrees are as follows.

$$\theta = \frac{2\pi \ell}{\lambda_g} \tag{2-66}$$

$$\ell = \frac{\theta \lambda_g}{360\sqrt{\varepsilon_{eff}}} \tag{2-67}$$

Example 2.11-2: Calculate the physical length of a microstrip transmission line for a given electrical length of the line.

Solution: In Genesys simply place an ideal transmission line element on a schematic and enter the desired impedance, electrical length, and frequency. Select the Advanced TLINE utility from the Schematic menu. The microstrip line will be created as shown in Figure 2-27.

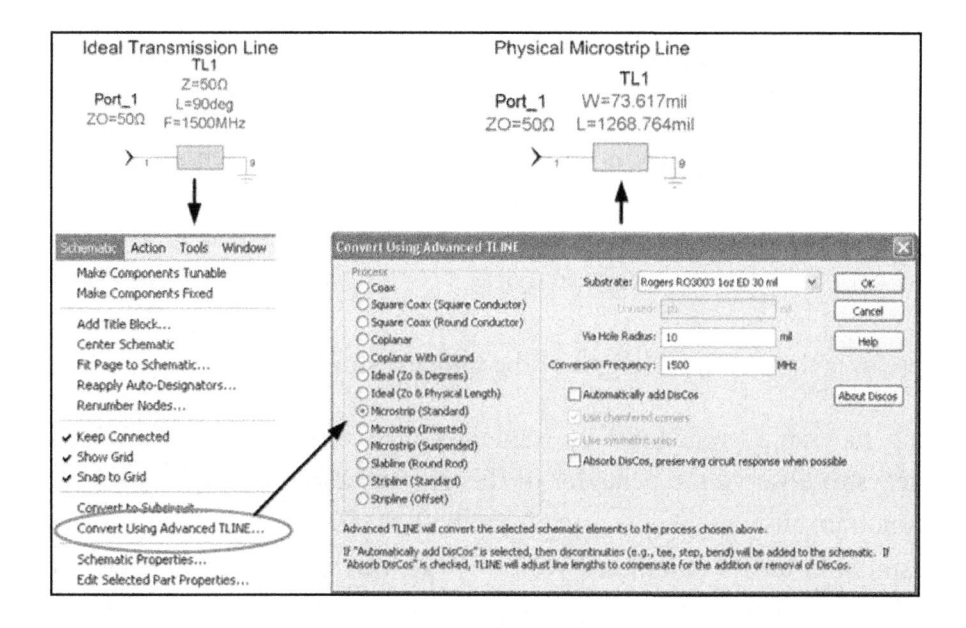

Figure 2-27 Advanced TLINE for calculation of microstrip line length

Figure 2-28 shows the correct method of modeling a microstrip short-circuited transmission line with a via hole. As the table inset of Figure 2-28 shows, the impedance of a shorted quarter-wave section is close to an open circuit. This type of line section could be used as a parallel resonant circuit.

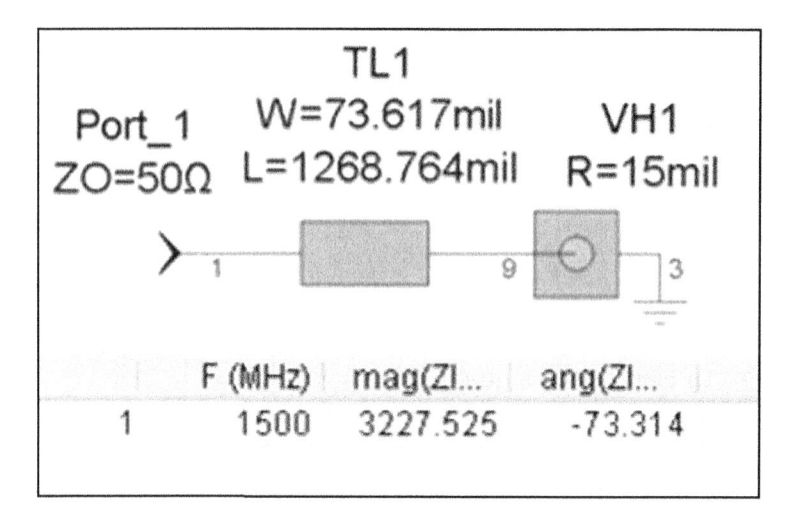

```
                      TL1
Port_1        W=73.617mil              VH1
ZO=50Ω        L=1268.764mil            R=15mil

      F (MHz)      mag(ZI...      ang(ZI...
  1    1500         3227.525       -73.314
```

Figure 2-28 Quarter wave short-circuited line schematic and impedance

2.11.3 Open-Circuited Transmission Line

Equation (2-39) showed that the input impedance of a lossless open circuited transmission line is a pure imaginary function; therefore, the input reactance is given by the following equation.

$$X_{in} = Z_O \cot\theta \qquad (2\text{-}68)$$

where: θ is the electrical length of the transmission line in degrees.

Using the same procedures of section 2.11.1, use a parameter sweep in Genesys to observe the behavior of this reactance as the length of the open circuit transmission line is varied from 0 to 360 degrees. Note that the transmission line is terminated with a 10^6 Ω load to emulate an open circuit termination on the transmission line in Figure 2-29. Comparing the open circuit reactance to the short-circuited line reactance we can see that a 90°, $\lambda_g/4$, offset is present.

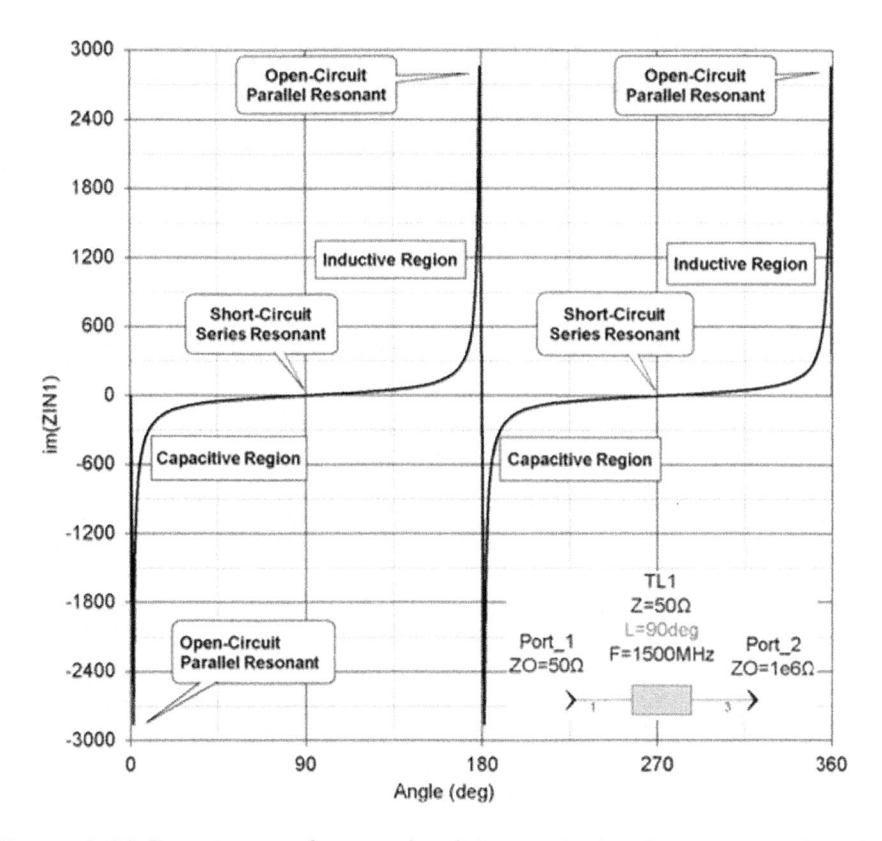

Figure 2-29 Reactance of open circuit transmission line versus electrical length

2.11.4 Modeling Open-Circuited Microstrip Lines

Care must be used when modeling the open circuit microstrip line due to the radiation effects from the end of the transmission line. The E fields that exist in the air space of the microstrip line add capacitance to the microstrip transmission line. On an open circuit microstrip line this fringing capacitance is referred to as an end effect. The end effect makes the line electrically longer than the physical length. This requires that the physical line length be shortened to achieve the desired reactance. The 'microstrip end' model accurately accounts for this end effect capacitance for the specified substrate. The end effect fringing field is visualized in Figure 2-30.

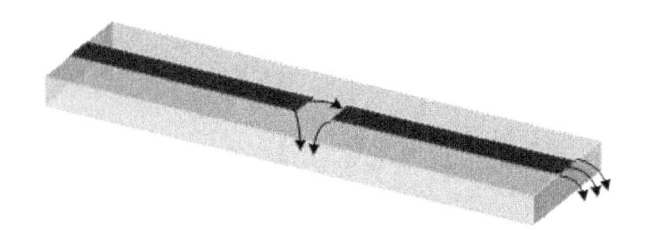

Figure 2-30 Fringing E fields in microstrip open circuit and gaps

Figure 2-31 shows the correct method of modeling a microstrip open circuit transmission line using an end effect. As this Figure shows, the impedance of a quarter-wave section of open circuit line is quite low, close to a short circuit. This type of line section could be used as a series resonant circuit.

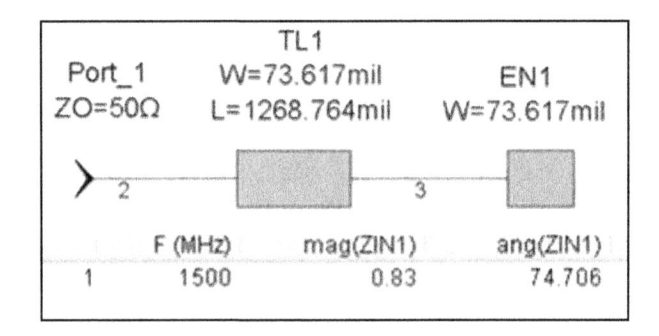

Figure 2-31 Quarter wave open circuited line schematic and impedance

2.11.5 Distributed Inductive and Capacitive Elements

Thus far we have dealt with only 50 Ω transmission lines. It is possible to synthesize series inductance by using short lengths of transmission lines that have considerably higher impedance than 50 Ω. It is possible to synthesize shunt capacitors by using short lengths of transmission lines that have considerably lower impedance than 50 Ω. Typical impedances would range from approximately 20 Ω for capacitive elements and 80 Ω for inductive elements. The actual impedance used is a compromise between the substrate height and dielectric constant and the ability to physically realize the distributed element. For example an 80 Ω line on a thin substrate or a high dielectric constant substrate may be too narrow to etch on a printed circuit board. In such a case it may be necessary to use a 70 Ω impedance to realize the inductive element. These elements can be used successfully in narrow bandwidth applications. These distributed elements are very useful in the design of filters; bias feed networks; and impedance matching networks.

2.11.6 Distributed Microstrip Inductance and Capacitance

For short lengths of high impedance transmission line use the following equation to calculate the length of microstrip line to synthesize a specific value of inductance [2].

$$Inductive\ Line\ Length = \frac{f \lambda_g L}{Z_L} \qquad (2\text{-}69)$$

$$Capacitive\ Line\ Length = f \lambda_g Z_C C \qquad (2\text{-}70)$$

where:
    f = frequency and which inductance is calculated
    L = nominal inductance value
    C = nominal capacitance value
    Z_L = impedance of inductive transmission line
    λ_g = wavelength using the effective dielectric constant

The Genesys Advanced TLINE calculator is quite useful to convert an ideal transmission line element to a physical microstrip line. Figure 2-32 shows the lumped element equivalent circuit to the cascade of a low impedance and high impedance microstrip line. The printed circuit layout shows the line width relationship among the 50 Ω, 20 Ω, and 80 Ω microstrip lines.

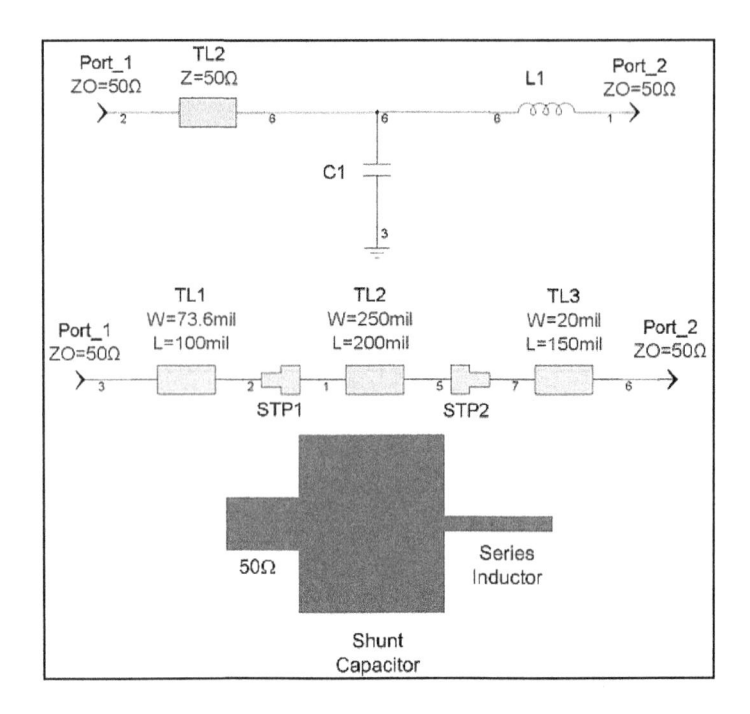

Figure 2-32 Distributed capacitive and inductive lines with PCB layout

2.11.7 Step Discontinuities

Examine the printed circuit board (PCB) layout of the transmission lines in Figure 2-32. Note the change in geometry as the impedance transitions from 50 Ω to 20 Ω and then to 80 Ω. These changes in geometry are known as discontinuities. Discontinuities in geometry result in fringing capacitance and parasitic inductance that will modify the frequency response of the circuit. At RF and lower microwave frequencies (up to about 2 GHz) the effects of discontinuities are minimal and sometimes neglected [4]. As the operation frequency increases, the effects of discontinuities can significantly alter the performance of a microstrip circuit. Genesys has several model elements that help to account for the effects of discontinuities. These include: T-junctions, cross junctions, open circuit end effects, coupling gaps, bends, and slits. A Microstrip Step element (STP) can be placed between series lines of abruptly changing geometry to account for the step discontinuity. The Genesys schematic of Figure 2-32 shows the proper placement of a microstrip step element between the low impedance and high impedance lines.

2.11.8 Microstrip Bias Feed Networks

Another useful purpose for high impedance and low impedance microstrip transmission lines is the design of bias feed networks. Often it is necessary to insert voltage and current to a device that is attached to a microstrip line. Such a device could be a transistor, MMIC amplifier, or diode. The basic bias feed or "bias decoupling network" consists of an inductor (used as an "RF Choke") and shunt capacitor (bypass capacitor). At lower RF frequencies (< 200 MHz) these networks are almost entirely realized with lumped element components. Even at these low frequencies it is very important to account for the parasitics in the components. Fig. 2-33 shows a typical series inductor, shunt capacitor, lumped element bias feed and its effect on a 50 Ω transmission line.

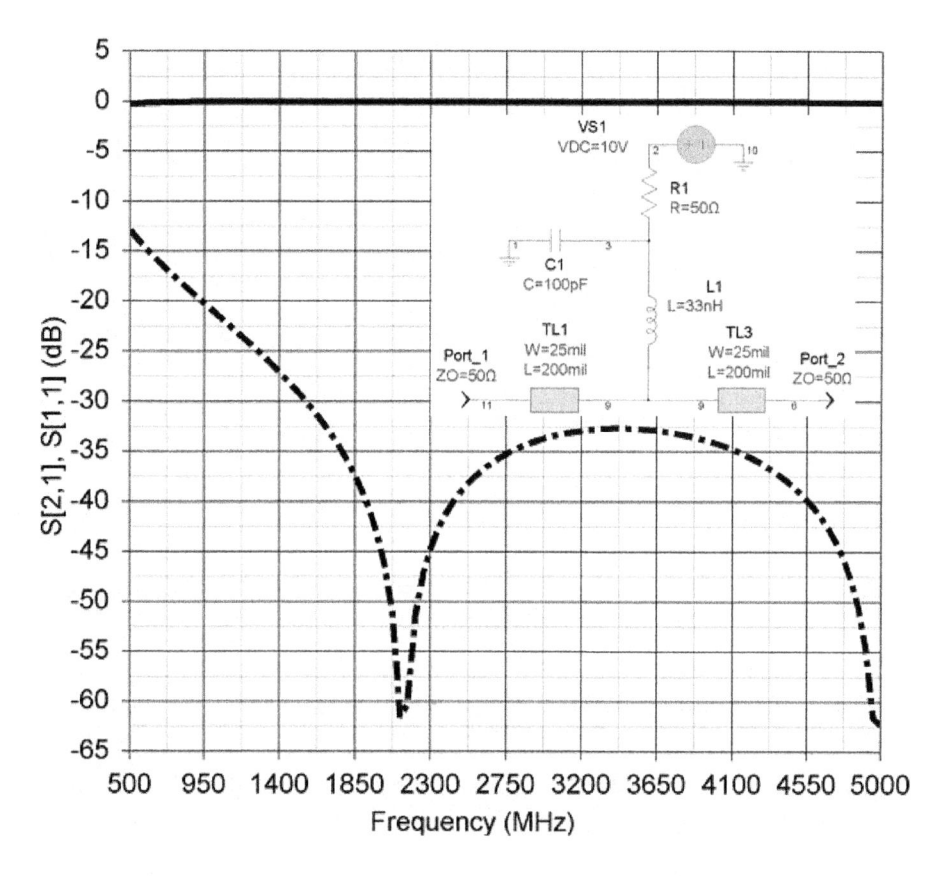

Figure 2-33 Typical inductor and bypass capacitor bias insertion network

2.11.9 Distributed Bias Feed

A high impedance microstrip line of $\lambda_g/4$ can be used to replace the lumped element inductor. Similarly a $\lambda_g/4$ of low impedance line can be used to model the shut capacitor.

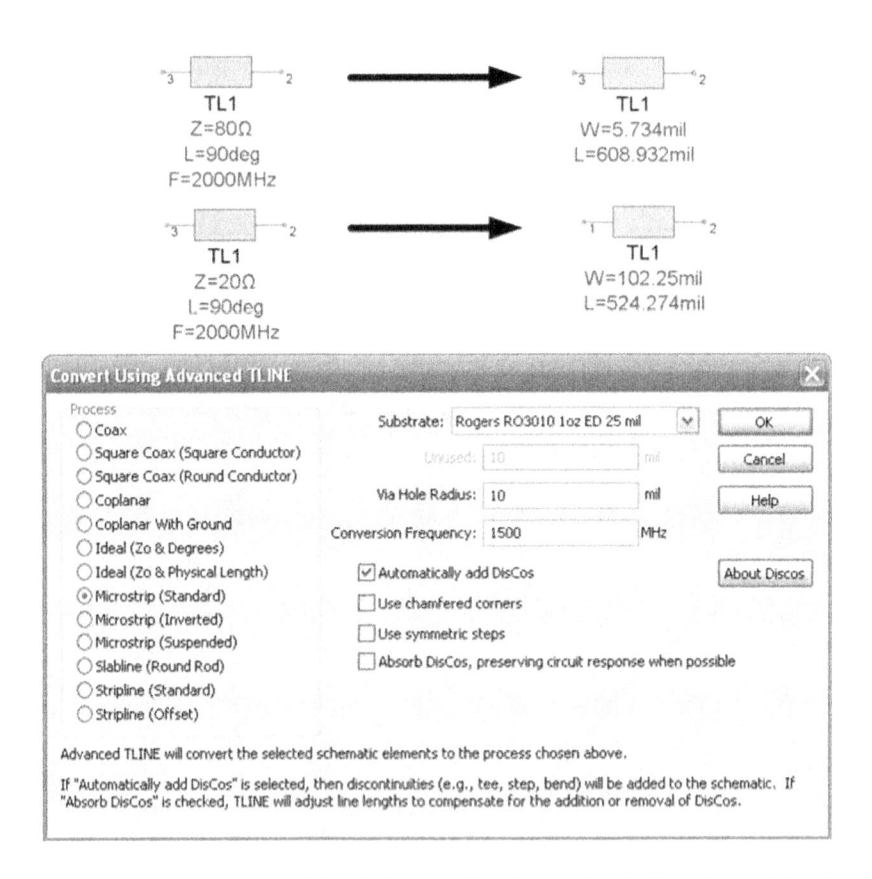

Figure 2-34 Advanced TLINE for 20 Ω and 80 Ω microstrip line synthesis

Example 2.11-3: Use the Advanced TLine utility to calculate the physical line length of the $\lambda_g/4$ sections of 80 Ω and 20 Ω microstrip lines at a frequency of 2 GHz. Create a schematic of a distributed bias feed network.

Solution: Use the 80 Ω high impedance quarter wave section and a shunt capacitance as shown in Figure 2-35. A microstrip taper, TP1, is used to connect the low impedance line to the high impedance line. Note the use of the microstrip tee junction, TE2. The tee junction accurately models the electrical length of the junction and includes all parasitic effects of the discontinuity. An end-effect element is used on the open circuit line. The response of the bias feed is characterized by the null in the return loss and very low insertion loss near the design frequency of 2 GHz. The return loss null occurs at 1.85 GHz suggesting that the high impedance line length should be decreased to center the design on 2 GHz.

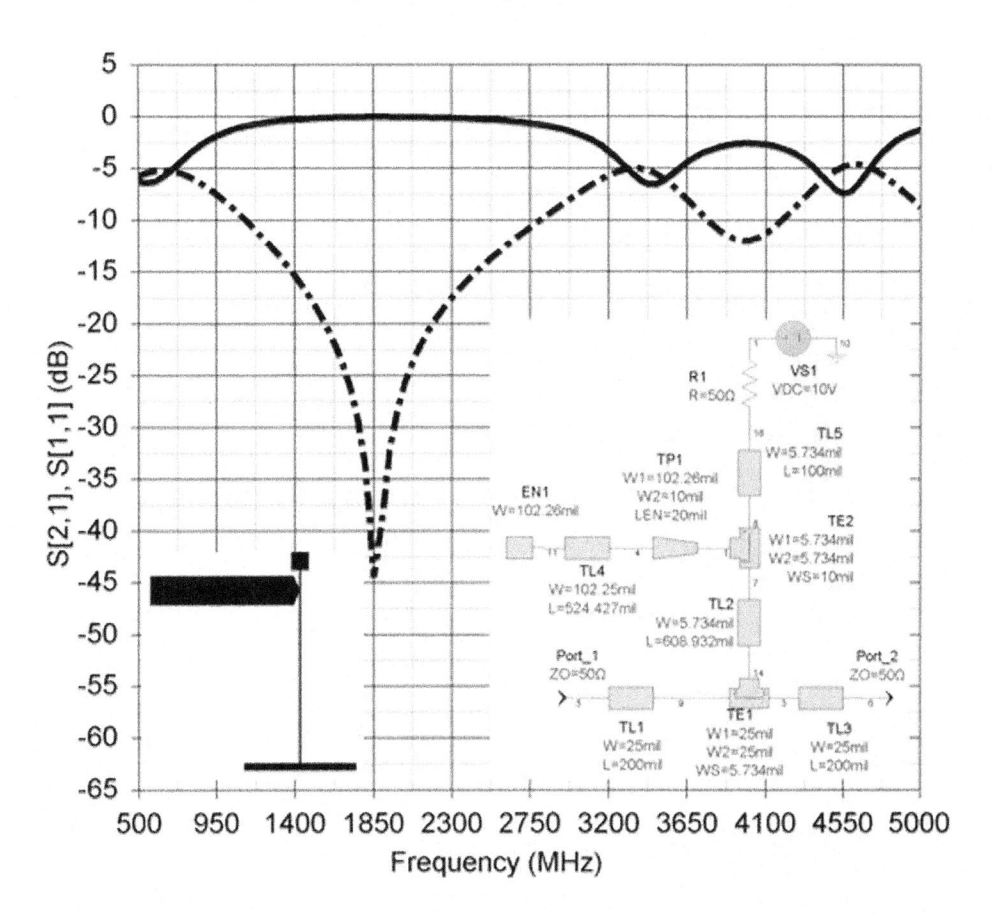

Figure 2-35 Bias feed modeled with distributed transmission line elements

A modified version of the open circuited transmission line is the radial stub. The radial stub can be used in applications where an open circuit transmission line is needed. Fig. 2-36 shows the use of the radial stub replacing the open circuit transmission line in the bias feed. Comparing the responses we can see that the network using the radial stub achieves a slightly wider bandwidth. This is one of the advantages of using the radial stub. The radial stub may also result in a slightly smaller PCB pattern. The PCB patterns are overlaid on the graphs of Figs. 2-35 and 2-36.

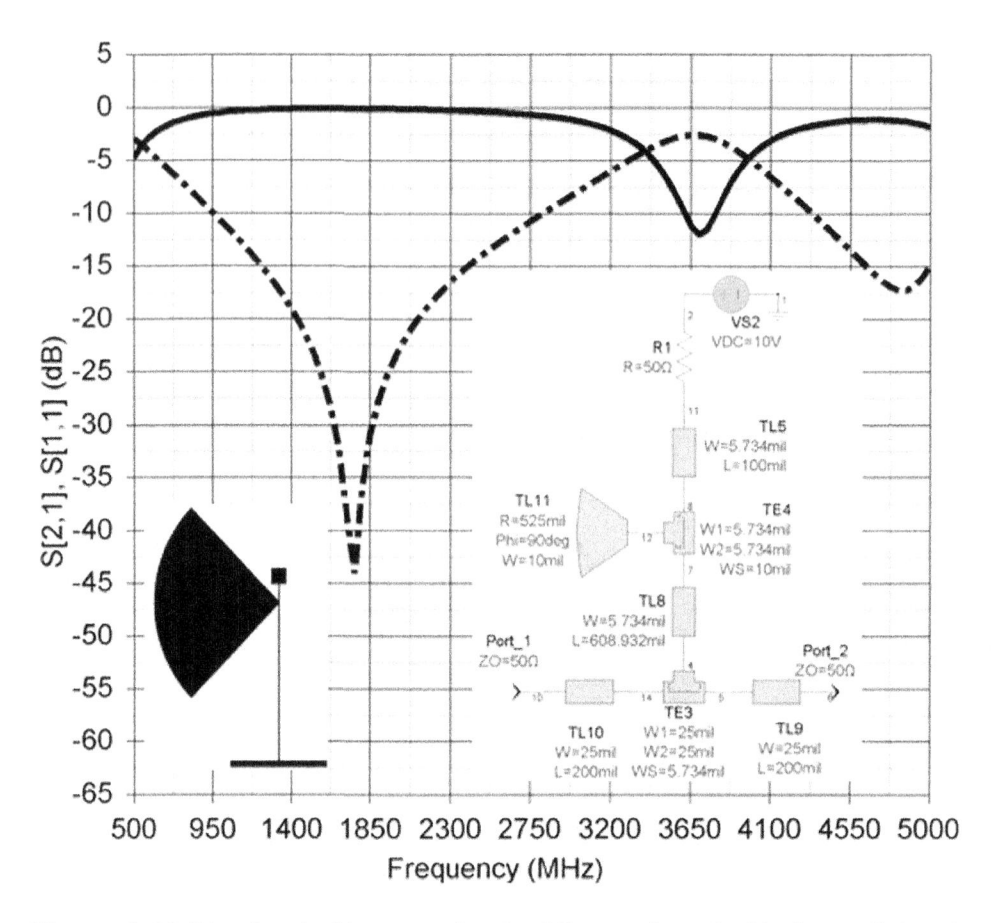

Figure 2-36 Bias feed with open circuited line replaced with the radial stub

2.12 Coupled Transmission Lines

There are three primary methods in which coupled lines are used in microwave circuit design. These are end-coupled, edge-coupled, and broadside coupled line structures. End coupled lines are often used to realize microstrip resonators and filters. Edge coupled lines are used in both coupler and filter designs. Broadside coupled lines are popular with various coupler designs.

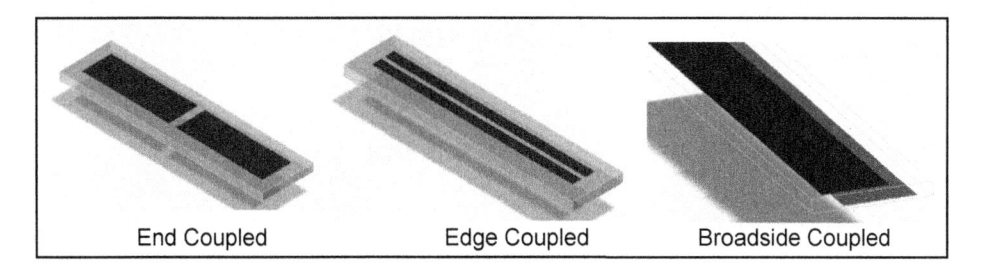

Figure 2-37 Types of coupled line structures

Current flow in the edge-coupled and broadside coupled sections can be difficult to quantify. Because of the coupling between the lines there exist two modes of impedance required to characterize the circuit. These are known as the even mode, Z_{oe}, and odd mode, Z_{oo}, impedance. Figure 2-38 shows the field distribution on the edge coupled lines for both conduction modes. In the even mode, a common displacement current flows from the conductors as shown by the E fields of the same polarity. In the odd mode, a component of the electric field is at opposite direction with respect to the two conductors. Therefore the magnitude of the even and odd mode impedance is strongly dependent on the separation between the lines which also determine the electrical coupling between the lines. The coupling between the lines, in dB, is defined by Equation (2-71).

$$C = 20 \log \left| \frac{Z_{oe} - Z_{oo}}{Z_{oe} + Z_{oo}} \right| \tag{2-71}$$

The even and odd mode impedances are then defined by the following equations [2].

$$Z_{oo} = Z_o \sqrt{\frac{1 - 10^{\left(\frac{-C}{20}\right)}}{1 + 10^{\left(\frac{-C}{20}\right)}}} \tag{2-72}$$

$$Z_{oe} = Z_o \sqrt{\frac{1 + 10^{\left(\frac{-C}{20}\right)}}{1 - 10^{\left(\frac{-C}{20}\right)}}} \tag{2-73}$$

$$Z = \sqrt{Z_{oo} Z_{oe}} \tag{2-74}$$

Even Mode   **Odd Mode**

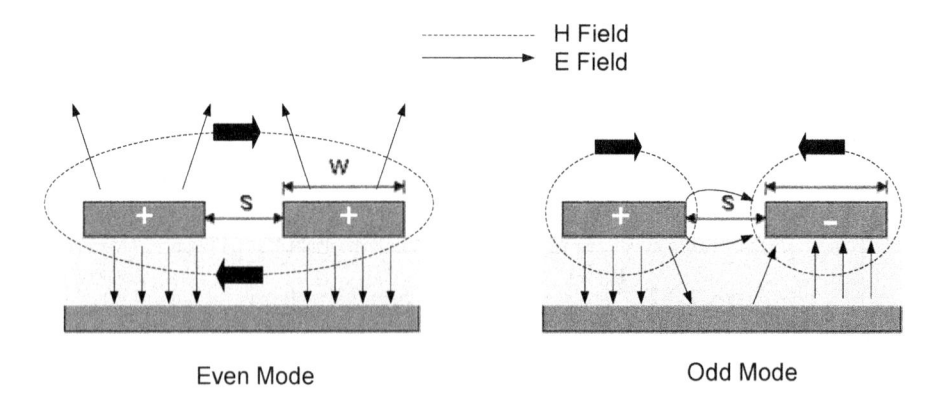

Figure 2-38 Edge coupled microstrip line field distribution

The Genesys TLINE utility has the capability to perform the calculations for coupled line impedance. The even and odd mode impedance can be calculated for a given characteristic impedance, Z_o, and coupling ratio in dB.

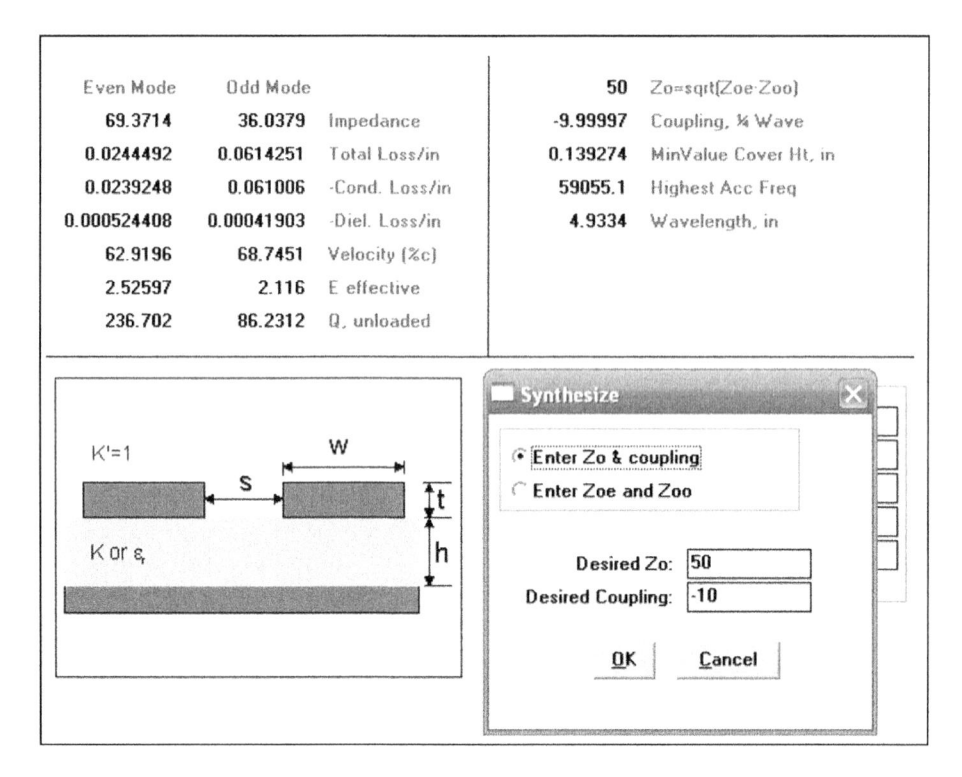

Even Mode	Odd Mode			
69.3714	36.0379	Impedance	50	Zo=sqrt(Zoe·Zoo)
0.0244492	0.0614251	Total Loss/in	-9.99997	Coupling, ¼ Wave
0.0239248	0.061006	-Cond. Loss/in	0.139274	MinValue Cover Ht, in
0.000524408	0.00041903	-Diel. Loss/in	59055.1	Highest Acc Freq
62.9196	68.7451	Velocity (%c)	4.9334	Wavelength, in
2.52597	2.116	E effective		
236.702	86.2312	Q, unloaded		

Synthesize
- Enter Zo & coupling
- Enter Zoe and Zoo

Desired Zo: 50
Desired Coupling: -10

OK   Cancel

Figure 2-39 TLINE calculation of microstrip edge coupled line impedance

2.12.1 Directional Coupler

One important use of coupled lines is the design of directional couplers. Directional couplers are useful components for sampling an RF signal without significantly loading or perturbing the input signal. Simple directional couplers are often used to provide a sample of an RF signal for measurement. High quality, precision, directional couplers can separate incident and reflected signals and are the fundamental component used for the measurement of VSWR and return loss. Two directional couplers can be placed back-to-back to form a dual directional coupler as shown in Figure 2-40. This type of coupler forms a four port network that can provide a sample of the forward power and reflected power between a source and load. Some basic properties of directional couplers include:

Insertion Loss:
Insertion Loss is simply the ratio of the output power at P2 to the input power at P1. Expressed in dB:

$$Insertion\ Loss\ (dB) = P2_{dBm} - P1_{dBm} \qquad (2\text{-}75)$$

Coupling:
The coupling factor is the ratio of the output power at P3 to the input power at P1. In microstrip and stripline circuits the coupled port is adjacent to the input line port. In waveguide couplers the coupled port is furthest from the input port.

$$Input\ Port\ Coupling\ (dB) = P1_{dBm} - P3_{dBm} \qquad (2\text{-}76)$$

Isolation:
Isolation is the ratio of the output power at P4 to the input power at P1.

$$Isolation\ (dB) = P1_{dBm} - P4_{dBm} \qquad (2\text{-}77)$$

Directivity:
Directivity is the difference between the isolation and the coupling when P2 is perfectly terminated in 50 Ω. Another way to think of a coupler's directivity is its ability to properly separate the forward and reflected waves.

$$Directivity\ (dB) = P3_{dBm} - P4_{dBm} \qquad (2\text{-}78)$$

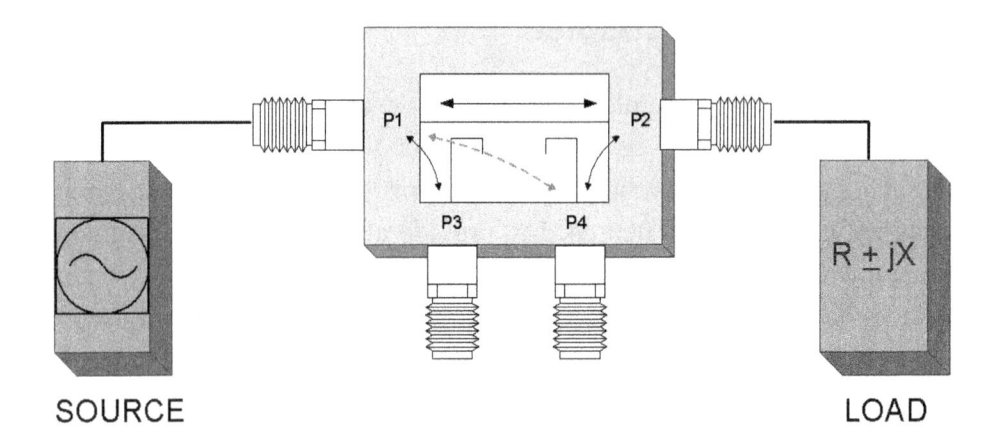

Figure 2-40 A dual directional coupler to measure VSWR, and return loss

A coupler will always have a finite amount of isolation. Ideally, all of the power input to P1 should be directed to P2 and P3. However some finite amount of power will show up at P4. This power will then add with any reflected power coming from P2 and being directed to P4. It is this finite isolation that limits the directivity of the directional coupler. Thus the power at P3 is the (Incident power at P1 – coupling factor) + the (Reflected power from P2 – isolation). For simple RF power sampling applications, the directivity is not that critical. But if we are using the directional coupler to measure VSWR, the directivity is very important. In VSWR measurement applications it is important to know the directivity of the directional coupler that is used to perform the measurement. A significant measurement error can exist when the coupler directivity is less than 40 dB. Figure 2-41 is a standard plot of measurement error vs. coupler directivity. The plot shows that when measuring a load that has an actual return loss of 20 dB with a directional coupler of 40 dB directivity, an error of +0.8 to -0.9 dB exists. This means that our instrument may read (-19.2 dB to -20.9 dB). We can readily see that if we used a directional coupler with directivity of 20 dB the resultant error could be approximately +2.3 dB to –3.0 dB. The asymmetry is due to whether the reflected signal is in phase with the forward signal or 180 degrees out of phase.

Transmission Lines

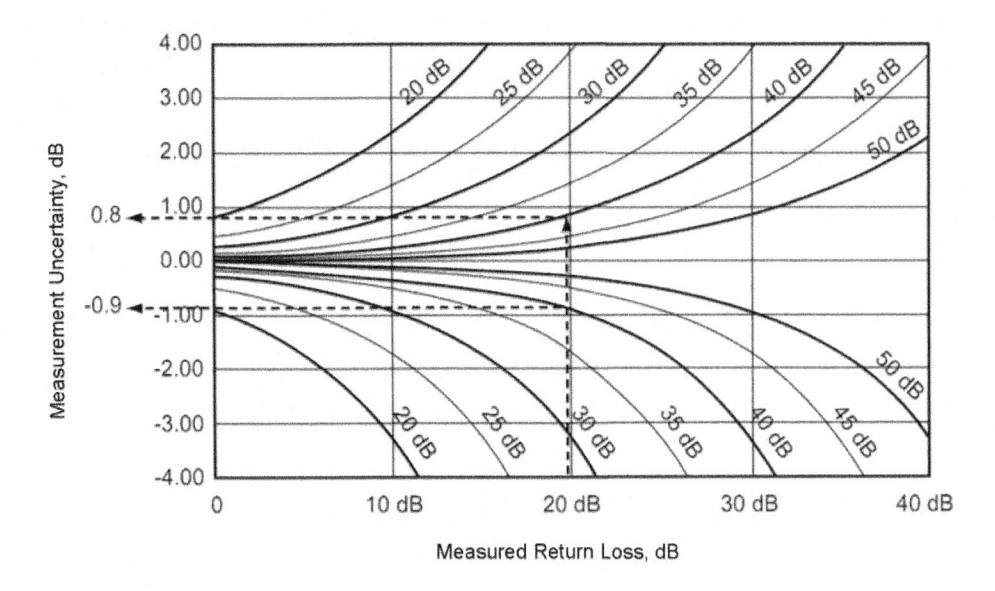

Figure 2-41 Error measurement versus coupler directivity (*Courtesy of Anritsu Corporation*)

2.12.2 Microstrip Directional Coupler Design

Example 2.12-1: Design a simple edge coupled microstrip directional coupler with a coupling factor of 10 dB at 5 GHz. Use Rogers RO3003 substrate with relative dielectric constant $\varepsilon_r = 3.0$ and 0.020-inch thickness.

Solution: Traditional coupler design required the computation of the even and odd mode impedance based on the characteristic impedance and coupling factor. Many references have tables and families of curves in which the line width and spacing could be obtained [2]. The Genesys TLINE utility provides an exact solution for the coupled line parameters. Select the coupled microstrip line calculator from the Rectangular transmission lines in the TLINE utility. Enter the dielectric constant (ε_r), the dielectric thickness (h), and conductor thickness (t) as shown in Figure 2-42. Select the Synthesis mode and choose to synthesize a coupled line pair based on the input characteristic impedance, Z_o, and the coupling in dB. The coupled line width, W, is then calculated be 38.86 mils and the line spacing, s, is 6.16 mils.

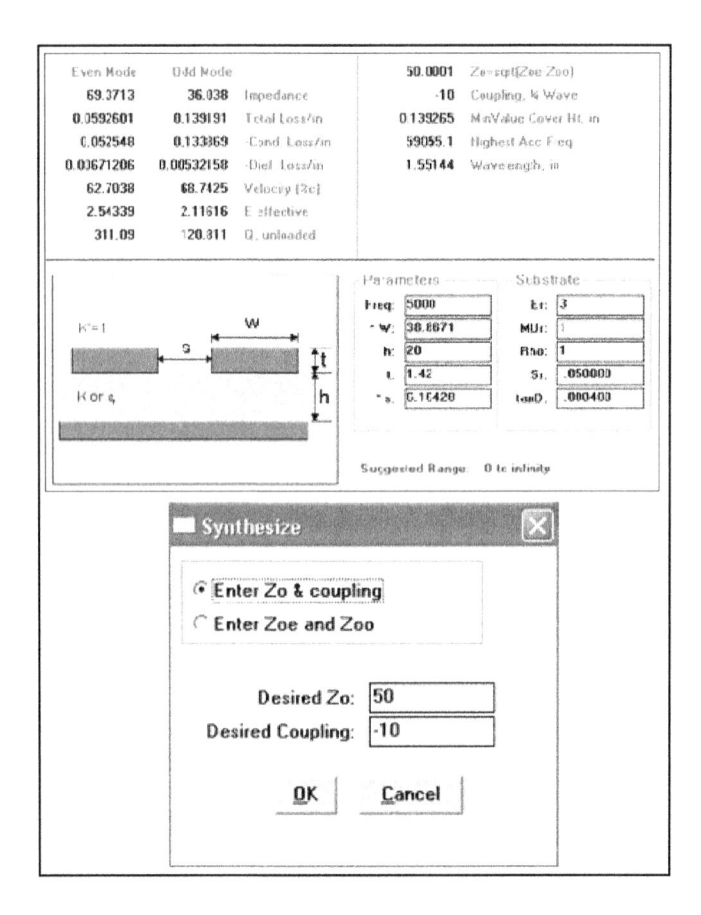

Figure 2-42 TLINE synthesis of a 10dB coupled line section

Add the RO3003, 0.020-inch thick substrate and create the coupler schematic using the Coupled Microstrip Line (Symmetrical) element. On the coupled port side add a Microstrip Bend with Optimal Miter at 90° from the main path. It is important to keep this side orthogonal from the main path so that no further parallel line coupling can occur. The optimal miter element automatically optimizes the miter for the least amount of discontinuity traversing the 90° bend. Lastly add a short section of line to each port to complete the circuit as shown in Figure 2-43. Use a Linear Analysis to sweep the coupler from 4500 MHz to 5500 MHz. Simulate the circuit and display the insertion loss (S21), coupled response (S31), isolation (S41), and the directivity. The directivity must be calculated by using Equation 2-78. It is convenient to implement simple mathematical operations directly in the Graph Properties window as opposed to using an

Transmission Lines 109

Equation block. The directivity is calculated by subtracting the coupled port response from the isolation.

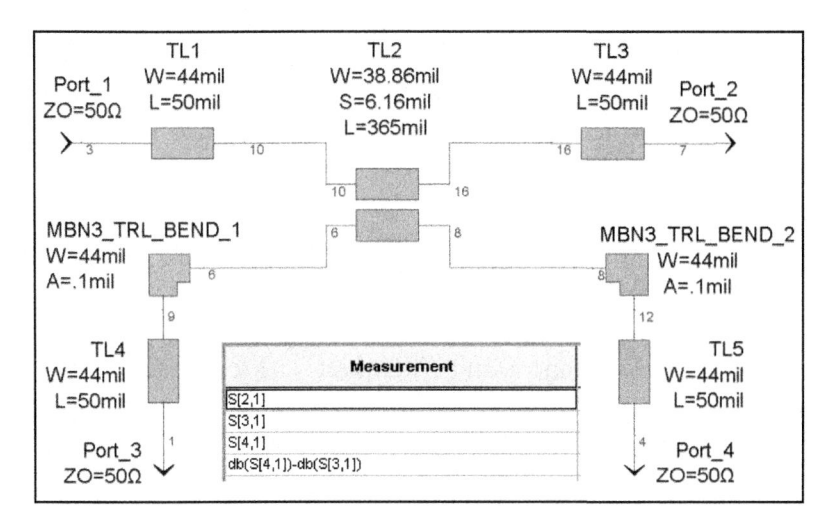

Figure 2-43 Schematic of the microstrip directional coupler

The simulated coupling factor is 10.157 dB which is close to the 10 dB design goal. The isolation is about 25 dB and directivity is about 15 dB. Clearly this coupler is not appropriate for VSWR measurement. It is useful for obtaining a sample of the input signal without disturbing or loading down the input signal.

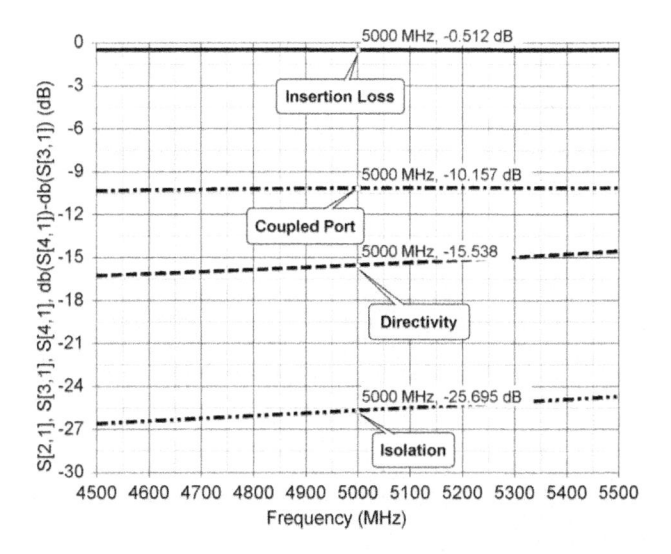

Figure 2-44 Simulated microstrip directional coupler response

References and Further Reading

[1] David M. Pozar, *Microwave Engineering*, Fourth Edition, John Wiley and Sons, Inc., 2012

[2] Ali A. Behagi and Stephen D. Turner, *Microwave and RF Engineering,* BT Microwave LLC, State College, PA, 2011

[3] Keysight Technologies, *Genesys 2014.03, Users Guide*, www.keysight.com.

[4] William Sinnema and Robert McPherson, Electronic Communications, Prentice-Hall Canada, Inc., Scarborough, Ontario, 1991

[5] UHF/Microwave Experimenters Manual, American Radio Relay League, Newington, CT.1990

[6] Reference: I. J. Bahl and D. K. Trivedi, "A Designer's Guide to Microstrip Line", Microwaves, May 1977, pp. 174-182.

[7] Microwave Handbook Volume 1, Radio Society of Great Britain, The Bath Press, Bath, U.K., 1989.

[8] Tatsuo Itoh, Planar Transmission Line Structures, IEEE Press, New York, NY, 1987

Problems

Determine the VSWR of a satellite antenna with a return loss of -11.4 dB.

2-1. The input reflection coefficient of a transistor is measured to be 0.22 at an angle of 32°. Determine the input VSWR of the device.

2-2. Determine the impedance of a quarter-wave transformer to match a 25 Ω load to a 50 Ω source.

2-3. Design the quarter-wave transformer from Problem 3 using a microstrip transmission line. The frequency of operation is 2.05 GHz. The dielectric constant is 3.0 with a thickness of 0.030 in. Determine the length and width of the microstrip line.

2-4. A radio transmitter is operating into a transmission line that measures a 3:1 VSWR. Determine the percentage of power that would be expected to reflect back into the transmitter.

2-5. A series RLC load, R = 75 Ω, L = 10 nH, C = 25 pF is connected to a 50 Ω transmission line. Setup a Linear Analysis in Genesys to sweep the frequency from 200 MHz to 2000 MHz in 200 MHz steps. Display the input reflection Coefficient, S11, and VSWR in a Table.

2-6. Create a simple schematic using the RG8 coaxial cable. Set the length to 50 ft. Calculate the insertion loss in a Table. Terminate the coaxial line with a 100 Ω resistor and display the input return loss and reflection coefficient in the same Table.

2-7. Calculate the cutoff frequency of the TE1,0 mode in a rectangular waveguide with a height of 0.200 inches and a width of 0.470 inches. Also calculate the waveguide wavelength, λ_g.

2-8. Design a distributed bias feed network for a C Band amplifier operating at 6.0 GHz. Use a microstrip substrate with a dielectric constant of 10.2 and a thickness of 0.025 inches. Plot the insertion loss and return loss from 2 GHz to 10 GHz.

2-9. Determine the physical length of a $\lambda_g/4$ open circuit microstrip transmission line with an impedance of 20 Ω. The frequency of operation is 10 GHz. Use a microstrip dielectric constant of 2.2 and a thickness of 0.010 inches. Determine whether an end-effect model element should be used.

2-10. Design a 35dB directional coupler to be used in a 100W transmitter operating at 4 GHz. Design the coupler in stripline. Use a dielectric constant of 3.0 and a thickness of 0.060 for each half of the stripline transmission media. Using Genesys determine the directivity of the coupler. Comment on coupler's ability to measure a 1.25:1 VSWR.

Chapter 3

Network Parameters and the Smith Chart

3.1 Introduction

At low frequencies below the VHF range, the terminal voltages V_1 and V_2 and the terminal currents I_1 and I_2 of a two-terminal network, shown in Fig. 3-1, can be related to each other by a different set of matrix parameters. The most common representations are the impedance matrix (Z parameters), the admittance matrix (Y parameters), the hybrid matrix (h parameters), and the transmission matrix (ABCD parameters).

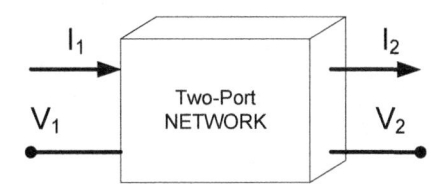

Figure 3-1 Low frequency two-port network

3.1.1 Z Parameters

The network representation with Z parameters, relating the input and output voltages to input and output currents, is given by the following equations.

$$V_1 = Z_{11}I_1 + Z_{12}I_2 \qquad (3\text{-}1)$$
$$V_2 = Z_{21}I_1 + Z_{22}I_2 \qquad (3\text{-}2)$$

The Z parameters, also known as impedance parameters, are determined by making the following open circuit measurements:

$$Z_{11} = \left.\frac{V_1}{I_1}\right|_{I_2=0} \quad \textit{(Requires open circuit measurement)}$$

$$Z_{12} = \left.\frac{V_1}{I_2}\right|_{I_1=0} \quad \textit{(Requires open circuit measurement)}$$

$$Z_{21} = \left.\frac{V_2}{I_1}\right|_{I_2=0} \quad \textit{(Requires open circuit measurement)}$$

$$Z_{22} = \left.\frac{V_2}{I_2}\right|_{I_1=0} \quad \textit{(Requires open circuit measurement)}$$

The Z parameters are very useful when two-terminal networks are connected in series. In this case the overall Z parameters are simply the algebraic sum of the individual Z parameters.

3.1.2 Y Parameters

Similarly, the input and output currents can be related to input and output voltages by Y parameters as shown in Equations (3-3) and (3-4).

$$I_1 = Y_{11}V_1 + Y_{12}V_2 \qquad (3\text{-}3)$$

$$I_2 = Y_{21}V_1 + Y_{22}V_2 \qquad (3\text{-}4)$$

The Y parameters, also known as admittance parameters, are determined by making the following short circuit measurements.

$$Y_{11} = \left.\frac{I_1}{V_1}\right|_{V_2=0} \quad \textit{(Requires short circuit measurement)}$$

$$Y_{12} = \left.\frac{I_1}{V_2}\right|_{V_1=0} \quad \textit{(Requires short circuit measurement)}$$

$$Y_{21} = \left.\frac{I_2}{V_1}\right|_{V_2=0} \quad \textit{(Requires short circuit measurement)}$$

$$Y_{22} = \left.\frac{I_2}{V_2}\right|_{V_1=0} \quad \textit{(Requires short circuit measurement)}$$

Network Parameters and the Smith Chart

The Y parameters are useful when two-terminal networks are connected in parallel. In this case the overall Y parameters are simply the algebraic sum of the individual Y parameters.

3.1.3 h Parameters

The network representation by Hybrid or h parameters, relating the input voltage and output current to input current and output voltage, is given by the following equations.

$$V_1 = h_{11}I_1 + h_{12}V_2 \qquad (3\text{-}5)$$
$$I_2 = h_{21}I_1 + h_{22}V_2 \qquad (3\text{-}6)$$

The h parameters can be determined from the following measurements.

$$h_{11} = \left.\frac{V_1}{I_1}\right|_{V_2=0} \quad \textit{(Requires short circuit measurement)}$$

$$h_{12} = \left.\frac{V_2}{V_1}\right|_{I_1=0} \quad \textit{(Requires open circuit measurement)}$$

$$h_{21} = \left.\frac{I_2}{I_1}\right|_{V_2=0} \quad \textit{(Requires short circuit measurement)}$$

$$h_{22} = \left.\frac{I_2}{V_2}\right|_{I_1=0} \quad \textit{(Requires open circuit measurement)}$$

The h parameters are often used to characterize the low frequency characteristics of transistor circuits. The parameter, h_{21}, defines the forward current gain while the parameter, h_{12}, defines the reverse voltage gain of the network. The parameter, h_{11}, is the input impedance and parameter, h_{22}, is the output admittance of the network.

3.1.4 ABCD Parameters

Another representation relating the input voltage and current to the output voltage and current, is by ABCD parameters.

$$V_1 = AV_2 - BI_2 \quad (3\text{-}7)$$
$$I_1 = CV_2 - DI_2 \quad (3\text{-}8)$$

The ABCD parameters are found by the following measurements.

$$A = \left.\frac{V_1}{V_2}\right|_{I_2=0} \quad \text{(Requires open circuit measurement)}$$

$$B = -\left.\frac{V_1}{I_2}\right|_{V_2=0} \quad \text{(Requires short circuit measurement)}$$

$$C = \left.\frac{I_1}{V_2}\right|_{I_2=0} \quad \text{(Requires open circuit measurement)}$$

$$D = -\left.\frac{I_1}{I_2}\right|_{V_2=0} \quad \text{(Requires short circuit measurement)}$$

The ABCD parameters in Equations (3-7) and (3-8) are often presented in matrix form and referred to as the transmission matrix [2].

$$\begin{bmatrix} V_1 \\ I_1 \end{bmatrix} = \begin{bmatrix} A & B \\ C & D \end{bmatrix} \cdot \begin{bmatrix} V_2 \\ -I_2 \end{bmatrix}$$

The ABCD parameters are very useful to characterize networks when individual circuits are cascaded in a chain fashion. They have also been popular in the design of telephone networks. In this case the overall ABCD matrix is found by the product of the individual ABCD matrices. For example, if two networks are connected in cascade the overall ABCD matrix is the product of individual ABCD matrices.

$$\begin{bmatrix} A & B \\ C & D \end{bmatrix} = \begin{bmatrix} A_1 & B_1 \\ C_1 & D_1 \end{bmatrix} \cdot \begin{bmatrix} A_2 & B_2 \\ C_2 & D_2 \end{bmatrix} \qquad (3\text{-}9)$$

At microwave frequencies, due to the difficulty in making short circuit and open circuit measurements, the two-terminal network representation by Z, Y, h, or ABCD parameters is not practical. Therefore, at microwave frequencies, a new representation known as Scattering or S parameters has been developed.

3.2 Development of Network S Parameters

At RF and microwave frequencies, where the wavelength of the voltage and current waveforms are comparable or smaller than the physical dimensions of the network, it is difficult to obtain perfect open and short circuit terminations. It is also difficult to measure voltage and current in high frequency circuits, therefore, the usefulness of the parameters in Section 3.1 diminish. In high frequency networks it is much easier to measure power than voltage or current. For RF and microwave networks, a form of the transmission matrix has been defined based on power measurements into the system's characteristic impedance. These parameters are known as S parameters, named after their scattering matrix form. Consider a two-port network, shown in Figure 3-2, where Z_{01} is the real characteristic impedance and V_1^+ and V_1^-, respectively, are the incident and reflected voltage waveforms at the input port. Similarly, Z_{02} is the real characteristic impedance and V_2^+ and V_2^-, respectively, are the incident and reflected voltage waveforms at the output port.

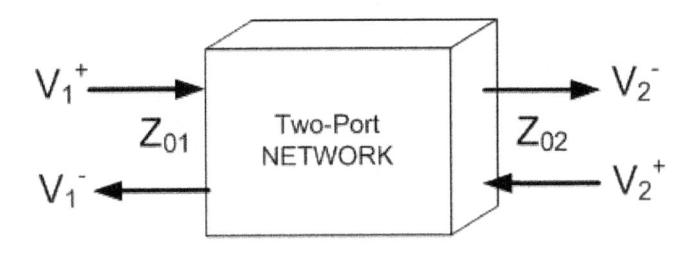

Figure 3-2 Two-port network with incident and reflected voltage waveforms

In order to obtain measurable power relations in terms of wave amplitudes, we need to define a new set of waveforms by normalizing the voltage amplitudes with respect to the square root of the respective characteristic impedances, namely:

$$a_1 = \frac{V_1^+}{\sqrt{Z_{01}}} \tag{3-10}$$

$$b_1 = \frac{V_1^-}{\sqrt{Z_{01}}} \tag{3-11}$$

$$a_2 = \frac{V_2^+}{\sqrt{Z_{02}}} \tag{3-12}$$

$$b_2 = \frac{V_2^-}{\sqrt{Z_{02}}} \tag{3-13}$$

The two-port network with normalized waveforms is shown in Figure 3-3.

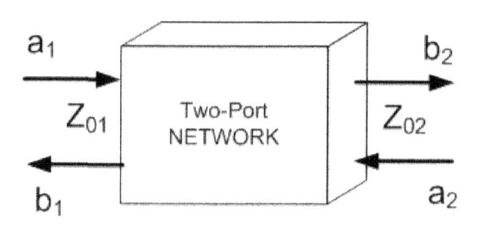

Figure 3-3 Two-port network with incident and reflected waveforms

Notice that:

$$|a_1|^2 = \frac{|V_1^+|^2}{Z_{01}} = \text{Incident power at the network input}$$

$$|b_1|^2 = \frac{|V_1^-|^2}{Z_{01}} = \text{Reflected power at the network input}$$

$$|a_2|^2 = \frac{|V_2^+|^2}{Z_{02}} = \text{Incident power at the network output}$$

$$|b_2|^2 = \frac{|V_2^-|^2}{Z_{02}} = \text{Reflected power at the network output}$$

The S parameters relate b_1 and b_2 to a_1 and a_2 by the following Equations [1].

$$b_1 = S_{11}a_1 + S_{12}a_2 \qquad (3\text{-}14)$$
$$b_2 = S_{21}a_1 + S_{22}a_2 \qquad (3\text{-}15)$$

In matrix form the scattering matrix is written as:

$$\begin{bmatrix} b_1 \\ b_2 \end{bmatrix} = \begin{bmatrix} S_{11} & S_{12} \\ S_{21} & S_{22} \end{bmatrix} \cdot \begin{bmatrix} a_1 \\ a_2 \end{bmatrix}$$

At RF and microwave frequencies the normalized voltage waveforms a_1, a_2, b_1, and b_2 represent vectors having both magnitude and phase. Terminating the output of the two-port network with a real load impedance that is equal to the real system characteristic impedance, $Z_{01} = Z_{02}$, forces $a_2=0$. Solving for the individual S parameters then gives the following relationships.

$$S_{11} = \frac{V_1^-}{V_1^+} = \left.\frac{b_1}{a_1}\right|_{a_2=0} \qquad (3\text{-}16)$$

$$S_{21} = \frac{V_2^-}{V_1^+} = \left.\frac{b_2}{a_1}\right|_{a_2=0} \qquad (3\text{-}17)$$

These measurements are better visualized using the network of Figure 3-4.

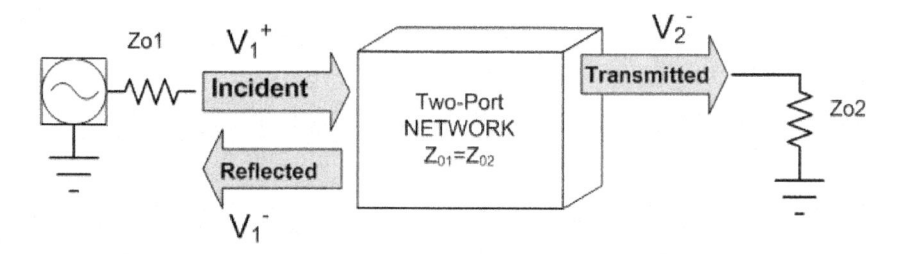

Figure 3-4 Measurement of S11 and S21 in a two-port network

Terminating the input of the two-port network with a real load impedance that is equal to the real system characteristic impedance, $Z_{01} = Z_{02}$, forces

$a_1 = 0$. The S parameters S_{22} and S_{12} can be solved by Equations (3-18) and (3-19) as demonstrated in Figure 3-5.

$$S_{22} = \frac{V_2^-}{V_2^+} = \frac{b_2}{a_2}\bigg|_{a_1=0} \quad (3-18)$$

$$S_{12} = \frac{V_1^-}{V_2^+} = \frac{b_1}{a_2}\bigg|_{a_1=0} \quad (3-19)$$

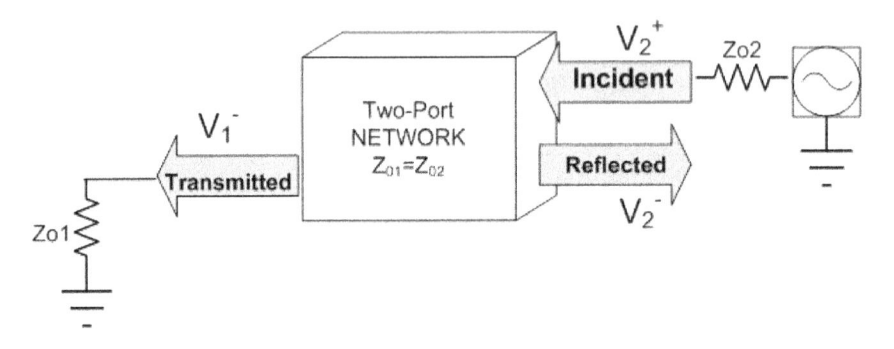

Figure 3-5 Measurement of S22 and S12 in a two-port network

S11 is often referred to as the input reflection coefficient and S22 as the output reflection coefficient of the network. S21 is the forward transmission and is often expressed as a gain or loss depending on whether S21, in dB, is positive or negative. S12 is the reverse transmission or isolation of the network. Because the S parameters are complex entities they must be measured with a Vector Network Analyzer, VNA, capable of measuring both amplitude and phase. A Scalar Network Analyzer is used to measure the magnitude of two-port networks only. Similar to the ABCD parameters, it is very easy to cascade S parameters to calculate the overall S parameters of a multi-stage network.

3.3 Using S Parameter Files in Genesys

Chapter one touched upon the use of S parameters to describe the network models of lumped components such as resistors, inductors, and capacitors. Any linear RF and microwave network can be described by its S parameters. As such, the S parameters find extensive use throughout linear circuit modeling and simulation. The S parameter file can serve as a substitute for any element model in a circuit design. The file is loaded into the simulation engine via a data element model such as the one shown in Figure 3-6. This schematic symbol represents a generic two port S parameter data file. In the Genesys parts selector there exists a variety of symbols that can represent S parameter files from one port to N-port scattering matrices. An N-port device will have N^2 S parameters. Genesys allows any schematic symbol to be applied to a data file element model. The generic data file element of Figure 3-6 can be changed to the symbol of a transistor, capacitor, or whatever element the S parameter data file represents. This is helpful for maintaining a more professional and meaningful schematic. The data file element can be edited so that the designer can browse to the location of the saved S parameter data file. The S Parameter file is simply an ASCII text file that can be edited with any text editor. For a two-port network the filename should use the .s2p extension. Similarly a one port network would use the .s1p file extension. Only one definition line is required and must begin with a (#) symbol. Any number of comment lines can be included and must begin with the (!) symbol. The only remaining requirement is the order in which the S parameters are entered into the text file. This order must be in the form of S_{11}, S_{21}, S_{12}, and S_{22}. Figure 3-7 shows the S parameter file of a C Band amplifier.

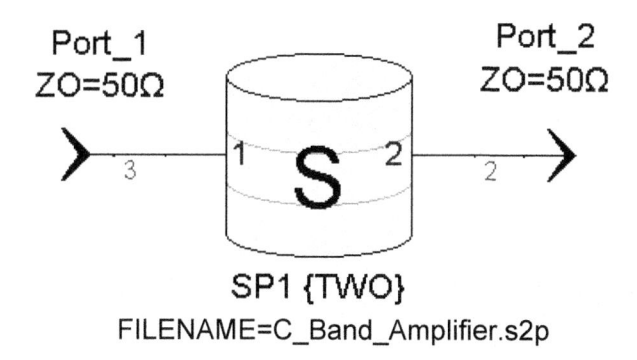

Figure 3-6 S Parameter data file in the Genesys schematic

```
!   S-Parameter File of 200W C-Band Amplifier
#  GHz  S   MA   R   50
!FREQ      |S11|              |S21|              |S12|              |S22|
5.800      0.070  -52.367     65.436   59.242    0.001   49.912      0.151  -69.520
5.804      0.067  -58.549     64.997   49.393    0.002  -91.508      0.159  -69.723
5.807      0.078  -57.215     64.574   39.402    0.003   15.689      0.150  -70.922
5.811      0.071  -62.285     64.270   29.318    0.004  -46.869      0.161  -68.117
5.814      0.062  -63.434     64.010   18.336    0.003  100.039      0.155  -72.297
5.818      0.068  -68.539     63.737    9.254    0.002   55.959      0.154  -73.473
5.821      0.069  -68.652     63.011   -1.009    0.001  -49.662      0.153  -72.922
5.825      0.067  -75.742     63.067  -11.077    0.000   72.992      0.155  -73.332
5.828      0.067  -79.371     63.110  -20.816    0.001  -46.926      0.149  -73.844
5.832      0.066  -80.000     62.883  -30.369    0.001  163.273      0.153  -76.746
5.835      0.064  -80.039     62.700  -40.990    0.002   -0.501      0.146  -75.992
5.839      0.064  -83.637     62.461  -50.729    0.001   14.107      0.146  -77.883
5.842      0.065  -93.328     62.278  -61.002    0.002  168.805      0.153  -75.441
5.846      0.063  -89.715     61.721  -70.473    0.004  -33.932      0.151  -80.746
5.849      0.066  -91.605     62.152  -80.652    0.002   84.551      0.151  -79.625
5.853      0.060  -96.484     61.416  -90.824    0.001   78.816      0.143  -83.137
```

Figure 3-7 C band amplifier S parameter data file

As Figure 3-7 shows the definition line contains four descriptive parameters for the data file. The available options for these parameters are summarized below [5]. Only a single space is required between each entry on a line of the S parameter file. For better visual presentation a tab space can be used between entries.

GHz: Units for the swept frequency data column. Frequency can be GHz, MHz, KHz, or Hz.

S: Defines network parameter type. Parameters can be S, Y, Z, or h parameters format.

MA: Magnitude-Angle format for the parameter data. Available formats are DB for dB-angle, MA for magnitude-angle, or RI for real-imaginary format.

R 50: Reference resistance. This is the characteristic impedance in which the parameters have been measured. The parameters will be normalized to this resistance.

Setup a linear analysis to analyze the amplifier's S parameter data file over a frequency range of 5800 MHz to 5820 MHz. Setup a tabular output and display each of the four S parameters. On the Table Properties window, set the units to absolute (Abs) and magnitude-angle format as shown in Figure

3-8. The resulting table is shown in Figure 3-9. Note that even though the S Parameters in the data file are recorded in increments greater than 1 MHz, Genesys has interpolated the values between each data point and can output the S parameters in less than 1 MHz increments.

Measurement	Label (Optional)	Units	Complex Format
S[1,1]		dB	Mag abs+Angle
S[2,1]		dB	Mag abs+Angle
S[1,2]		dB	Mag abs+Angle
S[2,2]		dB	Mag abs+Angle

Figure 3-8 Genesys Table for display of the S Parameters

	F (MHz)	mag(S[1,1]) (dB)	ang(S[1,1])	mag(S[2,1]) (dB)	ang(S[2,1])	mag(S[1,2]) (dB)	ang(S[1,2])	mag(S[2,2]) (dB)	ang(S[2,2])
1	5800	0.07	-52.367	65.436	59.242	1e-3	49.912	0.151	-69.52
2	5800.2	0.07	-52.676	65.414	58.75	1.05e-3	42.841	0.151	-69.53
3	5800.4	0.07	-52.985	65.392	58.257	1.1e-3	35.77	0.152	-69.54
4	5800.6	0.07	-53.294	65.37	57.765	1.15e-3	28.699	0.152	-69.55
5	5800.8	0.069	-53.603	65.348	57.272	1.2e-3	21.628	0.153	-69.561
6	5801	0.069	-53.913	65.326	56.78	1.25e-3	14.557	0.153	-69.571
7	5801.2	0.069	-54.222	65.304	56.287	1.3e-3	7.486	0.153	-69.581
8	5801.4	0.069	-54.531	65.282	55.795	1.35e-3	0.415	0.154	-69.591
9	5801.6	0.069	-54.84	65.26	55.302	1.4e-3	-6.656	0.154	-69.601
10	5801.8	0.069	-55.149	65.238	54.81	1.45e-3	-13.727	0.155	-69.611
11	5802	0.068	-55.458	65.217	54.318	1.5e-3	-20.798	0.155	-69.621
12	5802.2	0.068	-55.767	65.195	53.825	1.55e-3	-27.869	0.155	-69.632
13	5802.4	0.068	-56.076	65.173	53.333	1.6e-3	-34.94	0.156	-69.642
14	5802.6	0.068	-56.385	65.151	52.84	1.65e-3	-42.011	0.156	-69.652
15	5802.8	0.068	-56.694	65.129	52.348	1.7e-3	-49.082	0.157	-69.662
16	5803	0.068	-57.004	65.107	51.855	1.75e-3	-56.153	0.157	-69.672
17	5803.2	0.068	-57.313	65.085	51.363	1.8e-3	-63.224	0.157	-69.682
18	5803.4	0.067	-57.622	65.063	50.87	1.85e-3	-70.295	0.158	-69.693
19	5803.6	0.067	-57.931	65.041	50.378	1.9e-3	-77.366	0.158	-69.703
20	5803.8	0.067	-58.24	65.019	49.885	1.95e-3	-84.437	0.159	-69.713
21	5804	0.067	-58.549	64.997	49.393	2e-3	-91.508	0.159	-69.723
22	5804.2	0.068	-58.46	64.969	48.727	2.067e-3	-84.362	0.158	-69.803
23	5804.4	0.068	-58.371	64.941	48.061	2.133e-3	-77.215	0.158	-69.883
24	5804.6	0.069	-58.282	64.912	47.395	2.2e-3	-70.069	0.157	-69.963
25	5804.8	0.07	-58.193	64.884	46.729	2.267e-3	-62.922	0.157	-70.043

Figure 3-9 Genesys tabular display of the S Parameter data file

3.3.1 Scalar Representation of the S Parameters

When working with a two-port network, such as an amplifier, the engineer is usually more interested in the magnitude in (dB's) of the S parameters rather than the absolute units. This is referred to as the Scalar representation of the S parameters and is readily measured on scalar network analyzers. Because the S parameters are based on voltage waveforms they must be multiplied by 20log to convert to dB format.

$S_{11}(dB) = -20\log|S_{11}|$ *Input return loss*
$S_{21}(dB) = 20\log|S_{21}|$ *Insertion gain (+) or insertion loss (-)*
$S_{12}(dB) = 20\log|S_{12}|$ *Reverse gain (+) or isolation (-)*
$S_{22}(dB) = -20\log|S_{22}|$ *Output return loss*

With the C Band amplifier in Genesys, setup a rectangular graph and display the S Parameters in Scalar format. Assign each S parameter to the graph using the dB-magnitude format. The display is shown in Figure 3-10.

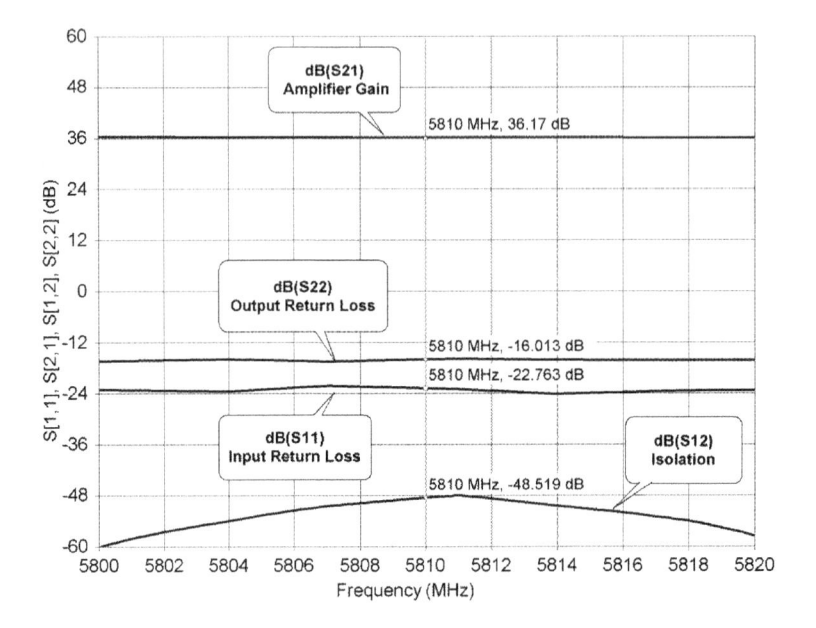

Figure 3-10 Scalar display of frequency swept amplifier S Parameters

3.4 Development of the Smith Chart

The most frequently used graphical tool used to visualize the vector properties of S parameters is the Smith Chart. The Smith Chart conformally maps the familiar rectangular impedance coordinates onto a polar plane. Essentially the reactive, normally the vertical axis, has been bent around in such a way that ± infinity is included within the boundary of the graph. Therefore any positive complex impedance can be plotted on the standard Smith Chart shown in Figure 3-11. Negative impedances or gain is outside of the standard Smith Chart. A compressed Smith Chart must be used to plot negative impedances. Figure 3-11 shows the Genesys standard Smith

Chart graph with impedance coordinates. A rectangular axis has been overlaid to show the relationship to the rectilinear grid system. The Smith Chart can be normalized to any characteristic impedance. The chart of Figure 3-11 is normalized to 50 Ω. The normalized impedance is a pure resistance that is a single point at the center of the chart. The purely real impedances exist along the horizontal axis from 0 Ω to infinity. Note the locations of the short circuit (0 Ω) and open circuit (infinity) on the real axis. The concentric circles that intersect the real axis are known as the constant resistance circles. The constant reactance circles appear as arcs on the standard Smith Chart. As shown in Figure 3-11 the reactance circles on the top half of the chart represent inductive reactance while the circles on the bottom half of the chart represent capacitive reactance. Any impedance defined by rectangular coordinates (R + jX) can be plotted as a point where the R value on the constant resistance circle intersects the X value on the reactance circle. Consider plotting the impedance 25 + j25 Ω on the standard Smith Chart. This impedance is shown in Figure 3-12.

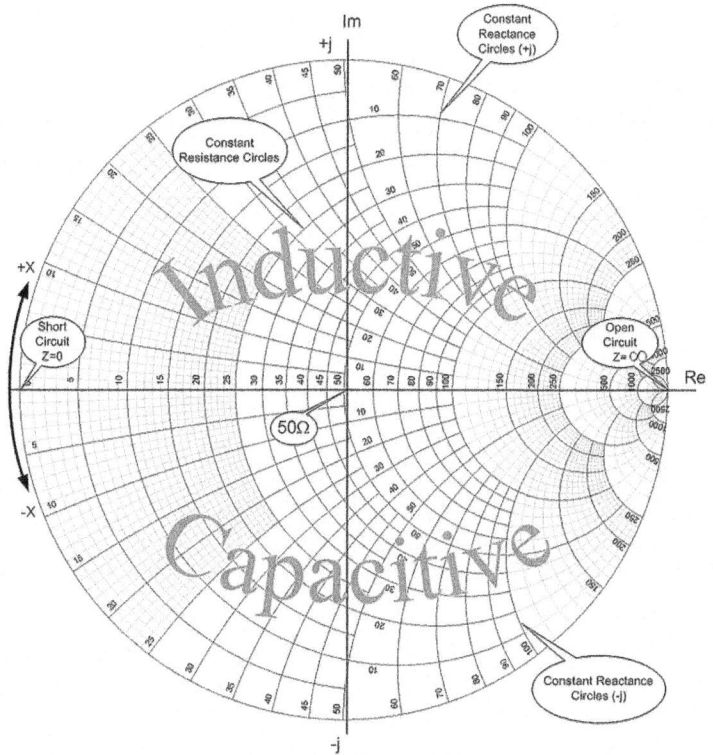

Figure 3-11 Standard Smith Chart with impedance coordinates in Genesys

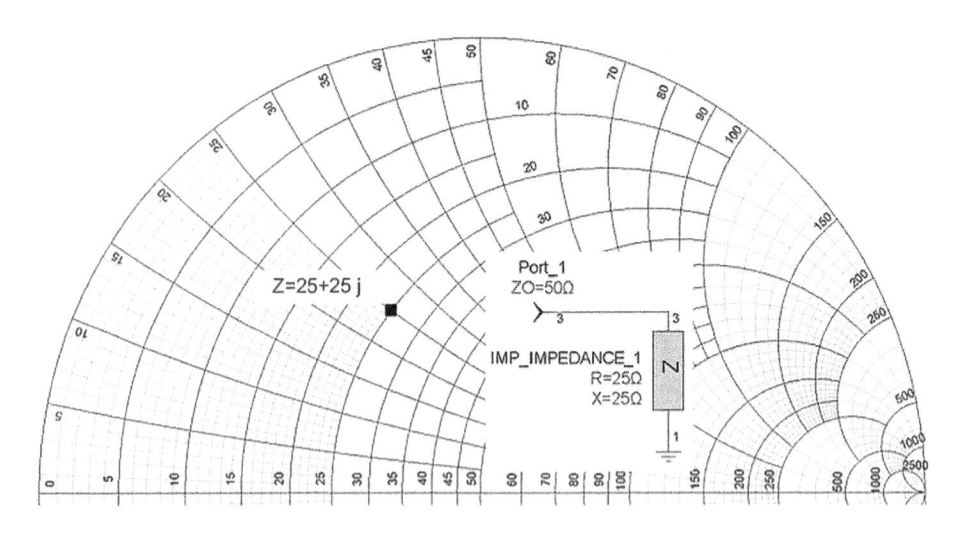

Figure 3-12 Plotting impedance on the Smith Chart

3.4.1 Normalized Impedance on the Smith Chart

The impedance on a Smith Chart is often presented in its normalized form. This means that the actual impedance is divided by the value of the characteristic impedance. The Genesys Smith Chart allows the selection of either normalized or actual impedance via the Properties window of the Smith Chart. Equation (3-20) shows how the reflection coefficient, Γ, is related to the normalized impedance on the Smith Chart.

$$\Gamma = \frac{(Z_{actual} - Z_o)}{(Z_{actual} + Z_o)} \tag{3-20}$$

Redefining Equation (3-20) in terms of normalized impedance where $z = \frac{(Z_{actual})}{(Z_o)}$, results in Equation (3-21).

$$\Gamma = \frac{(z-1)}{(z+1)} \tag{3-21}$$

The normalized impedance is shown on the Smith Chart of Figure 3-13. The normalized impedance is read as $z = 0.5 + j0.5\ \Omega$. The reflection coefficient on the Smith Chart is the vector from the center of the chart to the normalized impedance. The transmission coefficient is the vector from the origin (Z = 0) to the normalized impedance. The reflection S parameters, S_{11} and S_{22}, are measured as a reflection coefficient on the Smith Chart. The transmission S parameters, S_{21} and S_{12}, are measured as transmission parameters on the Smith Chart.

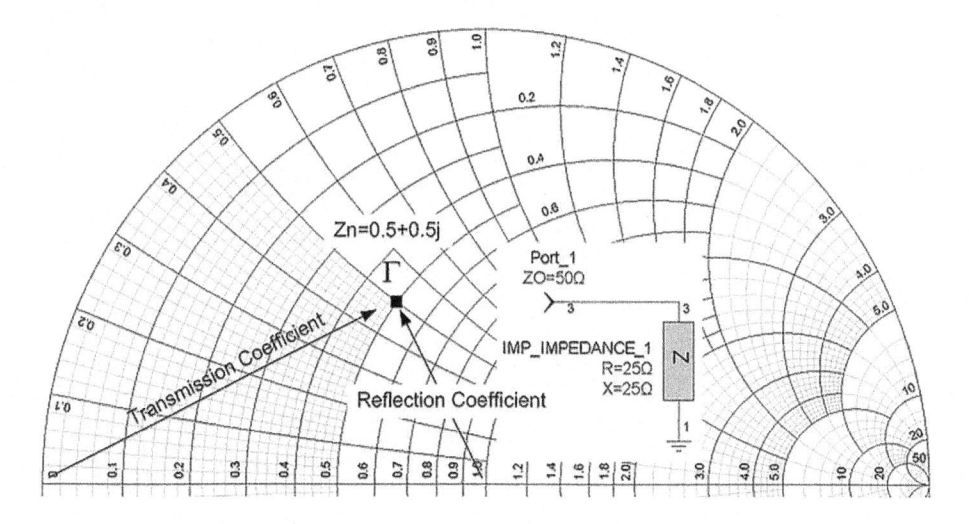

Figure 3-13 Impedance plotted with normalized impedance coordinates on the Smith Chart

Knowing that the reflection coefficients and S parameters are vector quantities, there must be a method to measure the angular portion of the vector. Figure 3-14 shows the angular measurement convention on the Smith Chart. Note that the reflection coefficient of a 50 Ω resistance (center of the chart) is equal to zero. A total reflection, like that due to a perfect short or open circuit, has a reflection coefficient equal to one. Therefore all positive impedances result in a reflection coefficient between 0 and 1. The reflection coefficient of the 25 + j25 Ω impedance can be determined in Genesys by displaying the magnitude and angle of the S parameter, S11. As F5igure 3-14 shows the reflection coefficient of this impedance is 0.447 at an angle of +116.565 degrees.

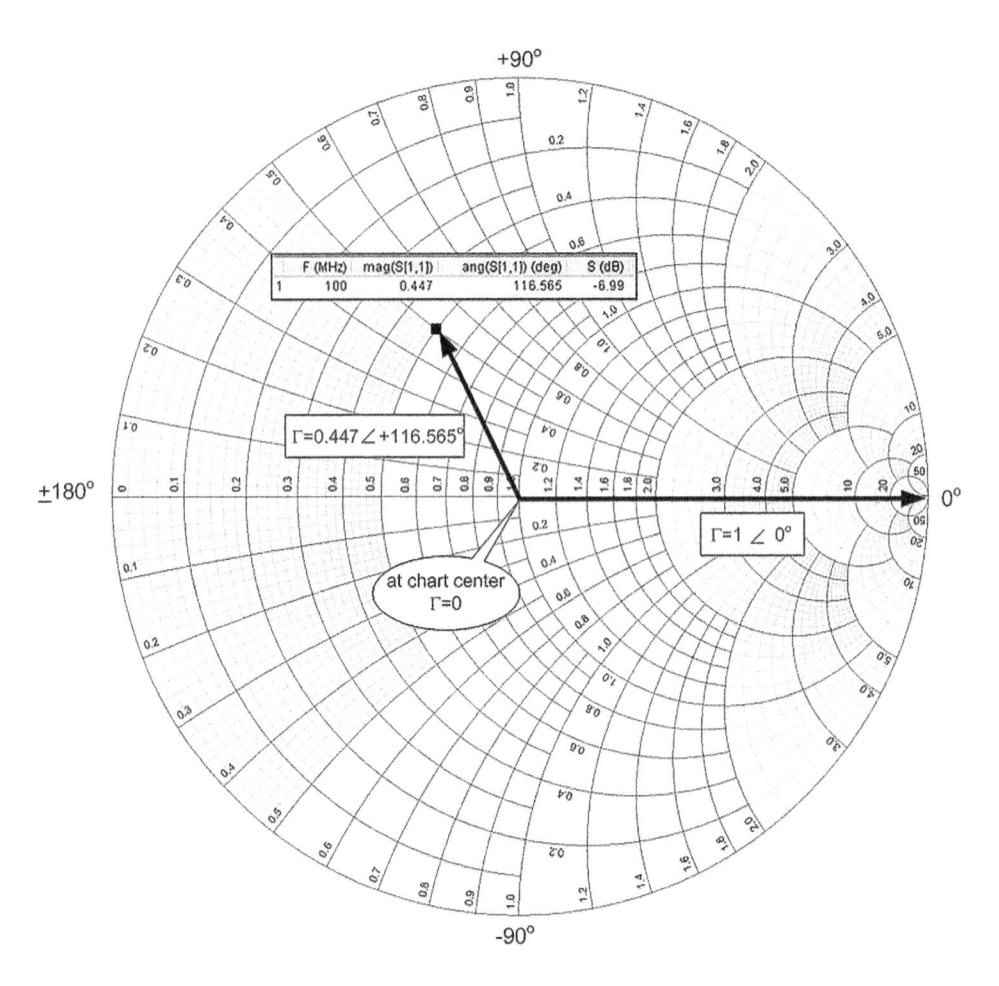

Figure 3-14 Angular measurement of reflection coefficients on the Smith Chart

3.4.2 Admittance on the Smith Chart

Admittance circles can also be displayed on the Smith Chart. The admittance circles can be enabled on the Genesys Smith Chart by their selection on the graph's Properties page. The admittance circles consist of constant conductance and susceptance circles which are inverted from the impedance circles. Subsequent chapters dealing with the subject of impedance matching will make frequent use of the admittance parameters. Having both impedance and admittance parameters displayed on the Smith Chart makes it very easy to design impedance matching networks that include both series and parallel (shunt) elements. It also becomes very easy

Network Parameters and the Smith Chart 129

to convert series impedance to its parallel admittance equivalent. The admittance of a network is the inverse of the impedance as given by Equation (3-22).

$$Y = \frac{1}{Z} = G \pm jB \qquad (3\text{-}22)$$

where,
G = conductance in mhos
B = susceptance in mhos

The equivalent admittance of a network can be read directly from the admittance circles on the Smith Chart. For the normalized 0.5 + j0.5 Ω series impedance the admittance is read directly from the chart as 1.0 - j1.0 mho. It is important to remember that there is an inversion in the sign of the imaginary component when converting from impedance to admittance or vice versa.

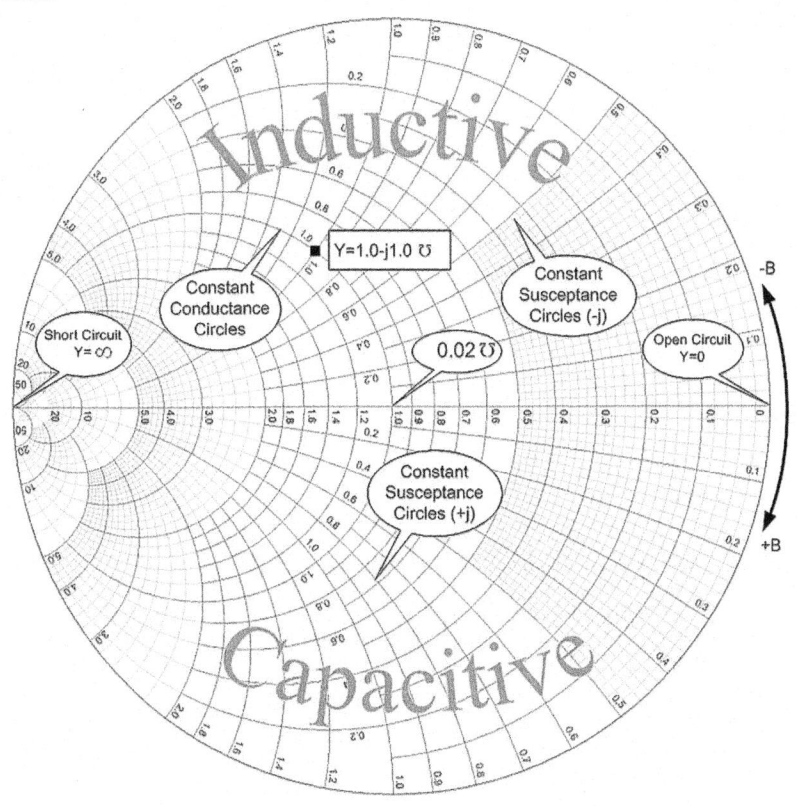

Figure 3-15 Admittance circles on the Smith Chart

3.5 Lumped Element Movements on the Smith Chart

Lumped element movements on the Smith Chart form the basis for impedance matching. The Smith Chart is a wonderfully intuitive tool, for the visualization of moving from one impedance to another, without involving circuit synthesis mathematics. Understanding the basic movements around the Smith Chart will build a foundation for the circuit designs covered in this text. It is helpful to display both the impedance and admittance coordinates simultaneously on the Smith Chart.

3.5.1 Adding a Series Reactance to an Impedance

Adding a series reactance to an impedance point on the Smith Chart causes the resulting impedance to move along the constant resistance circle in which the impedance intersects. A series inductance will move the impedance in a clockwise direction while a series capacitance will move the resulting impedance in a counter-clockwise direction on the constant resistance circle. The reactance that is added to the impedance by the series element can be read from the Smith Chart by finding the difference between the lines of reactance that intersect the start point and end point on the constant resistance circle. This reactance can then be converted to a capacitance or inductance value using Equations (3-23) and (3-24) [4].

$$C(series) = \frac{1}{\omega X n} \qquad (3\text{-}23)$$

$$L(series) = \frac{X n}{\omega} \qquad (3\text{-}24)$$

where,
  X = reactance measured along the arc length
  ω = the design frequency, $2\pi f$
  n = impedance normalizing value (50 Ω)

Example 3.5-1: Measure the amount of reactance required to move the impedance Z = 25 + j25 Ω from point A to point B on the Smith Chart, as shown in Figure 3-16.

Solution: The amount of reactance required in the inductor can be measured from the reactance lines that intersect the start point (A) and end point (B). As Figure 3-16 shows the reactance is approximately 0.364 Ω. Using a design frequency of 1000 MHz and Equation (3-22) the inductance is calculated to as 2.86 nH.

$$L(series) = \frac{0.364 \cdot (50)}{2 \cdot \pi \cdot 1000 \cdot 10^6} = 2.86 \; nH$$

Add the inductor to the impedance element in Genesys to verify that a 2.86 nH inductor moves the impedance to point B on the Smith Chart. Make the inductor tunable and set the initial value to 0.1 nH and tune the value of inductance to move the impedance from point A to point B.

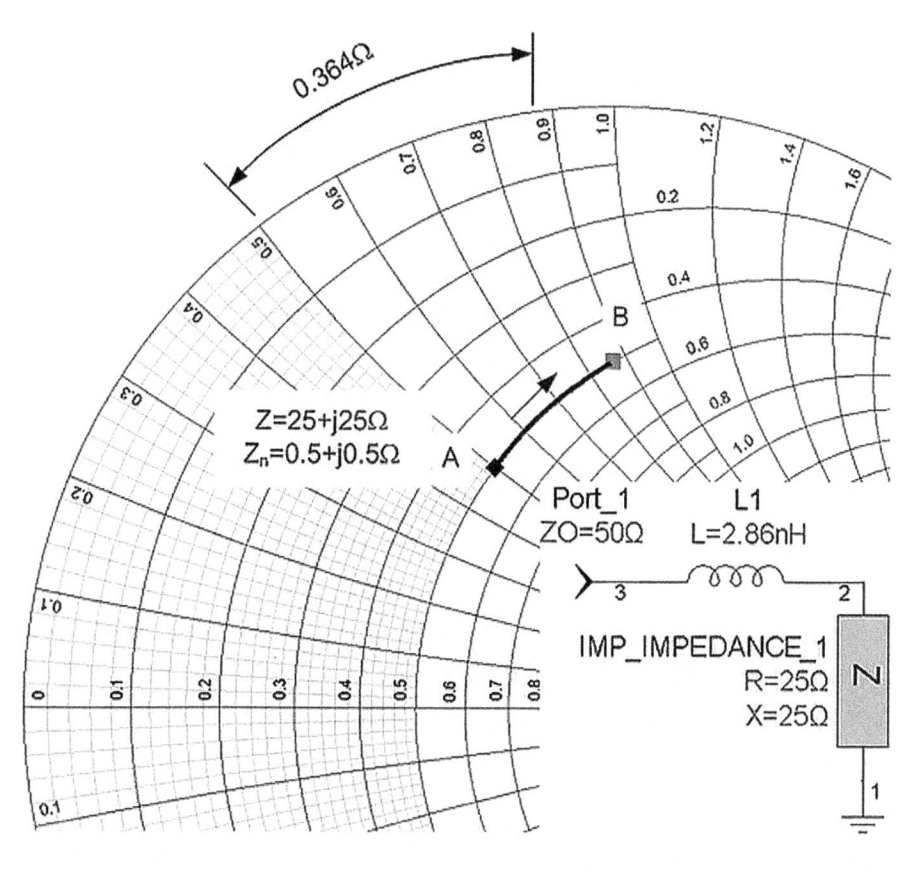

Figure 3-16 Adding a series inductance to 25 + j25 Ω impedance

3.5.2 Adding a Shunt Reactance to an Impedance

Adding a shunt element to an impedance point on the Smith Chart causes the resulting impedance to move along the constant conductance circle in which the impedance intersects. A shunt inductance will move the impedance in a counter-clockwise direction while a shunt capacitance will move the resulting impedance in a clockwise direction on the constant conductance circle. The susceptance that is added to the impedance by the shunt element can be read from the Smith Chart by finding the difference between the lines of susceptance that intersect the start point and end point on the constant conductance circle. This susceptance can then be converted to a capacitance or inductance value using Equations (3-25) and (3-26) [4].

$$L(shunt) = \frac{n}{\omega B} \qquad (3\text{-}25)$$

$$C(shunt) = \frac{B}{\omega n} \qquad (3\text{-}26)$$

where
  B = susceptance measured along the arc length
  ω = the design frequency $2\pi f$
  n = impedance normalizing value (50 Ω)

Example 3.5-2: Continuing with the previous example measure the susceptance required to move from point B to point C on the real axis.

Solution:. Add a shunt capacitance to move the impedance from point B to point C as shown in Figure 3-17. When adding a shunt element switch from the impedance grid to the admittance coordinates. The admittance follows the constant conductance circle in which the point lies by the difference between the intersecting susceptance lines. The susceptance as measured on the perimeter of the chart is 0.865 mhos. Use Equation (3-26) to calculate the capacitance.

$$C\ (shunt) = \frac{B}{\omega n} = \frac{0.865}{2 \cdot \pi \cdot 1000 \cdot 10^6 \cdot (50)} = 2.75\ pF$$

Alternatively you can add a shunt capacitor to the circuit and make the capacitance value tunable. Start with a very low value of approximately 0.1 pf and increase the value of capacitance until the admittance is moved from point B to point C.

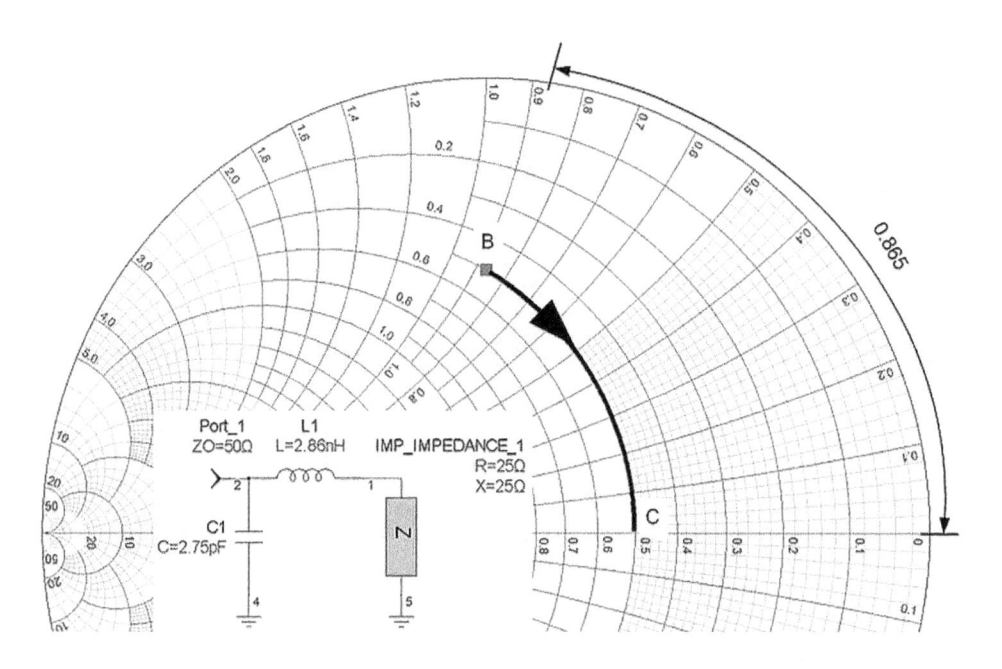

Figure 3-17 Adding the shunt capacitance to the network

The resulting admittance at point C is 0.5 mhos. This is the resulting normalized admittance looking into port one of the network shown in Figure 3-17. The actual admittance is found by multiplying the normalized admittance by the Smith Chart's characteristic admittance (1/50 Ω). Therefore the admittance at point C is 0.01 mhos or an impedance of 100Ω. These techniques form the basis for performing impedance transformations using the Smith Chart. In this example a complex impedance of 25 + j25 Ω has been transformed to a pure resistance of 100 Ω. The lumped element movement directions are summarized in Figure 3-18. By iterative addition of series and shunt elements an impedance point can be moved from one location to another on the Smith Chart for a given frequency. This process forms the basis of the topic of impedance matching which is fundamental to RF and microwave engineering and will be used throughout the remainder of the text.

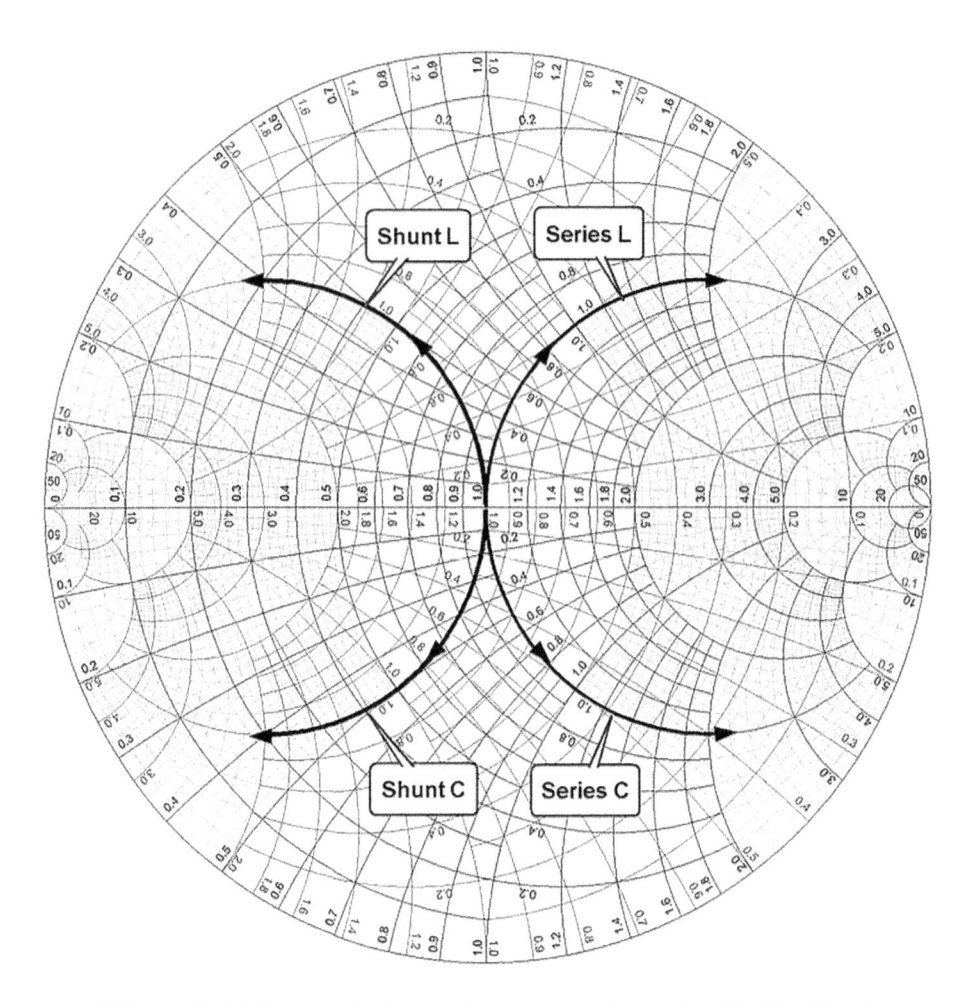

Figure 3-18 Lumped element movements on the Smith Chart

3.6 VSWR Circles on the Smith Chart

Equation (2-38) demonstrated that the VSWR of a network is related to the magnitude of the reflection coefficient, independent of the angle. From plotting the reflection coefficient on the Smith Chart we know that the origin of the vector is located in the center of the chart. This suggests that as the reflection coefficient vector rotates 360 degrees around the chart with a constant magnitude, the VSWR will remain constant. This locus of points around the center of the Smith Chart is known as the constant VSWR circle [3]. The constant VSWR circles can be displayed in Genesys by using the manual S parameter model element and a Parameter sweep. The manual S parameter model is shown in Figure 3-19. This model is similar to the impedance element used in Fig. 3-12 in that it is frequency independent. This means that the impedance or reflection coefficient will not change with frequency. The magnitude of S_{11}, which is the magnitude of the reflection coefficient, is assigned a variable named mag1. An Equation Editor block can be used to enter the desired VSWR and calculate the reflection coefficient. The reflection coefficient is then assigned to the variable "mag1".

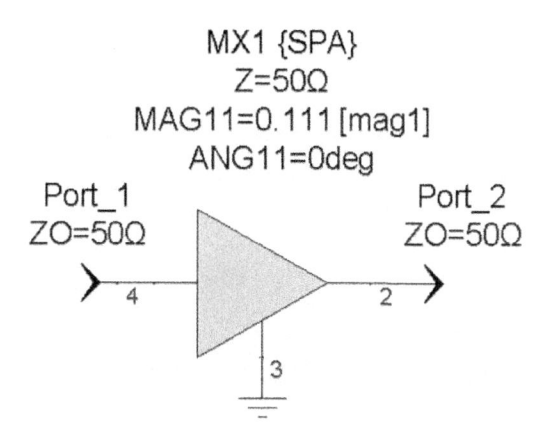

```
            MX1 {SPA}
              Z=50Ω
          MAG11=0.111 [mag1]
            ANG11=0deg
   Port_1                    Port_2
   ZO=50Ω                    ZO=50Ω
```

Figure 3-19 Manual S Parameter model in Genesys

Make the angle of S_{11} on the S parameter model a tunable variable. Then create a parameter sweep that will be used to sweep the angle of S_{11} from 0 to 360 degrees in 2 degree steps. Create a Linear Analysis to analyze the network at a fixed frequency as shown in Figure 3-20. Any frequency can be entered as the model is frequency independent. The equation editor block is presented in Figure 3-21. Enter the desired VSWR value in the Equation block and simulate the circuit. A family of circles is shown in Figure 3-22 for VSWR values of: 1.25, 2.0, 4.0, and 10.0. Any reflection coefficient intersecting the circle has a VSWR equal to the value of the circle. If the reflection coefficient is inside the circle, the VSWR is less than the value of the circle. Likewise if a reflection coefficient is outside of the circle, the VSWR is greater than the value of the circle. Figure 3-22 gives intuitive feel for relating reflection coefficients on the Smith Chart to their VSWR value.

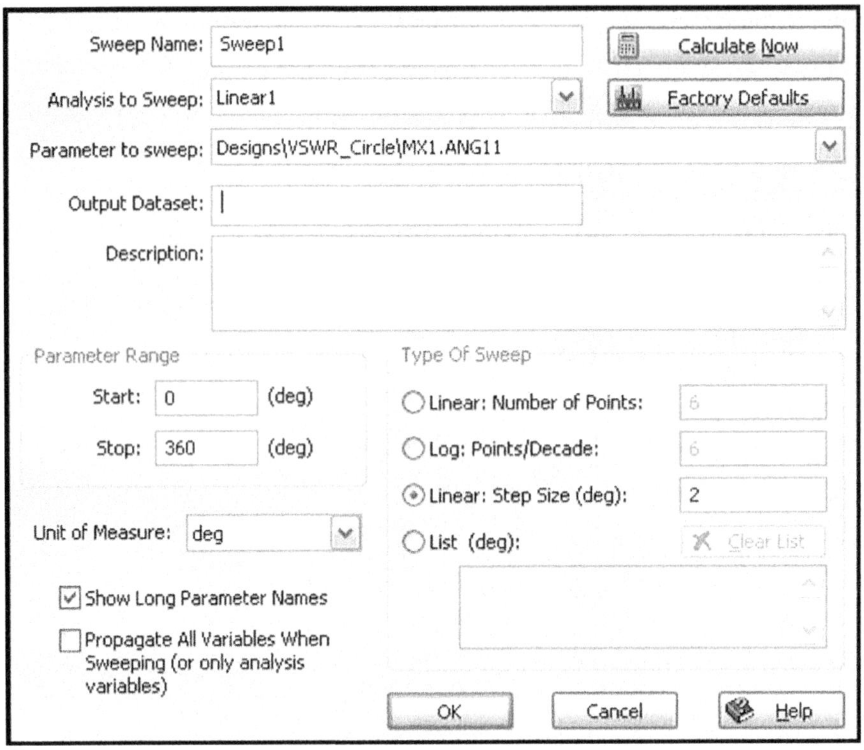

Figure 3-20 Parameter sweep for the VSWR circle

```
1    'Enter Sweep Parameters
2    vswr1=1.25
3    mag1=(vswr1-1)/(vswr1+1)
4    angle=Sweep1_Data.MX1_ANG11SWP_F
5    Gamma1=Sweep1_Data.S(1,1)
6    ⇒setindep("Gamma1","angle")
7
```

Figure 3-21 Generation of the VSWR circle on the Smith Chart

From the plot of the VSWR circles in Figure 3-22 we can see that the value of the VSWR is equal to the magnitude of the reflection coefficient as labeled on the horizontal axis to the right of the center of the chart. Conversely the horizontal axis values on the left-hand side of the chart represent $1/\Gamma$.

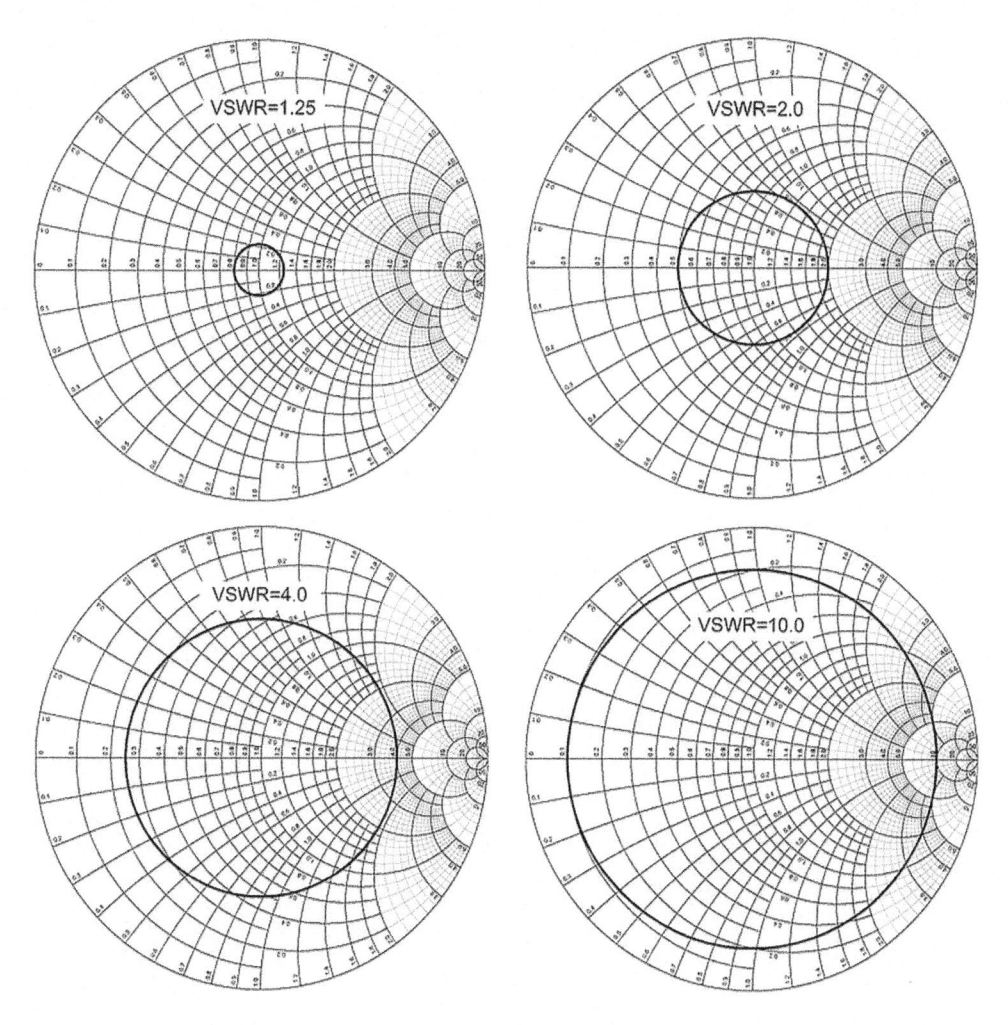

Figure 3-22 VSWR circles shown for VSWR values of 1.25, 2.0, 4.0, and 10.0

3.7 Adding a Transmission Line in Series with an Impedance

We have seen that adding a reactance in series with an impedance point causes the impedance to follow the constant resistance circles. Adding a transmission line of the same impedance as the Smith Chart's normalized impedance, in series with an impedance point causes the resulting impedance to follow the constant VSWR circle in which the impedance lies. The impedance moves in a clockwise direction on the constant VSWR circle.

Example 3.7-1: Revisiting the 25+j25 Ω impedance in Genesys, calculate the electrical length of a series transmission line moving the impedance at point A to point B on the Smith Chart, as shown in Figure 3-23.

Solution: Add a series transmission line to the impedance and make the electrical length tunable. Increase the line length to move the impedance to point B. The electrical line length is 58.45 degrees. The impedance on the real axis (zero reactance) on the right hand side of the Smith Chart would represent a point of maximum voltage-minimum current along the transmission line.

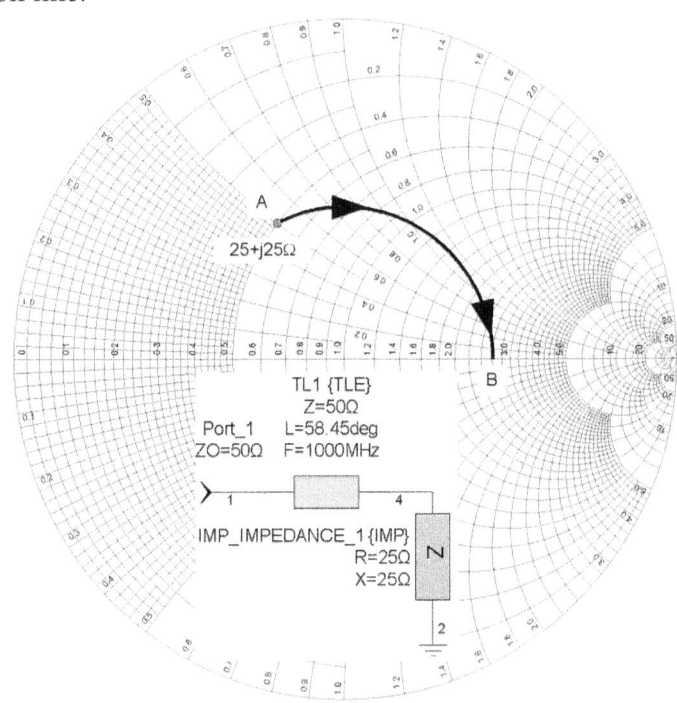

Figure 3-23 Series transmission line moves impedance to a maximum voltage point

Transmission line lengths are sometimes referred to in terms of fractional wavelengths. Because one wavelength is equal to 360°, a 58.45° electrical length represents (58.45/360) 0.162λ. Continue to add electrical length to the transmission line to reach point C on the Smith Chart. As Figure 3-24 shows, the real impedance (zero reactance) on the left side of the horizontal

Network Parameters and the Smith Chart 139

axis on the Smith Chart represents a minimum voltage-maximum current point along the transmission line.

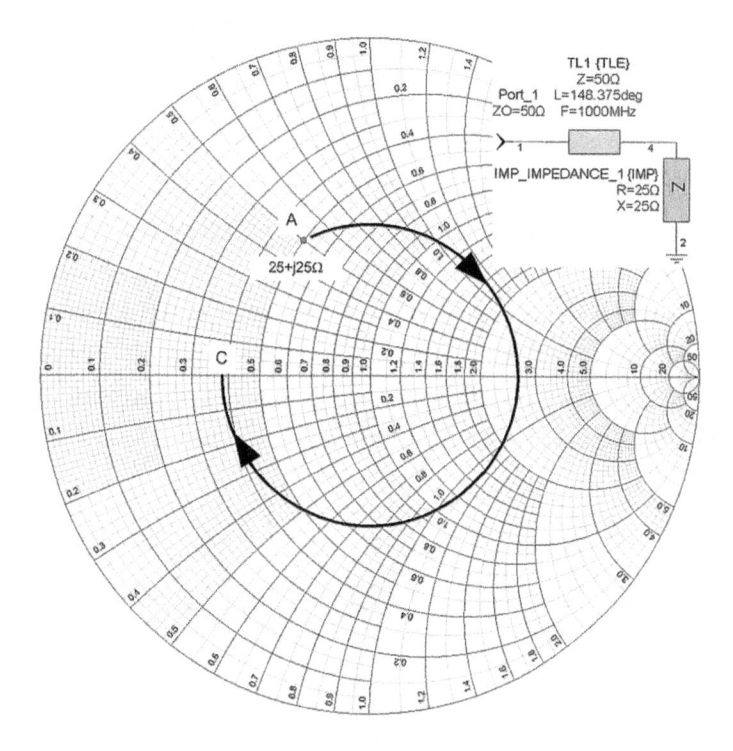

Figure 3-24 Series transmission line moves impedance to a minimum voltage point

Further increasing the length of the transmission line we find that we arrive back at point A at 180 degrees of electrical length. Therefore the electrical distance around the Smith Chart is 180° or $\lambda/2$ wavelength. The points of maximum voltage and minimum voltage will repeat every $\lambda/2$ wavelength.

3.8 Adding a Transmission Line in Parallel with an Impedance

In Chapter 2 section 2.11 we have seen that the open and short-circuited transmission lines could take on the equivalence of an inductor, capacitor, or series and parallel resonant circuits depending on the electrical length of the line. Therefore the shunt transmission line will behave more like the lumped element movements on the Smith Chart.

3.8.1 Short Circuit Transmission Lines

At DC and low frequencies, a short circuit is a very low inductance but this is not the case at higher RF and microwave frequencies. Figure 3-25 shows that a transmission line with 0° length (perfect short) appears at the short circuit point on the Smith Chart. As 45° of electrical length is added to the short circuit, the impedance moves clockwise along the outer circumference of the Smith Chart to the position shown at the top of the chart. As the line length is increased to 90° we see that the short circuit has been transformed to an open circuit. At 180° line length the impedance will travel completely around the Smith Chart and appear as a short circuit again. Therefore depending on the line length the short circuit transmission line can be transformed into a shunt capacitor or shunt inductor.

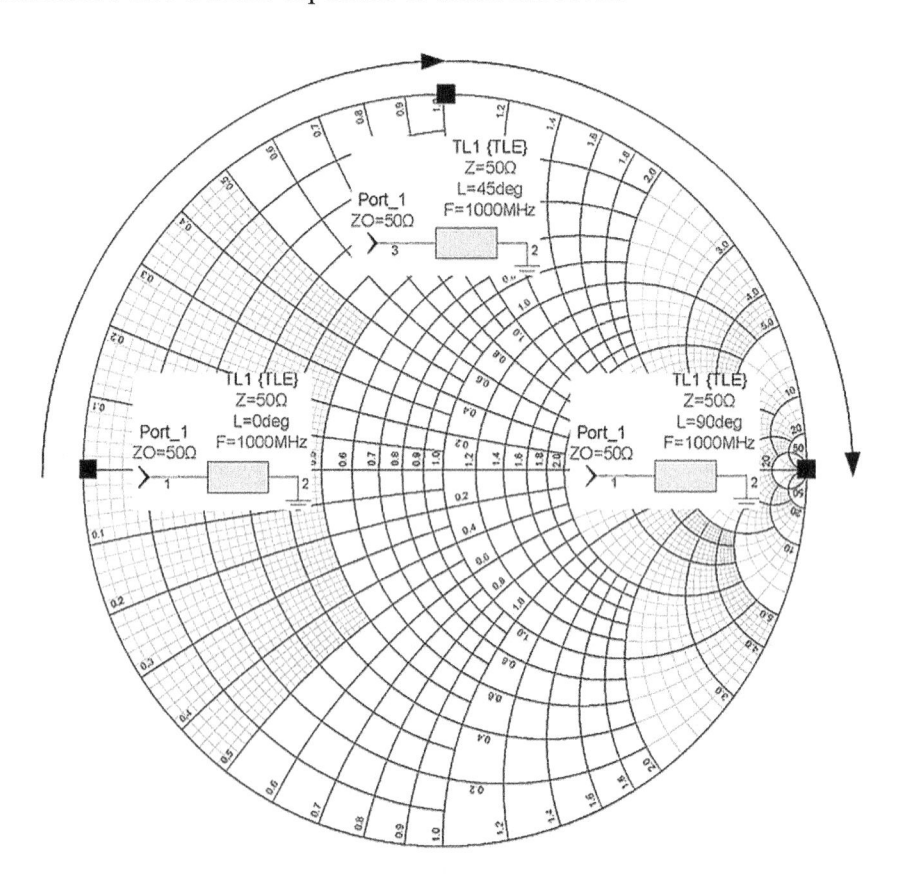

Figure 3-25 Short circuit transmission line impedance at electrical lengths: 0°, 45°, and 90°

3.8.2 Open Circuit Transmission Lines

Figure 3-26 shows the characteristic of the open circuit transmission line. At 0° electrical line length it appears as a perfect open circuit. As the electrical length is increased, the impedance moves clockwise around the circumference to the 45° position at the bottom of the chart. At 90° electrical length the open circuit now appears as a short circuit. This property of transforming open circuits to short circuits and vice versa is one that is used frequently throughout microwave circuit design.

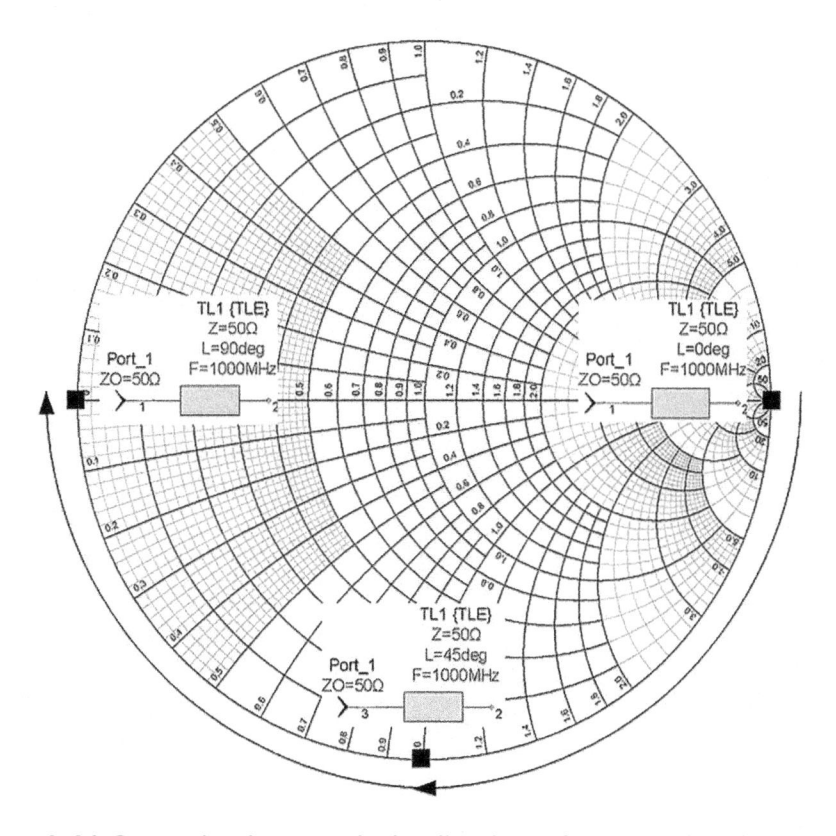

Figure 3-26 Open circuit transmission line impedance at electrical lengths: 0°, 45°, and 90°

3.9 Open and Short Circuit Shunt Transmission Lines

For small fractional wavelength transmission lines the open circuit shunt transmission line acts as a shunt capacitor.

Example 3.9-1: Measure the electrical length of a shunt transmission line or the amount of shunt capacitance to move the impedance $Z = 25 + j25\ \Omega$ from point A to the center of Smith Chart, as shown in Figure 3-27.

Solution: Plotting the impedance on the Smith Chart with the admittance circles shows that the impedance lies directly on the unit conductance circle. Therefore an open circuit shunt transmission line can move this impedance directly to 50 Ω. As the schematic inset of Figure 3-27 shows, a 45.38° electrical length of a 50 Ω shunt transmission line moves the impedance to the center of the Smith Chart. The second schematic inset of Figure 3-27 shows that this circuit is equivalent to a 3.2 pF capacitor in shunt with the load impedance.

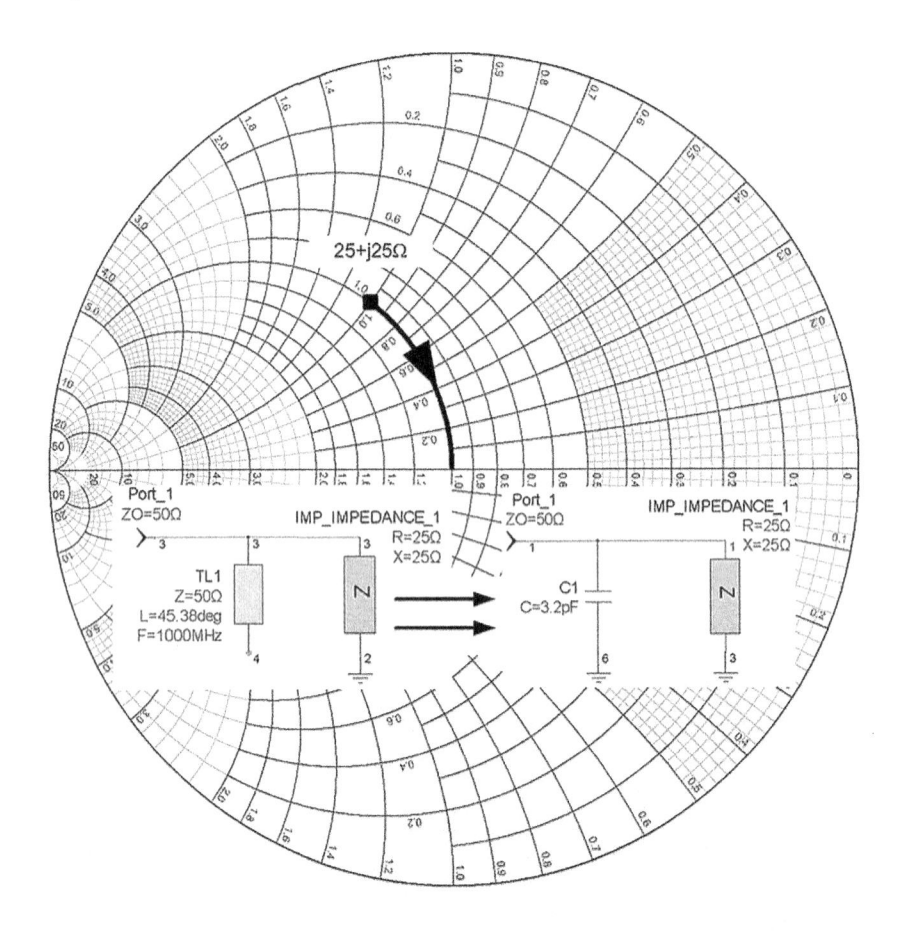

Figure 3-27 Open circuit shunt transmission line added to 25+j25 Ω impedance

Network Parameters and the Smith Chart 143

For small fractional wavelength transmission lines the short circuit shunt transmission line acts as a shunt inductor. Consider the 4.3 –j14 Ω impedance as shown in Figure 3-28. This impedance lies on the unit conductance circle on the bottom half of the Smith Chart. A 50 Ω shunt transmission line added to the impedance moves along the constant conductance circle to the center of the Smith Chart. Similarly a 2.4 nH shunt inductor has the same effect at a frequency of 1000 MHz. These movements form the basis for distributed network impedance matching which is covered in detail in Chapter 6.

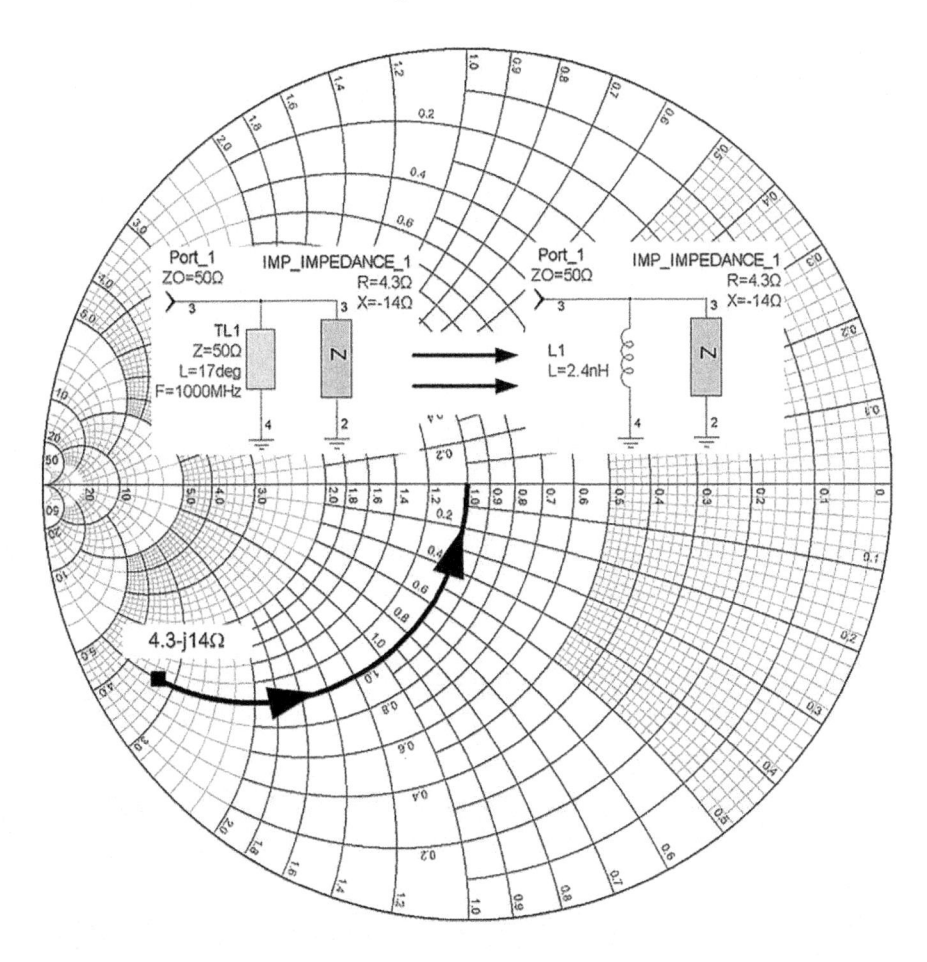

Figure 3-28 Short circuit shunt transmission line added to 4.3-j14 Ω impedance

References and Further Reading

[1] David M. Pozar, *Microwave Engineering*, Fourth Edition, John Wiley and Sons, Inc., 2012

[2] Ali A. Behagi and Stephen D. Turner, *Microwave and RF Engineering*, BT Microwave LLC, State College, PA, 2011

[3] William Sinnema, Electronic Transmission Technology, Prentice-Hall, Inc, Englewood Cliffs, New Jersey 07632, 1979

[4] Chris Bowick, RF Circuit Design, Second Edition, Newnes, Elsevier, 2008

[5] Keysight Technologies, *Genesys 2014.03, Users Guide*, www.keysight.com

Problems

3-1. Place a 20 + j30 Ω impedance at point A on the Smith Chart. Add a series inductance to move the impedance along the constant resistance circle to point B having the impedance 20 + j50 Ω. Using a design frequency of 1000MHz, calculate the inductance that this reactance represents.

3-2. Continuing with Problem 3-1 enable the admittance coordinates on the Smith Chart to add a shunt element. Add a shunt capacitance to move the impedance to the real axis as shown in Figure 3-16. Measure the susceptance required to move to the real impedance axis by the difference between the intersecting susceptance lines.

3-3. Use the Genesys Equation Editor to calculate the magnitude of reflection coefficient for a desired VSWR = 2.

3-4. Using Genesys, create a constant VSWR circle for a VSWR=20. Comment on the VSWR value required to place the VSWR circle on the circumference of the Smith Chart.

3-5. A load of 75 + j20 Ω is connected to a 50 Ω transmission line. Calculate the load admittance and the input impedance if the line is 0.2 wavelengths long.

3-6. For the load impedance in Problem 3-4, determine the reflection coefficient and the transmission coefficient.

3-7. For the load impedance in Problem 3-4, determine the normalized value of the load impedance if the impedance is normalized to 75 Ω.

3-8. A series RLC load, R = 100 Ω, L = 20 nH, C = 25 pF is connected to a 50 Ω transmission line. Calculate the VSWR and reflection coefficient at the load at 100 MHz.

3-9. Using the RLC load impedance of problem 3-8 determine the impedance with a series transmission line of characteristic impedance of 50 Ω and electrical length of 180 degrees.

3-10. Determine the input impedance of a network that has a reflection coefficient of 0.5 at an angle of 112°.

3-11. Create a one-port S parameter text file with the impedance of Problem 3-10 at a frequency of 1 GHz. Plot the S parameter, S11, on the Smith Chart in Genesys.

3-12. Using the S parameter file of the C Band amplifier of Figure 3-8, plot the input return loss, S11, and output return loss, S22, on the Smith Chart. Determine the worst case input and output VSWR for this amplifier.

Chapter 4

Resonant Circuits and Filters

4.1 Introduction

The first half of this chapter examines resonant circuits. Lumped element resonant circuits and the lumped equivalent networks of mechanical and distributed resonators are considered. Resonant circuits are used in many applications, such as filters, oscillators, tuners, tuned amplifiers, and microwave communication networks. The analysis of basic lumped element series and parallel RLC resonant circuits is implemented in the Genesys software. The discussion turns to microwave resonators with an analysis of the Q factor measurement of transmission line resonators. Using the Genesys software a robust technique is demonstrated for the evaluation of Q factor from the measured S parameters of a resonant circuit. The second half of the chapter is an introduction to the vast subject of filter networks. The design of lumped element filters is introduced and followed by an introduction to distributed element filters. The chapter concludes with an introduction to Electromagnetic (EM) simulation of distributed filters.

4.2 Resonant Circuits

Near resonance, RF and microwave resonant circuits can be represented either as a lumped element series or parallel RLC network.

4.2.1 Series Resonant Circuits

In this section we analyze the behavior of the resonant circuit in Genesys.

Example 4.2-1: Consider the one port resonator that is represented as a series RLC circuit of Figure 4-1. Analyze the circuit, with R = 10 Ω, L = 10 nH, and C = 10 pF.

Solution: The plot of the resonator's input impedance in Figure 4-1 shows that the resonance frequency is about 503.3 MHz and the input impedance at resonance is 10 Ω, the value of the resistor in the network.

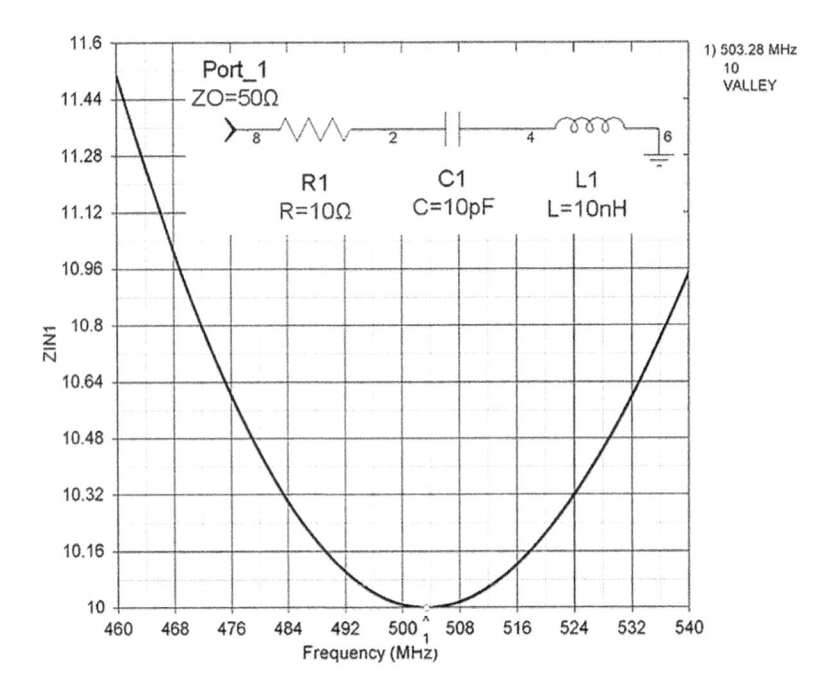

Figure 4-1 One-port series RLC resonator circuit and input impedance

The input impedance of the series RLC resonant circuit is given by,

$$Z_{in} = R + j\omega L - j\frac{1}{\omega C}$$

where, $\omega = 2\pi f$ is the angular frequency in radian per second.

If the AC current flowing in the series resonant circuit is I, then the complex power delivered to the resonator is

$$P_{in} = \frac{|I|^2}{2}Z_{in} = \frac{|I|^2}{2}\left(R + j\omega L - j\frac{1}{\omega C}\right) \quad (4\text{-}1)$$

At resonance the reactive power of the inductor is equal to the reactive power of the capacitor. Therefore, the power delivered to the resonator is equal to the power dissipated in the resistor

$$P_{in} = \frac{|I|^2 R}{2} \quad (4\text{-}2)$$

4.2.2 Parallel Resonant Circuits

Example 4.2-2: Analyze a rearrangement of the RLC components of Figure 4-1 into the parallel configuration of Figure 4-2. The schematic of Figure 4-2 represents the lumped element representation of the parallel resonant circuit.

Solution: The plot of the magnitude of the input impedance shows that the resonance frequency is still 503.3 MHz where the input impedance is R = 10 Ω. Again this shows that the impedance of the inductor cancels the impedance of the capacitor at resonance. In other words, the reactance, X_L is equal to the reactance, X_C, at the resonance frequency.

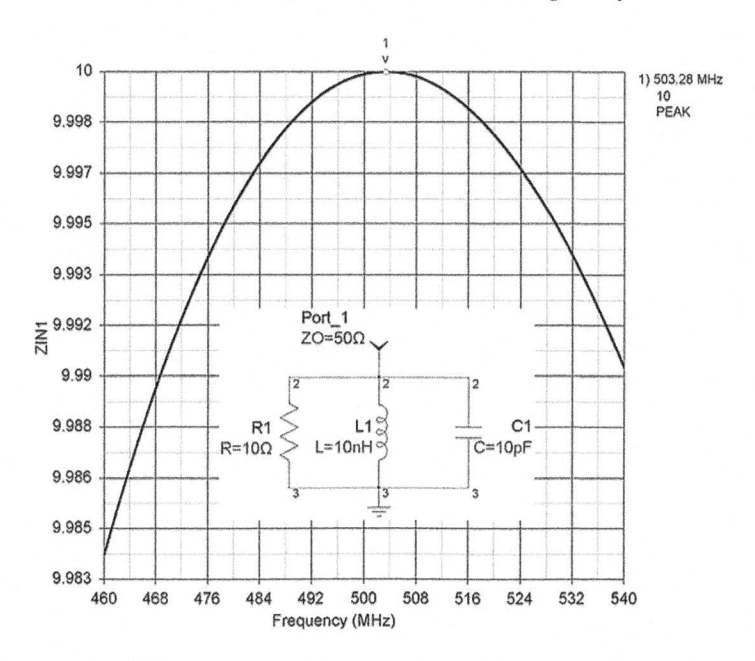

Figure 4-2 One-port parallel RLC resonant circuit and input impedance

The input admittance of the parallel resonant circuit is given by:

$$Y_{IN} = \frac{1}{R} + j\omega C - j\frac{1}{\omega L}$$

If the AC voltage across the parallel resonant circuit is V, then the complex power delivered to the resonator is

$$P_{in} = \frac{|V|^2}{2} Y_{in} = \frac{|V|^2}{2}\left(\frac{1}{R} + j\omega C - j\frac{1}{\omega L}\right) \qquad (4\text{-}3)$$

At resonance the reactive power of the inductor is equal to the reactive power of the capacitor. Therefore, the power delivered to the resonator is equal to the power dissipated in the resistor

$$P_{in} = \frac{|V|^2}{2R} \qquad (4\text{-}4)$$

The resonance frequency for the parallel resonant circuit as well as the series resonant circuit is obtained by setting $\omega_0 C = \dfrac{1}{\omega_0 L}$ or:

$$\omega_o = 2\pi f_o = \frac{1}{\sqrt{LC}} \qquad (4\text{-}5)$$

where, ω_0 is the angular frequency in radian per second and f_0 is equal to the frequency in Hertz.

4.2.3 Resonant Circuit Loss

In Figure 4-1 and 4-2 the resistor, R1, represents the loss in the resonator. It includes both the losses in the capacitor as well as the inductor. The Q factor can be shown to be a ratio of the energy stored in the inductor and capacitor to the power dissipated in the resistor as a function of frequency [6]. For the series resonant circuit of Figure 4-1 the Q factor is defined by:

$$Q_u = \frac{X}{R} = \frac{\omega_o L}{R} = \frac{1}{\omega_o RC} \qquad (4\text{-}6)$$

The Q factor of the parallel resonant circuit is simply the inverse of the series resonant circuit.

$$Q_u = \frac{R}{X} = \frac{R}{\omega_o L} = \omega_o RC \qquad (4\text{-}7)$$

Resonant Circuits and Filters 151

Notice that as the resistance increases in the series resonant circuit, the Q factor decreases. Conversely as the resistance increases in the parallel resonant circuit, the Q factor increases. The Q factor is a measure of loss in the resonant circuit. Thus a higher Q corresponds to lower loss and a lower Q corresponds to a higher loss. It is usually desirable to achieve high Q factors in a resonator as it will lead to lower losses in filters or lower phase noise in oscillators. Note that the resonator Q of Equation (4-6) and (4-7) is defined as the unloaded Q of the resonator. This means that the resonator is not connected to any source or load impedance and as such is unloaded. The measurement of Q_u requires that the resonator be attached (coupled) to a signal source or load of some finite impedance. Equations (4-6) and (4-7) would then have to be modified to include the source and load resistance. We might also surmise that any reactance associated with the source or load impedance may alter the resonant frequency of the resonator. This leads to two additional definitions of Q factor; the loaded Q and external Q.

4.2.4 Loaded Q and External Q

Example 4.2-3: Analyze the parallel resonator that is attached to a 50 Ω source and load as shown in Figure 4-3.

Solution: Using Equation (4-7) to define the Q factor for the circuit requires that we include the source and load resistance which is 'loading' the resonator. This leads to the definition of the loaded Q, Q_L, for the parallel resonator as defined by Equation (4-8).

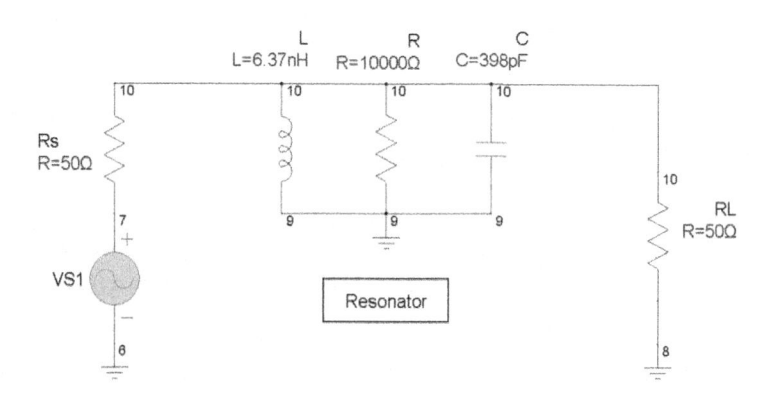

Figure 4-3 Parallel resonator with source and load impedance attached

$$Q_L = \frac{R_S + R + R_L}{\omega_o L} \tag{4-8}$$

Conversely we can define a Q factor in terms of only the external source and load resistance. This leads to the definition of the external Q, Q_E.

$$Q_E = \frac{R_S + R_L}{\omega_o L} \tag{4-9}$$

The three Q factors are related by the inverse relationship of Equation (4-10).

$$\frac{1}{Q_L} = \frac{1}{Q_E} + \frac{1}{Q_U} \tag{4-10}$$

At RF and microwave frequencies it is difficult to directly measure the Q_u of a resonator. We may be able to calculate the Q factor based on the physical properties of the individual inductors and capacitors as we seen in chapter 1. This is usually quite difficult and the Q factor is typically measured using a Vector Network Analyzer, VNA. Therefore, the measured Q factor is usually the loaded Q, Q_L. External Q is often used with oscillator circuits that are generating a signal. In this case the oscillator's load impedance is varied so that the external Q can be measured. The loaded Q of the network is then related to the fractional bandwidth by Equation (4-11) [8].

$$Q_L = \frac{f_o}{BW} \tag{4-11}$$

where, BW is the 3 dB bandwidth in Hertz and f_0 is equal to the resonant frequency in Hertz.

4.3 Lumped Element Parallel Resonator Design

Example 4.3-1: In this example we design a lumped element parallel resonator at a frequency of 100 MHz. The resonator is intended to operate between a source resistance of 100 Ω and a load resistance of 400 Ω.

Solution: Best accuracy would be obtained by using S parameter files or Modelithics models for the inductor and capacitor. However a good first order model can be obtained by using the Genesys inductor and capacitor models that include the component Q factor. These models save us the work of calculating the equivalent resistive part of the inductor and capacitor model. Use the Q factors shown in the schematic of Figure 4-4.

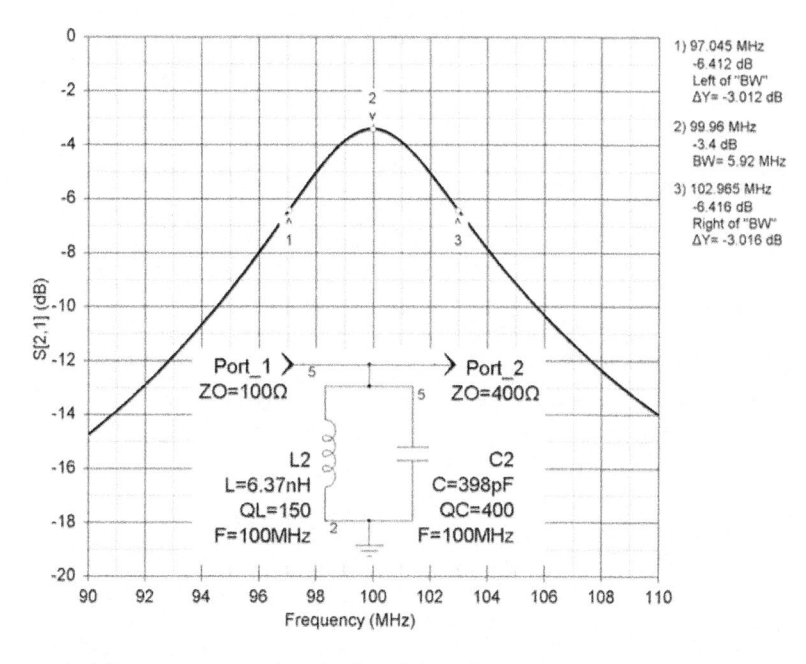

Figure 4-4 Resonator using inductor and capacitor with assigned Q values

Note that markers have been placed on the plot of the insertion loss, S21 that gives a direct readout of the 3 dB bandwidth. The loaded Q, Q_L can be calculated using Equation (4-11).

$$Q_L = \frac{f_o}{BW} = \frac{99.955\,MHz}{5.923\,MHz} = 16.87$$

The designer must use caution when sweeping resonant circuits in Genesys. Particularly high Q band pass networks require a large number of discrete frequency steps in order to achieve the necessary resolution required to accurately measure the 3dB bandwidth. In this example the Linear Analysis is set up to sweep the circuit from 90 MHz to 110 MHz using 2000 points.

Place a marker anywhere on the trace and double click to open the Marker Properties window. Enter any name for the marker, select Bandwidth (Tracks Peak), and make sure that -3.01dB is entered as the relative offset. As Figure 4-4 shows the 3 dB bandwidth is automatically calculated as 5.923 MHz. The frequency peak and bandwidth label next to marker 2 is then used to calculate Q_L.

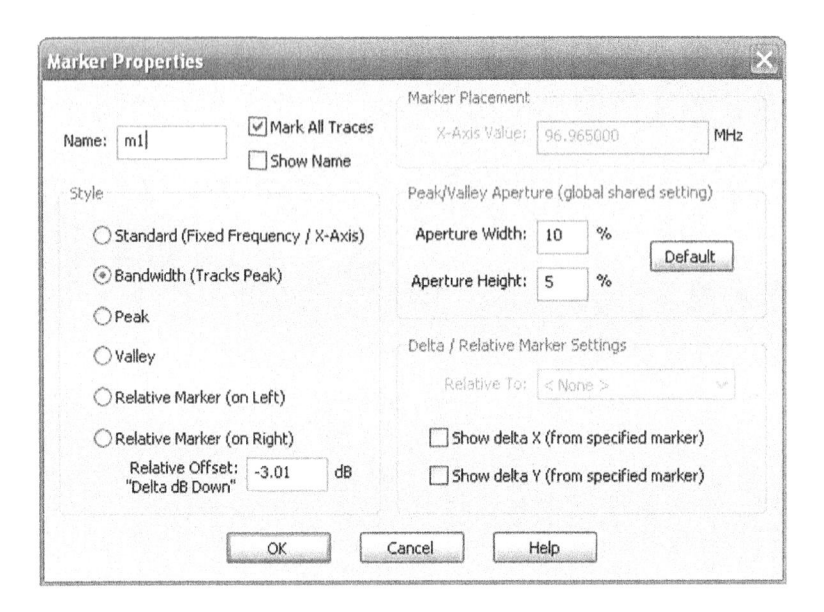

Figure 4-5 Bandwidth marker settings for measurement of 3dB bandwidth

4.3.1 Effect of Load Resistance on Bandwidth and Q_L

In RF circuits and systems the impedances encountered are often quite low, ranging from 1 Ω to 50 Ω. It may not be practical to have a source impedance of 100 Ω and a load impedance of 400 Ω.

Example 4.3-2: Using the previous parallel LC circuit example, change the load resistance from 400 Ω to 50 Ω and re-examine the circuit's 3 dB bandwidth and Q_L.

Solution: The 3 dB bandwidth is now 12.926 MHz resulting in a loaded Q factor of 7.73. The loaded Q factor has decreased by nearly half of the original value. We have increased the bandwidth or de-Q'd the resonator.

Resonant Circuits and Filters 155

This can also be thought of as tighter coupling of the resonator to the load.

$$Q_L = \frac{f_o}{BW} = \frac{99.965\,MHz}{12.926\,MHz} = 7.73$$

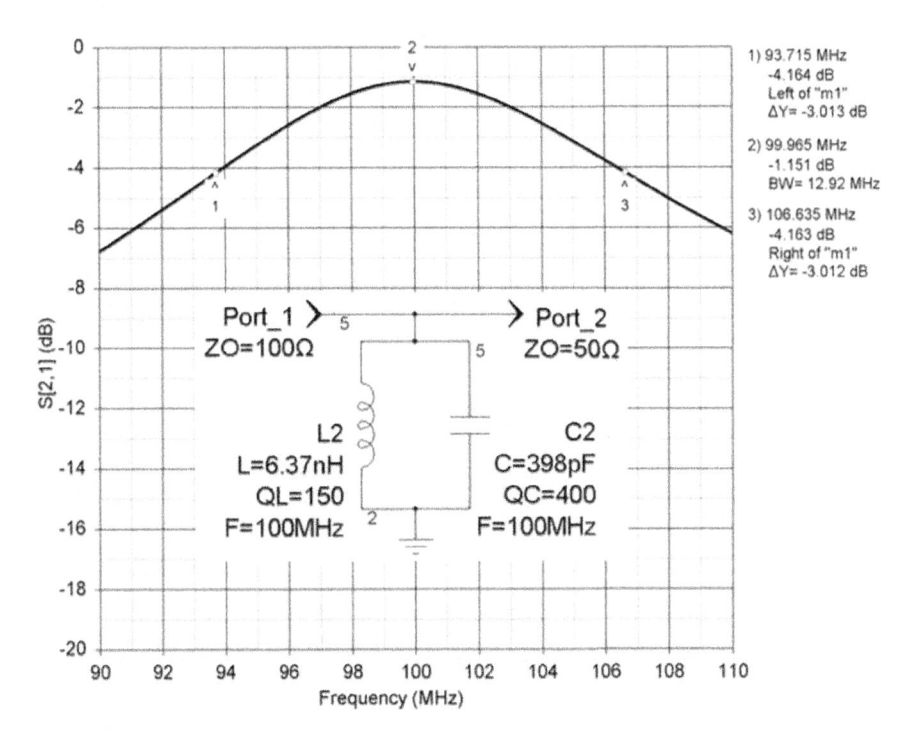

Figure 4-6 A parallel resonance circuit showing the 3dB bandwidth and insertion loss

4.4 Lumped Element Resonator Decoupling

To maintain the high Q of the resonator when attached to a load such as 50 Ω, it is necessary to transform the low impedance to high impedance presented to the load. The 50 Ω impedance can be transformed to the higher impedance of the parallel resonator thereby resulting in less loading of the resonator impedance. This is referred to as loosely coupling the resonator to the load. The tapped-capacitor and tapped-inductor networks can be used to accomplish this Q transformation in lumped element circuits.

4.4.1 Tapped Capacitor Resonator

Example 4.4-1: Consider rearranging the parallel LC network of Figure 4-6 with the tapped capacitor network shown in Figure 4-7. Re-examine the circuit's 3 dB bandwidth and Q_L.

Solution: The new capacitor values for C1 and C2 can be found by the simultaneous solution of the following equations.

$$C_T = \frac{C1 \cdot C2}{C1+C2} \qquad (4\text{-}12)$$

$$R_{L1} = R_L \left(1+\frac{C1}{C2}\right)^2 \qquad (4\text{-}13)$$

R_{L1} is the higher, transformed, load resistance. In this example substitute $R_{L1} = 400\ \Omega$, the original load resistance value. C_T is simply the original capacitance of 398 pf. The capacitor values are found to be: C1 = 1126.23 pF and C2 = 616.1 pF. The new resonator circuit is shown in Figure 4-7. Sweeping the circuit we see that the response has returned to the original performance of Figure 4-6. The 3 dB bandwidth has returned to 5.923 MHz making the Q_L equal to:

$$Q_L = \frac{99.925}{5.923} = 16.87$$

The 50 Ω load resistor has been successfully decoupled from the resonator. The tapped capacitor and inductor resonators are popular methods of decoupling RF and lower microwave frequency resonators. It is frequently seen in RF oscillator topologies such as the Colpitts oscillator in the VHF frequency range.

Resonant Circuits and Filters

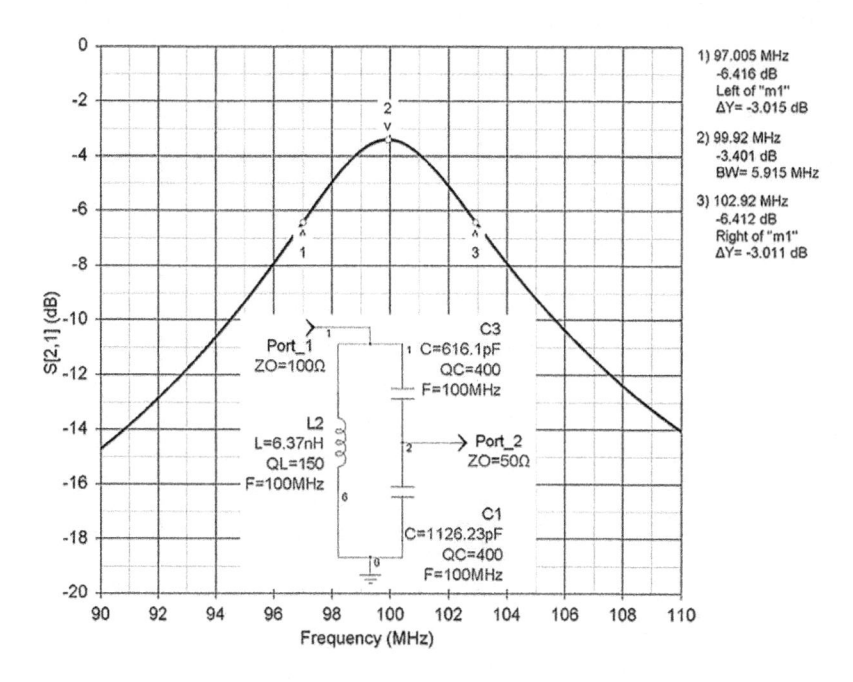

Figure 4-7 Parallel LC resonator using a tapped capacitor and response

4.4.2 Tapped Inductor Resonator

Example 4.4-2: Similarly design a tapped inductor network to decouple the 50 Ω source impedance from loading the resonator.

Solution: Replace the 100 Ω source impedance with a 50 Ω source and use a tapped inductor network to transform the new 50 Ω source to 100 Ω. Modify the circuit to split the 6.37 nH inductor, L_T, into two series inductors, L1 and L2. The inductor values can then be calculated by solving the following equation set simultaneously. R_{S1} is the higher, transformed, source resistance. In this example substitute $R_{S1} = 100$ Ω,

$$Rs1 = Rs \left(\frac{n}{n_1} \right)^2 \tag{4-14}$$

$$n = n_1 + n_2 \tag{4-15}$$

Solving the equation set results in values of L1=4.5 nH and L2=1.87 nH. The resulting schematic and response is shown in Figure 4-8. The new response is identical to the plot of Figure 4-5. Therefore we now have a source and load resistance of 50 Ω and have not reduced the Q of the resonator from what we had with the original source resistance of 100 Ω and a load resistance of 400 Ω.

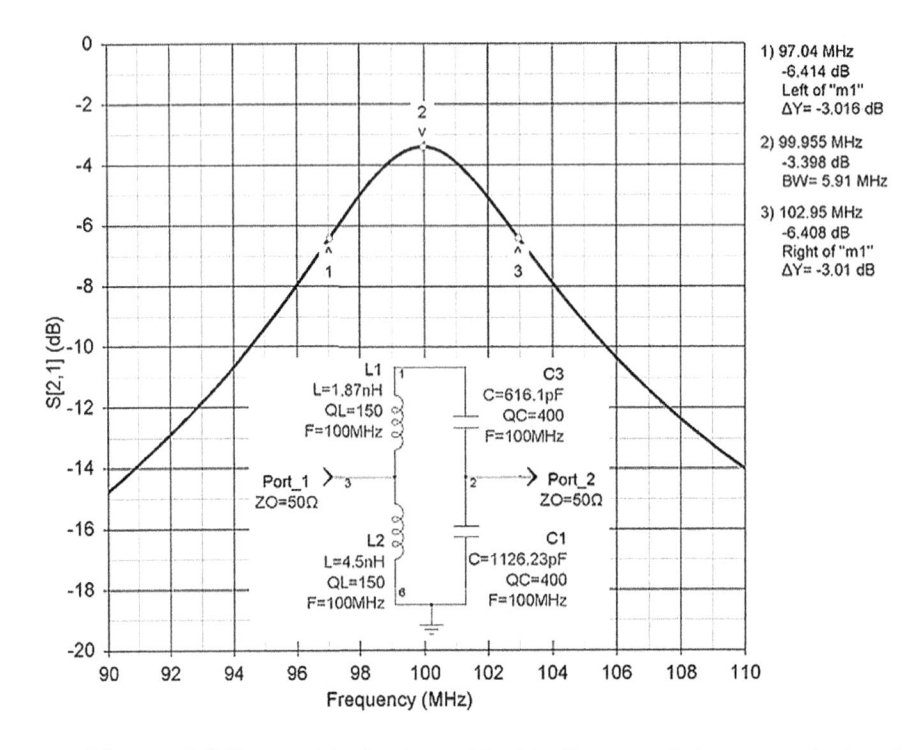

Figure 4-8 Tapped-inductor added to the parallel resonant circuit

4.5 Practical Microwave Resonators

At higher RF and microwave frequencies resonators are seldom realized with discrete lumped element RLC components. This is primarily due to the fact that the small values of inductance and capacitance are physically unrealizable. Even if the values could be physically realized we would see that the resulting Q factors would be unacceptably low for most applications. Microwave resonators are realized in a wide variety of physical forms. Resonators can be realized in all of the basic transmission line forms that were covered in Chapter 2. There are many specialized resonators such as ceramic dielectric resonator pucks that are coupled to a microstrip transmission line as well as Yittrium Iron Garnet spheres that are loop coupled to its load. These resonators are optimized for very high Q factors and may be tunable over a range of frequencies.

Figure 4-9 Ceramic dielectric resonator (puck) and rectangular coaxial resonator

4.5.1 Transmission Line Resonators

From Fig. 2-26 we have seen that a quarter-wave short-circuited transmission line results in a parallel resonant circuit. Similarly Figure 2-29 showed that a half-wave open circuited transmission line results in a parallel resonant circuit. Such parallel resonant circuits are often used as one port resonators. Near the resonant frequency, the one port resonator behaves as a parallel RLC network as shown in Figure 4-2. As the frequency moves further from resonance the equivalent network becomes more complex typically involving multiple parallel RLC networks. One port resonators are coupled to one another to form filter networks or directly to a transistor to form a microwave oscillator. Knowing the losses due to the physical and electrical parameters of the transmission line, one can calculate the Q_u of the transmission line resonator. The microstrip resonator Q_u is comprised of losses due to the conductor metal, the substrate dielectric, and radiation losses. The Q_u is often dominated by the conductor Q. Unfortunately it can be quite difficult to accurately determine the conductor losses in a microstrip resonator. T. C. Edwards has developed a set of simplified expressions for the conductor losses [4]. Equation (4-16) is an approximation of the conductor losses that treats the transmission line as a perfectly smooth surface.

$$\alpha_c = 0.072 \frac{\sqrt{f}}{W_e Z_o} \lambda_g \quad \text{dB/inch} \quad (4\text{-}16)$$

where,
f = the frequency in GHz
W_e = the effective conductor width (inches)
Z_o = the characteristic impedance of the line
α_c = Conductor loss in dB/inch
λ_g = wavelength in dielectric in inches

A microstrip conductor is actually not perfectly smooth but exhibits a certain roughness. The surface roughness exists on the bottom of the microstrip conductor where it contacts the dielectric. This can be seen by magnifying the cross section of a microstrip line's contact with the dielectric material. The surface roughness is usually specified as an r.m.s. value.

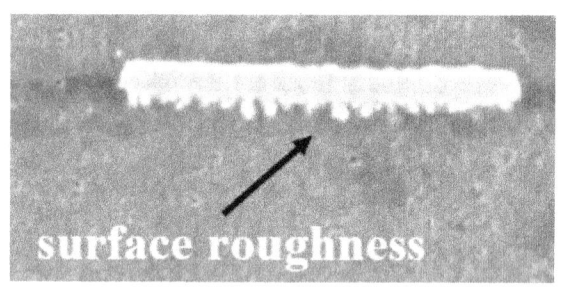

Figure 4-10 Cross section of microstrip line showing surface roughness at the conductor to dielectric interface (courtesy of Tektronix)

Edwards modified Equation (4-16) to include the effects of the surface roughness as given by Equation (4-17).

$$\alpha_c' = \alpha_c \left[1 + \frac{2}{\pi} \tan^{-1} \left(1.4 \left(\frac{\Delta}{\delta_s} \right)^2 \right) \right] \quad \text{dB/inch} \quad (4\text{-}17)$$

where, Δ is the r.m.s. surface roughness and δ_s is the conductor skin depth.

Resonant Circuits and Filters

The corresponding Q factor related to the conductor is then given by:

$$Q_c = \frac{27.3\sqrt{\varepsilon_{eff}}}{\alpha_c \lambda_o} \quad (4\text{-}18)$$

The dielectric loss is determined by the dielectric constant and loss tangent. It is calculated using Equation (4-19).

$$\alpha_d = 27.3 \frac{\varepsilon_r (\varepsilon_{eff}-1)\tan\delta}{\sqrt{\varepsilon_{reff}}(\varepsilon_r - 1)\lambda_o} \quad \text{dB/inch} \quad (4\text{-}19)$$

where:
ε_r = the substrate dielectric constant
ε_{reff} = the effective dielectric constant
$\tan\delta$ = the loss tangent of the dielectric
λ_o = wavelength in inches

The corresponding Q factor due to the dielectric is then given by:

$$Q_d = 27.3 \frac{\sqrt{\varepsilon_{eff}}}{\alpha_d \lambda_o} \quad (4\text{-}20)$$

We know that a microstrip line will also have some radiation of energy from the top side of the line. The open circuit stub will also experience some radiation effect from the open circuited end. On low dielectric constant substrates, $\varepsilon_r \leq 4.0$, the radiation losses are more significant for high impedance lines. Conversely for high dielectric constant substrates, $\varepsilon_r \geq 10$, low impedance lines experience more radiation loss [10]. The radiation Q factor is presented as Equation (4-21) [9].

$$Q_r = \frac{Z_o(f)}{480\pi \left(\frac{h}{\lambda_o}\right)^2 \left\{\left(\frac{\varepsilon_{eff(f)}+1}{\varepsilon_{eff(f)}}\right) - \left[\frac{(\varepsilon_{eff(f)}-1)^2}{2(\varepsilon_{eff(f)})^{3/2}} \ln\left(\frac{\sqrt{\varepsilon_{eff(f)}}+1}{\sqrt{\varepsilon_{eff(f)}}-1}\right)\right]\right\}} \quad (4\text{-}21)$$

where, h = substrate thickness in cm.

Note that in Equation (4-21) the line impedance and effective dielectric constant are defined as functions of frequency. This includes the dispersion or frequency dependent effect of Z_o and ε_{eff}. Dispersion tends to slightly increase the ε_{eff} as the frequency increases. This dispersive Z_o and ε_{eff} are given in Equations (4-22) and (4-23).

$$\varepsilon_{eff(f)} = \frac{\varepsilon_r - \varepsilon_{eff}}{1 + \left[(0.6 + 0.009 Z_o) \left(\frac{f}{Z_o / 8\pi(h - 2t)} \right)^2 \right]} \quad (4\text{-}22)$$

where,
    h = substrate thickness in mils
    t = conductor thickness in mils

$$Z_{o(f)} = Z_o \sqrt{\frac{\varepsilon_{eff}}{\varepsilon_{eff(f)}}} \quad (4\text{-}23)$$

Finally the resultant overall unloaded Q factor, Q_u, of the microstrip line can be determined by the reciprocal relationship of Equation (4.24).

$$\frac{1}{Q_u} = \frac{1}{Q_c} + \frac{1}{Q_d} + \frac{1}{Q_r} \quad (4\text{-}24)$$

4.5.2 Microstrip Resonator Example

Example 4.5-1: Consider a 5 GHz half wavelength open circuit microstrip resonator. The resonator is realized with a 50 Ω microstrip line on Roger's RO3003 dielectric. Calculate the unloaded Q factor of the resonator. The substrate parameters are defined as:

    Dielectric constant $\varepsilon_r = 3$
    Substrate height   h = 0.030 in.
    Conductor thickness   t = .0026 in.
    Line Impedance $Z_o = 50$ Ω
    Conductor width   w = 0.075 in.
    Loss tangent $\tan\delta = 0.0013$

Resonant Circuits and Filters

Solution: Using the simplified expression of Equation (4-16) for a smooth microstrip line the loss and conductor Q factor is calculated as:

$$\alpha_c = 0.072 \frac{\sqrt{5}}{0.077(50)} \cdot 1.52 = 0.063 \; dB/inch$$

$$Q_c = \frac{27.3\sqrt{2.41}}{(0.063)(2.36)} = 283.1$$

The dielectric loss and Q factor are then calculated from Equation (4-19) and (4-20).

$$\alpha_d = 27.3 \frac{(3)(2.41-1)(0.0013)}{\sqrt{2.41}(3-1)(2.36)} = 0.021 \, dB/inch$$

$$Q_d = 27.3 \frac{\sqrt{2.41}}{(0.021)(2.36)} = 875.2$$

For simplicity the radiation Q factor will be omitted. We will model the resonator using the Genesys Linear simulator. Linear simulators often do not model the radiation effects of the microstrip line. Therefore the overall unloaded Q factors then becomes.

$$\frac{1}{Q_u} = \frac{1}{283.1} + \frac{1}{875.2} = \frac{1}{213.9}$$

$$Q_u = 213.9$$

4.5.3 Genesys Model of the Microstrip Resonator

The half wave open circuit microstrip resonator is modeled in Genesys as shown in Fig 4-11. Note that the source and load impedance has been increased to 5000 Ω to avoid loading the impedance of the parallel resonant circuit. Perform a linear sweep of the resonator using 4001 points from 4500 MHz to 5400 MHz.

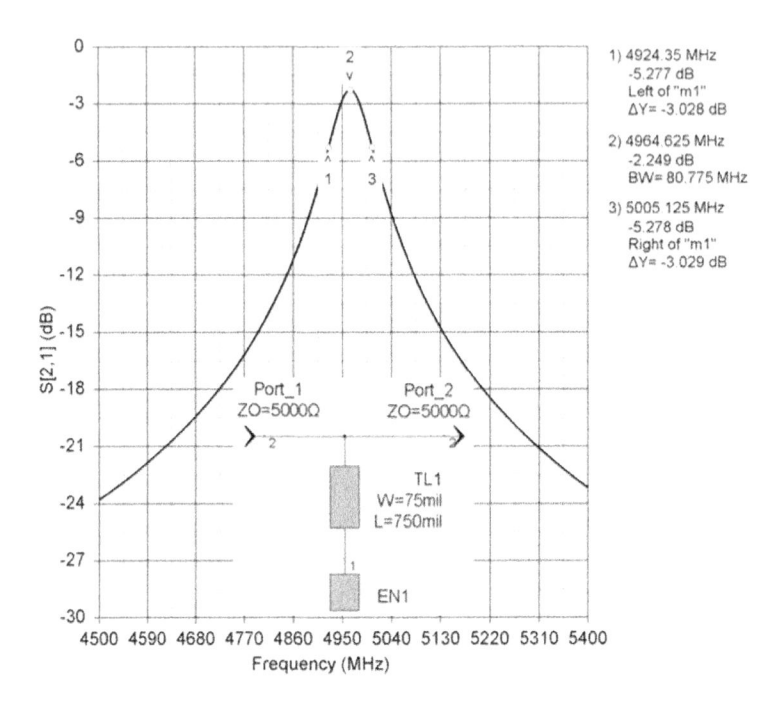

Figure 4-11 Half-wave open circuit microstrip resonator

Using the techniques of section 4.2.3 and Equation (4-10), the 3dB bandwidth is measured to determine the loaded Q of the resonator.

$$Q_L = \frac{4964.625}{80.775} = 61.5$$

The insertion loss at the resonant frequency can be used to relate the Q_L to the Q_u as shown by Equation (4-25). The Q_u as simulated by Genesys is within 20% of the value calculated using the substrate physical and electrical characteristics.

$$InsertionLoss(dB) = 20\log\frac{Q_u}{Q_u - Q_L} \qquad (4\text{-}25)$$

$$Q_u = \frac{\left(10^{IL(dB)/20}\right)Q_L}{\left(10^{IL(dB)/20}\right) - 1} = \frac{\left(10^{2.249/20}\right)(61.5)}{\left(10^{2.249/20}\right) - 1} = 269.5$$

It is also interesting to note that the Q_L of the resonator is related to the group delay through the two-port network.

$$Q_L = 2\pi f\left(\frac{t_d}{2}\right) \qquad (4\text{-}26)$$

where:
  t_d is the group delay in seconds
  f is the frequency in Hertz

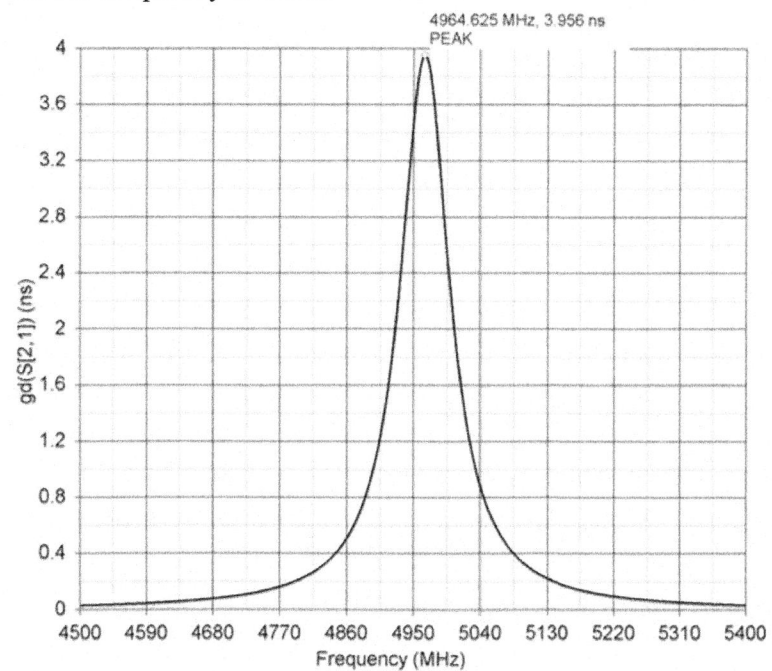

Figure 4-12 Group delay across the half-wave open circuit microstrip resonator

$$Q_L = 2\pi \left(4964.4 \cdot 10^6\right)\left(\frac{3.955 \cdot 10^{-9}}{2}\right) = 61.6$$

4.6 Resonator Series Reactance Coupling

To reduce the loading on the half wave resonator of Figure 4-11, the source and load impedances of 5000 Ω were used in Genesys. In practice the resonator is typically coupled to lower impedance circuits. If we attempt to examine the resonator on a network analyzer, most modern test equipment will have 50 Ω impedance levels. Such resonators are often coupled to the circuit by a highly reactive circuit element. This reactive element can be realized as a series capacitor or inductor. The resonator is then analyzed as a one port network. As the frequency is swept over a narrow frequency range around the resonant frequency, a circle is formed on the Smith Chart. This trace is known as the Q circle of the resonator [5]. Figure 4-13 shows the scalar plot and the Smith Chart plot of the resonator's input reflection coefficient, S11. Three plots are shown each with a different value of coupling capacitance. We can see that as the coupling capacitance changes, the resonant frequency of the circuit also changes. The series capacitance acts to decrease the overall resonance frequency of the circuit. This new resonance frequency is known as the loaded resonance frequency, f_L. Because the series capacitance lowers the frequency, the length of the resonator was decreased to 0.727 inches to return the resonant frequency close to 5 GHz. With the coupling capacitance set at 0.0485 pF the Q circle passes through the center of the Smith Chart at the resonant frequency. This is known as critical coupling and is characterized by having the lowest return loss on the scalar plot of Figure 4-13. With the capacitance increased to 0.108 pF the resonator is more strongly coupled to the 50 Ω load. The scalar plot shows that the resonance frequency is decreased. The Smith Chart shows a larger Q circle which is a characteristic of an over coupled resonator. With the coupling capacitor set to 0.025 pF the resonance frequency increases. The Smith Chart shows that the Q circle becomes much smaller thus under coupling the resonator. As Figure 4-13 shows, the value of the coupling capacitor also has an impact on the size of the Q circle. The diameter of the Q circle is dependent on the coupling of the resonator to the 50 Ω source.

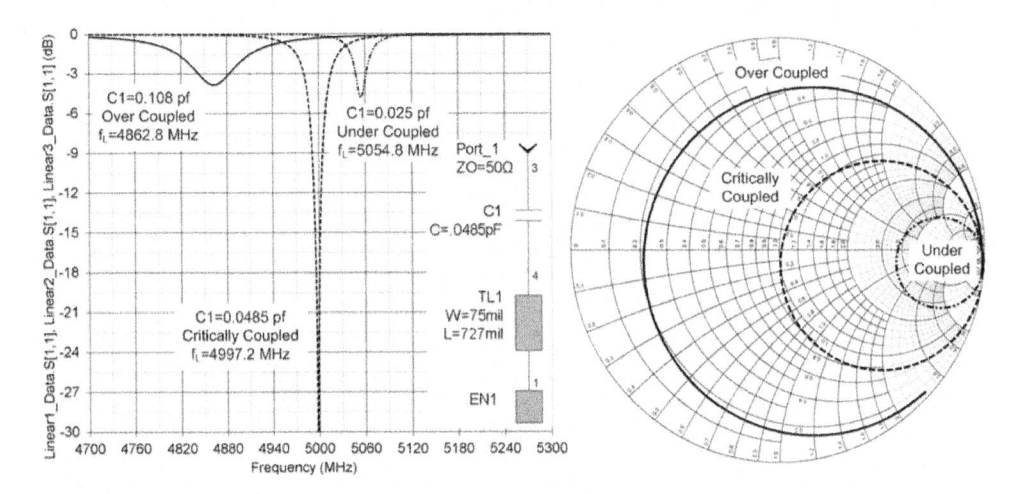

Figure 4-13 Capacitive coupled half-wave microstrip resonator and input reflection S11

4.6.1 One Port Microwave Resonator Analysis

The microstrip half wave resonator was fairly easy to model and analyze in Genesys. Many microwave resonators are not as easy to model. High Q microwave resonators are often realized as metallic cavities or dielectric resonators for which there are no native models in Genesys. The reactive coupling of the resonator to the circuit can be even more difficult to model. The coupling usually occurs by magnetic or electric coupling by a probe or loop inserted into the cavity. An E field probe coupled to a coaxial cavity resonator is shown in Figure 4-14. A dielectric resonator is coupled to a microstrip line by flux linkage in air as shown in Figure 4-15. Again there is no model in Genesys to directly model this coupling mechanism to the resonator. The designer is left to develop approximate models based on a lumped RLC equivalent models and couple the resonator to the circuit using an ideal transformer model. Linear simulation can still be of value in the design and evaluation process if we have a measured S parameter file of the resonator's reflection coefficient. Just as we have used S parameter models to represent capacitors and inductors we can also use the measured S parameters of a resonator. All modern vector network analyzers have the ability to save an S parameter data file for any measurement that can be made by the instrument. This section will show how we can use Genesys to analyze the S parameter file of a microwave resonator.

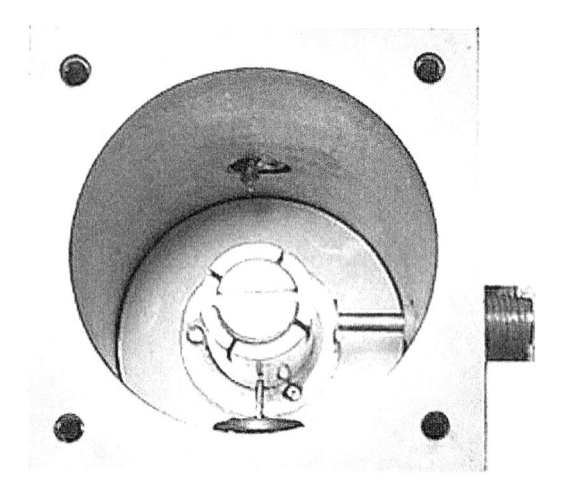

Figure 4-14 Cylindrical coaxial cavity with E field probe coupled to center conductor

The coupling of microwave resonators is often characterized by a coupling coefficient k. The coupling coefficient is the ratio of the power dissipated in the load to the power dissipated in the resonator [5].

$$k = \frac{P_{load}}{P_{resonator}} = \frac{Q_o}{Q_{ext}} \qquad (4\text{-}27)$$

where:
    Q_o is the unloaded Q of the resonator
    Q_{ext} is the external Q of the resonator

When P_{load} is equal to $P_{resonator}$, k = 1 and the critical coupling case exists. Substituting the reciprocal Q factor relationship of Equation (4-10) into Equation (4-27) we can relate the coupling coefficient to the loaded Q_L and unloaded Q_o of the resonator.

$$Q_L = \frac{Q_o}{1+k} \qquad (4\text{-}28)$$

Resonant Circuits and Filters

Figure 4-15 Dielectric resonator coupled to microstrip transmission line

Now the unloaded Q_o of the resonator can be calculated if the Q_L and k can be measured. Because the resonator is a one port device we cannot pass a signal through the device and measure the 3dB bandwidth as was done in section 4.5.3. Kajfez [5] has described a technique to extract the coupling coefficient k and Q_L values from the Q circle of the resonator. Consider the Q circle on the Smith Chart of Figure 4-16. A line that is projected from the center of the Smith Chart to intersect the Q circle with minimum length will intersect the circle at the loaded resonance frequency, f_L. The length of this vector is labeled as $|\Gamma_L|$. As the line projects along a path of the diameter of the circle it intersects the circle near the circumference of the Smith Chart at a point defined as $|\Gamma_d|$. The input reflection coefficient of the Q circle can be defined using the following empirical equation [5].

$$\Gamma_i = \Gamma_d \left[1 - \frac{2k}{1+k} \cdot \frac{1}{1+jQ_L 2 \frac{\omega - \omega_L}{\omega_o}} \right] \quad (4\text{-}29)$$

Lines that are projected from Γ_d through the Q circle at the angles $\pm\phi$ are related to the loaded Q by Equation (4-30).

$$Q_L = \frac{f_L}{f_1 - f_2} \tan\phi \quad (4\text{-}30)$$

If we set $\phi = 45°$ then Equation (4-30) reduces to the straightforward definition of Q_L given by (4-31).

$$Q_L = \frac{f_L}{f_1 - f_2} \quad (4\text{-}31)$$

In the previous section we saw that the diameter of the Q circle is directly related to the coupling coefficient. The diameter can be measured from:

$$|\Gamma_d| - |\Gamma_L| = d \quad (4\text{-}32)$$

The coupling coefficient is then derived from the diameter of the Q circle.

$$k = \frac{d}{2-d} \quad (4\text{-}33)$$

Finally the unloaded resonator Q_o is then calculated from Equation (4-28). We can also find the unloaded resonance frequency directly from the Q circle. Follow the reactive line on the Smith Chart that intersects the Q circle at Γ_d to the next Q circle intersection. The frequency at this Q circle intersection is the unloaded resonance frequency, f_o.

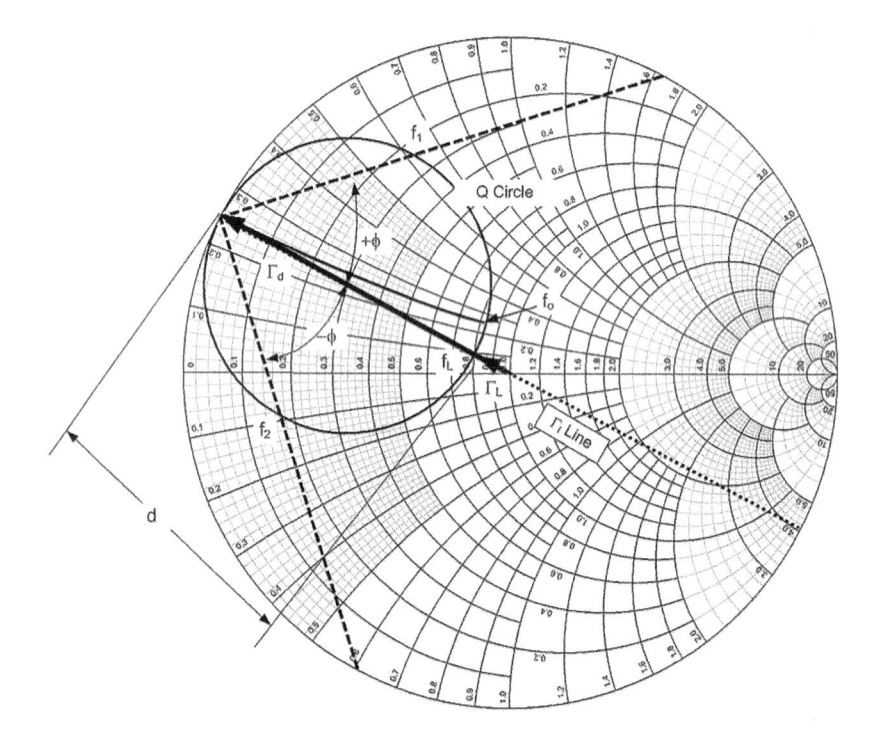

Figure 4-16 Resonator Q measurement from the resonator Q circle

4.6.2 Smith Chart Q_o Measurement of the Microstrip Resonator

Example 4.6-1: Use the Smith Chart technique to measure the Q_o of the half wave microstrip resonator of Figure 4-13. Solve for Q_o for all three coupling cases: under-coupled, over-coupled, and critically coupled cases.

Solution: The graphical technique requires that three overlays be placed on the Smith Chart.

    I. Γ_i Line
    II. Q_L Lines
    III.   Ideal Q Circle

In the Genesys workspace a separate schematic and linear analysis is used to model a circuit that generates each of these overlays. The first schematic and analysis creates a line the passes through the center of the Smith Chart and extends to the circumference. A second schematic and analysis combination generates the line pair at an angle of $\pm 45°$ from the Γ_i line. The third schematic produces an ideal Q circle. The ideal Q circle is overlaid on the S parameter data of the resonator under test. It helps to align the Γ_i line and Q_L lines. It is especially helpful when the measured data of a resonator may not form a full circle. The analysis that accompanies these designs can be simulated at any arbitrary frequency range. Only the schematic for the resonator under test needs to be swept over the actual measurement frequency range. The output from each Data Set can be plotted to the same Smith Chart so that the Γ_i line and Q_L lines are effectively an overlay on the Smith Chart. The schematics used to generate the Smith Chart overlays are given in Appendix B. Therefore the complete workspace is a collection of four schematics and linear analysis. The three overlays can be rotated around the chart by tuning the 'angle' variable. A 'circle-diameter' variable is used to vary the size of the ideal Q circle. The 'coupling-loss' variable moves the ideal Q circle toward the center of the chart as the resonator coupling loss increases. An Equation Editor is used to make these variables common among the three overlay schematics. The Equation Editor is also used to calculate the coupling coefficient (k), Q_L, and Q_o based on measured parameters on the Smith Chart. The procedure for using the Q measurement overlays is summarized. Figure 4-17 shows the Genesys schematic of the microstrip resonator. The measurement process is summarized as:

1. Move the cursor over the Q circle to determine the minimum reflection coefficient, $|\Gamma_L|$. Place a marker on this point and enter the frequency, f_L, in the Equation Editor.
2. Adjust the "angle" of the Γ_i control using the slider control so that the Γ_i line intersects the $|\Gamma_L|$ marker. Note that the Q_L lines move along with the Γ_i line.
3. Iterate between the ideal Q circle diameter and coupling loss to get the best fit over the resonator's Q circle.
4. Using the cursor measure the $|\Gamma_d|$ and $|\Gamma_L|$ and enter the values in the Equation Editor.
5. Using the cursor measure f_1 and f_2 at the intersection of the Q lines and the Q circle and enter their value in the Equation Editor.

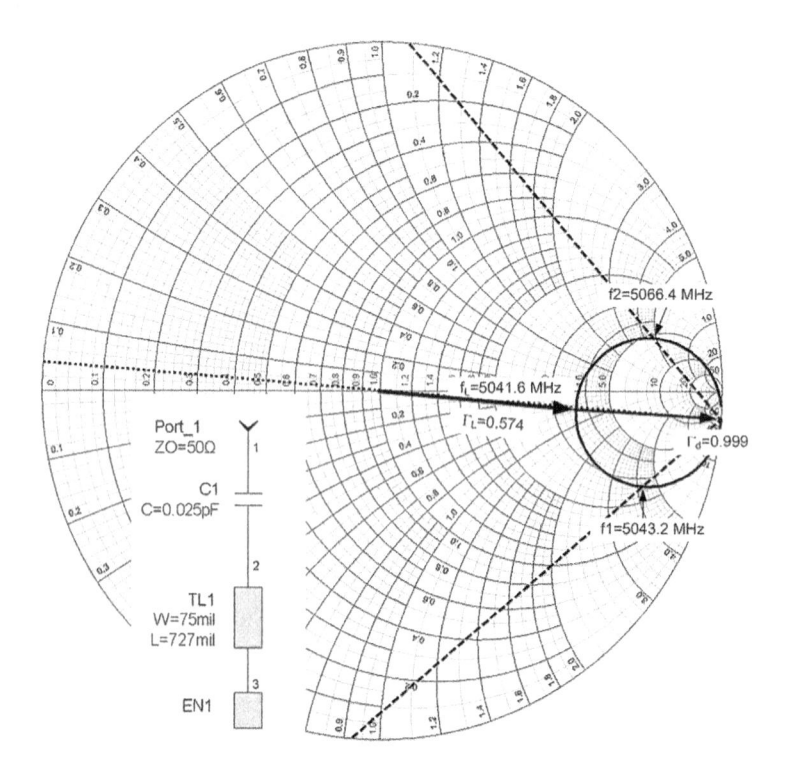

Figure 4-17 Under-coupled resonator Q circle and overlays for Q measurement

Resonant Circuits and Filters 173

```
1   'Ideal Q Circle Overlay Parameters
2   angle=175.348
3   circlediameter=7440.36
4   couplingLoss=0.02
5   circleangle=-angle/2
6   'Calculation of Qo
7   'Enter measurement from Smith Chart
8   GammaD=0.999      'Enter magnitude of GammaD
9   GammaL=0.574      'Enter magnitude of GammaL
10  fL=5054.8
11  f1=5066.4
12  f2=5043.2
13  d=GammaD-GammaL   'Under and critically coupled
14  'd=GammaD+GammaL  'For over coupled
15  k=d/(2-d)
16  QL=fL/abs(f1-f2)
17  Qo=QL*(1+k)
```

Figure 4-18 Calculation of Q_L and Q_o for the under-coupled resonator

Repeat steps 1 through 5 for the critically coupled resonator of Figure 4-19.

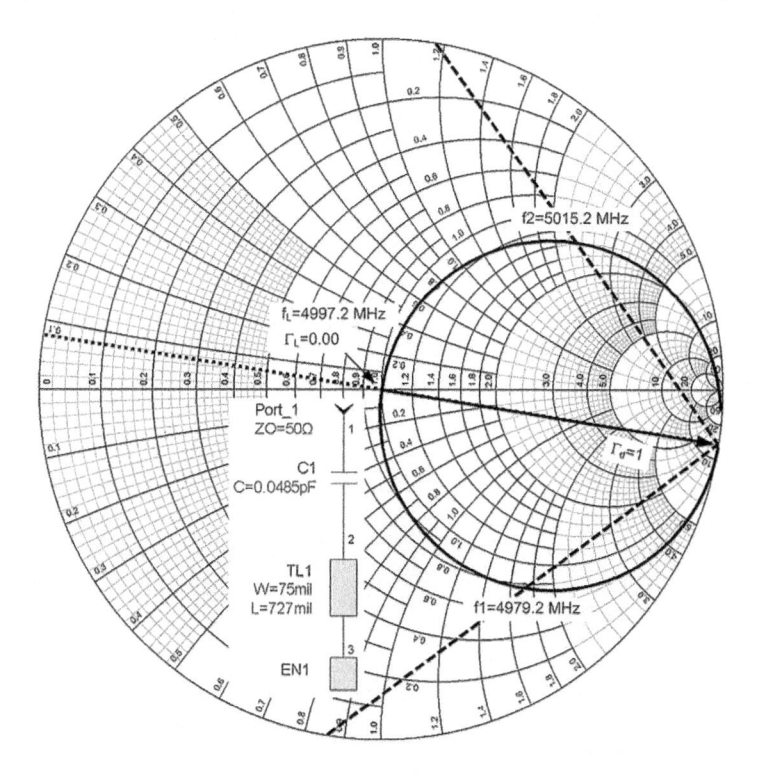

Figure 4-19 Critically-coupled Q circle and overlays for Q measurement

```
1   'Ideal Q Circle Overlay Parameters
2   angle=171.046
3   circlediameter=2025.53
4   couplingLoss=0.008
5   circleangle=-angle/2
6   'Calculation of Qo
7   'Enter measurement from Smith Chart
8   GammaD=1      'Enter magnitude of GammaD
9   GammaL=0      'Enter magnitude of GammaL
10  fL=4997.2
11  f1=5015.2
12  f2=4979.2
13  d=GammaD-GammaL    'Under and critically coupled
14  'd=GammaD+GammaL   'For over coupled
15  k=d/(2-d)
16  QL=fL/abs(f1-f2)
17  Qo=QL*(1+k)
```

Figure 4-20 Calculation of Q_L and Q_o for the critically-coupled resonator

Repeat steps 1 through 5 for the over coupled resonator of Figure 4-21.

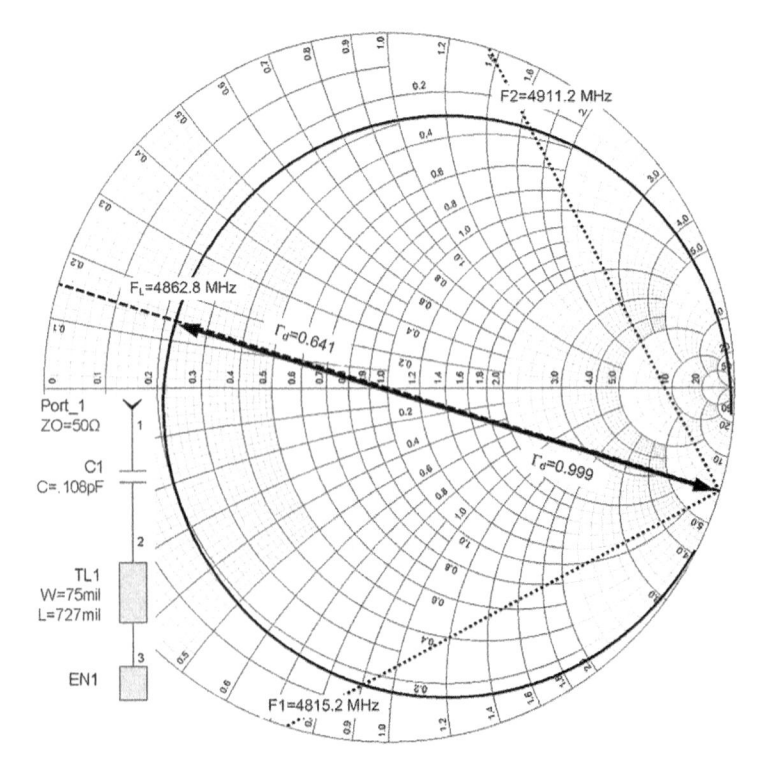

Figure 4-21 Over-coupled resonator Q circle and overlays

Resonant Circuits and Filters

```
1    'Ideal Q Circle Overlay Parameters
2    angle=162.754
3    circlediameter=460.37
4    couplingLoss=0.008
5    circleangle=-angle/2
6    'Calculation of Qo
7    'Enter measurement from Smith Chart
8    GammaD=0.999      'Enter magnitude of GammaD
9    GammaL=0.641      'Enter magnitude of GammaL
10   fL=4862.8
11   f1=4911.2
12   f2=4815.2
13   d=GammaD+GammaL   'Under and critically coupled
14   'd=GammaD+GammaL  'For over coupled
15   k=d/abs(2-d)
16   QL=fL/abs(f1-f2)
17   Qo=QL*(1+k)
```

Figure 4-22 Calculation of Q_L and Q_o for the over-coupled resonator

The three cases of the half wave microstrip resonator reveal the usefulness of the Smith Chart overlay technique for the measurement of reflection based one port resonator, Q_o. Even though the loaded Q's were varied from 50.6 to 217.8 the unloaded Q_o was calculated to within 2% error.

Case	Coupling Coefficient, k	Measured Q_L	Calculated Q_o
Under Coupled	0.270	217.8	276.6
Critical Coupled	1.000	138.8	277.6
Over Coupled	4.556	50.6	281.4

Table 4-1 Comparison of Q_o calculation at various coupling coefficients

4.7 Filter Design at RF and Microwave Frequency

In Section 4.3 we have seen that it is possible to change the shape of the frequency response of a parallel resonant circuit by choosing different source and load impedance values. Likewise multiple resonators can be coupled to one another and to the source and load to achieve various frequency shaping responses. These frequency shaped networks are referred to as filters.

4.7.1 Filter Topology

The subject of filter design is a complex topic and the subject of many dedicated texts [1-2]. This section is intended to serve as a fundamental primer to this vast topic. It is also intended to set a foundation for successful filter design using the Genesys software. The four most popular filter types are: Low Pass, High Pass, Band Pass, and Band Stop. The basic transmission response of the filter types is shown in Figure 4-23. The filters allow RF energy to pass through their designed pass band. RF energy that is present outside of the pass band is reflected back toward the source and not transmitted to the load. The amount of energy present at the load is defined by the S21 response. The amount of energy reflected back to the source is characterized by the S11 response.

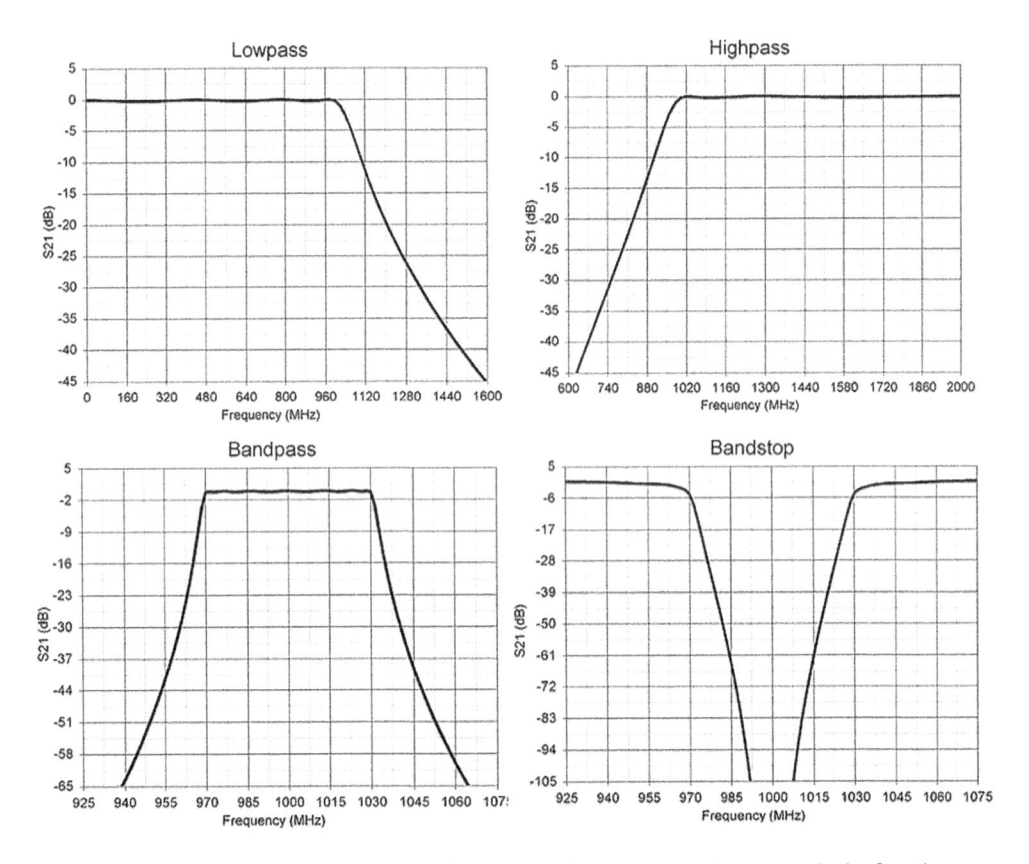

Figure 4-23 Transmission (S21) versus frequency characteristic for the basic filter types

4.7.2 Filter Order

The design process for all of the major filter types is based on determination of the filter pass band, and the attenuation in the reject band. The attenuation in the reject band that is required by a filter largely determines the slope needed in the transmission frequency response. The slope of the filter's response is related to the order of the filter. The steeper the slope or 'skirt' of the filter; the higher is the order. The term order comes from the mathematical transfer function that describes a particular filter. The highest power of s in the denominator of the filter's Laplace transfer function is the order of the filter. For the simple low pass and high pass filters presented in this chapter the filter order is the same as the number of elements in the filter. However this is not the case for general filter networks. In more complex types of lowpass and highpass filters as well as bandpass and bandstop filters the filter order will not be equal to the number of elements in the filter. In the general case the filter order is the total of the number of transmission zeros at frequencies:

- $F = 0$ (DC)
- $F = \infty$
- $0 < F < \infty$ (specific frequencies between DC and ∞)

Transmission zeros block the transfer of energy from the source to the load. In fact the order of a filter network can be solved visually by adding up the number of transmission zeros that satisfy the above criteria. Figure 4-24 shows the relationship between the filter order and slope of the response for a Low Pass filter. Each filter of Figure 4-24 has the same cutoff frequency of 1000 MHz. The third order filter has an attenuation of about 16 dB at a rejection frequency of 2000 MHz. The fifth order filter shows an attenuation of 39 dB and the seventh order filter has more than 61 dB attenuation at 2000 MHz. It is therefore clear that the order of the filter is one of the first criteria to be determined in the filter design. It is dependent on the cutoff frequency of the pass band and the amount of attenuation desired at the rejection frequency.

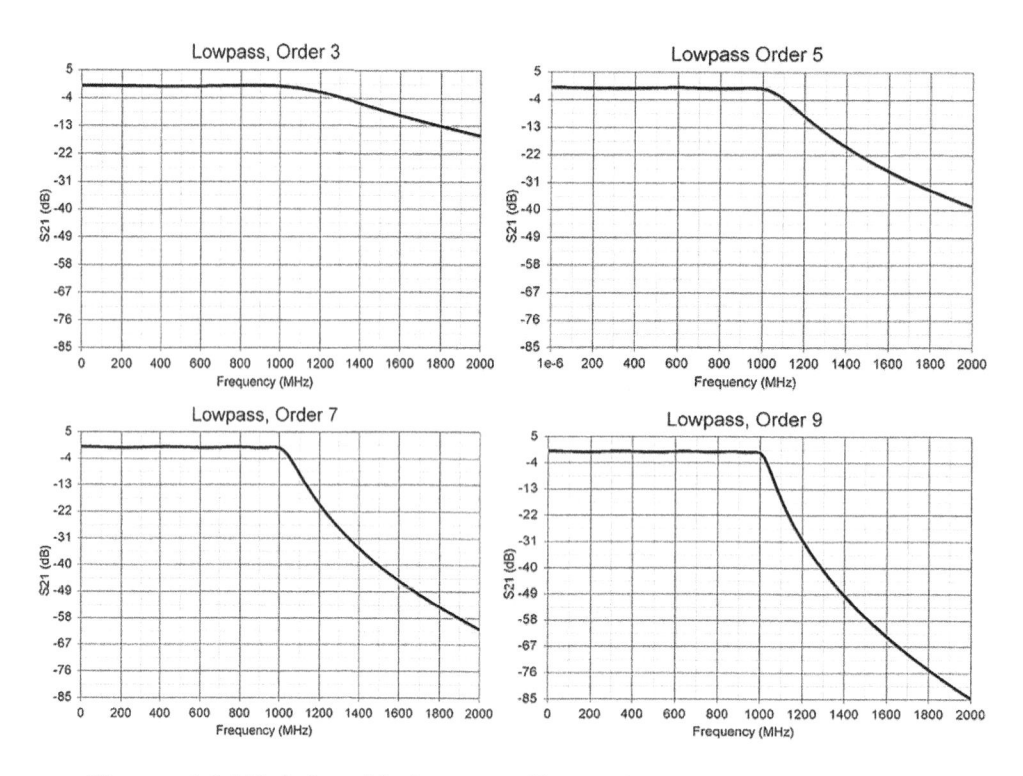

Figure 4-24 Relationship between filter order and the slope of S21

4.7.3 Filter Type

The shape of the filter passband and attenuation skirt can take on different shape relationships based on the coupling among the various reactive elements in the filter. Over the years several polynomial expressions have been developed for these shape relationships. Named after their inventors, some of the more popular passive filter types include: Bessel, Butterworth, Chebyshev, and Cauer. Figure 4-25 shows the general shape relationship among these filter types for a given seventh order filter. The Bessel filter type is a low Q filter and does not exhibit a steep roll off compared to its counterparts. The benefit of the Bessel filter is its linear phase or flat group delay response. This means that the Bessel filter can pass wideband signals while introducing little distortion. The Butterworth is a medium Q filter that has the flattest pass band of the group. The Chebyshev response is a higher Q filter and has a noticeably steeper skirt moving toward the reject band.

Resonant Circuits and Filters 179

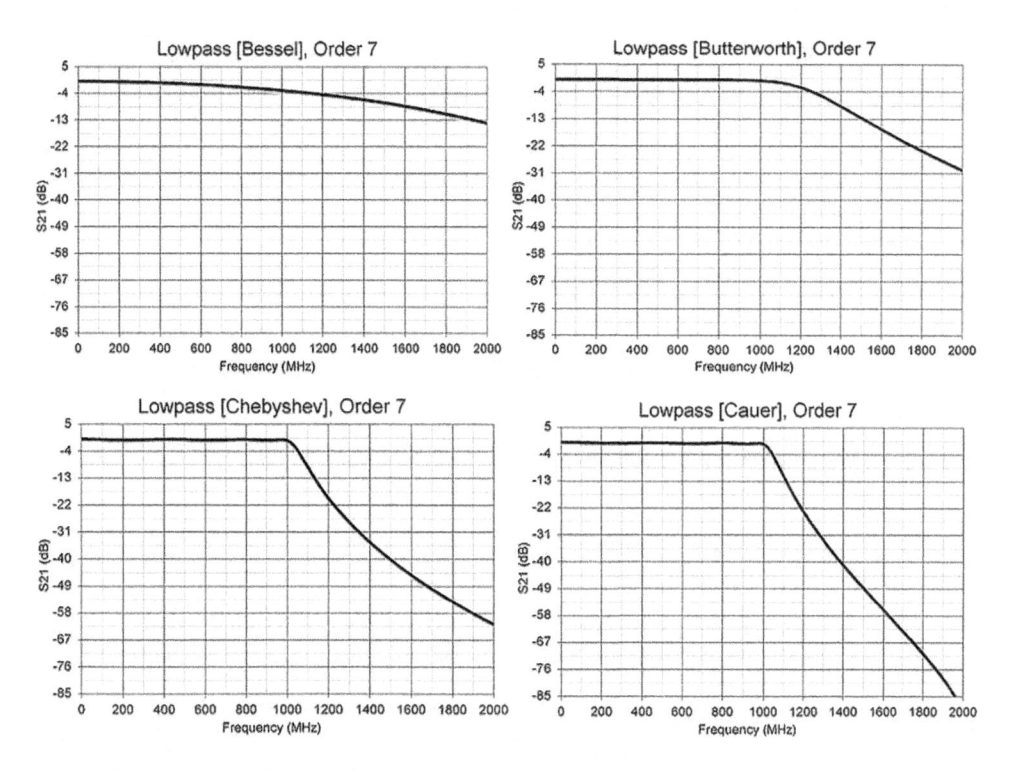

Figure 4-25 General characteristic shape of Bessel, Butterworth, Chebyshev, and Cauer filters

As a result it exhibits more transmission ripple in the pass band. The Cauer filter has the steepest slope of all of the four filter types. The Cauer filter is also known as an elliptic filter. Odd order Chebyshev and Cauer filters can be designed to have an equal source and load impedance. The even order Chebyshev and Cauer filters will have a different output impedance from the specified input impedance. Another interesting characteristic of the Cauer filter is that it has the same ripple in the rejection band as it has in the pass band. The Butterworth, Chebyshev and Cauer filters differ from the Bessel filter in their phase response. The phase response is very nonlinear across the pass band. This nonlinearity of the phase creates a varying group delay. The group delay introduces varying time delays to wideband signals which, in turn, can cause distortion to the signal. Group delay is simply the derivative, or slope, of the transmission phase and defined by Equation (4-34). Figure 4-26 shows the respective set of filter transmission characteristics with their corresponding group delay. Note the relative values of the group delay on the right hand axis.

$$\tau_g = -\frac{d\phi}{d\omega} \qquad (4\text{-}34)$$

where, ϕ is the phase shift in radians and ω is the frequency in radians per second

From the group delay plots of Figure 4-26 it is evident that the group delay peaks near the corner frequency of the filter response. The sharper cutoff characteristic results in greater group delay at the band edge.

4.7.4 Filter Return Loss and Passband Ripple

The Bessel and Butterworth filters have a smooth transition between their cutoff frequency and rejection frequency. The forward transmission, S21, is very flat vs. frequency. The Chebyshev and Cauer filters have a more abrupt transition between their cutoff and rejection frequencies. This makes these filter types very popular for many filter applications encountered in RF and microwave engineering. It is important to note however that the steeper filter skirt results as a certain amount of impedance mismatch between the source and load impedance.

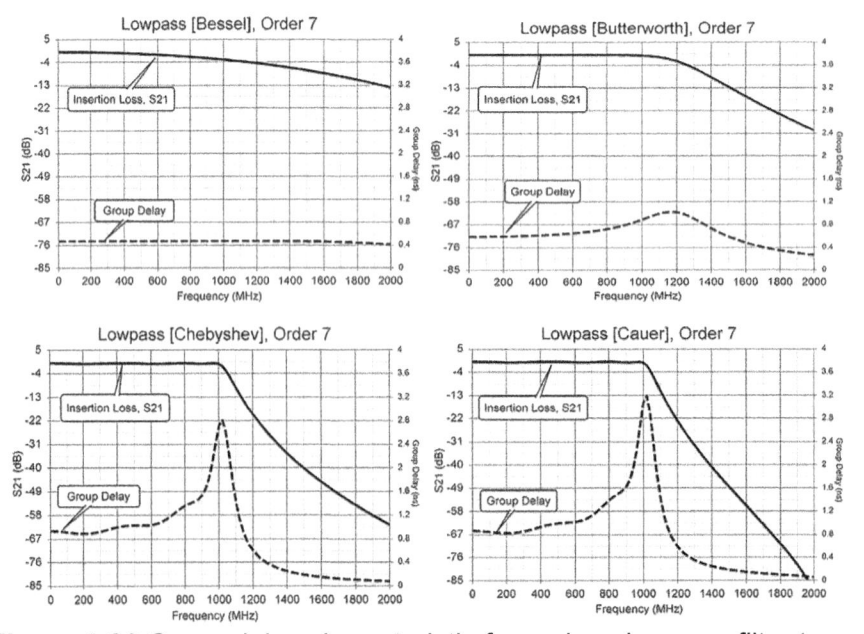

Figure 4-26 Group delay characteristic for various lowpass filter types

Resonant Circuits and Filters

The Chebyshev and Cauer filter types have ripple in the forward transmission path, S21. The amount of ripple is caused by the degree of mismatch between the source and load impedance and thus the resulting return loss that is realized by these filter types. For a given Chebyshev or Cauer filter order, the roll off of the filter response is also steeper for greater values of passband ripple. The cutoff frequency of the filters that have passband ripple is then defined as the passband ripple value. For all-pole filters such as the Butterworth, the cutoff frequency is typically defined as the 3 dB rejection point. Figure 4-27 shows the passband ripple of a fifth order low pass filter for ripple values of 0.01, 0.1, 0.25, and 0.5 dB. Note that the ripple shown is produced by ideal circuit elements. In practice the finite unloaded Q or losses in the inductors and capacitors will tend to smooth out this ripple.

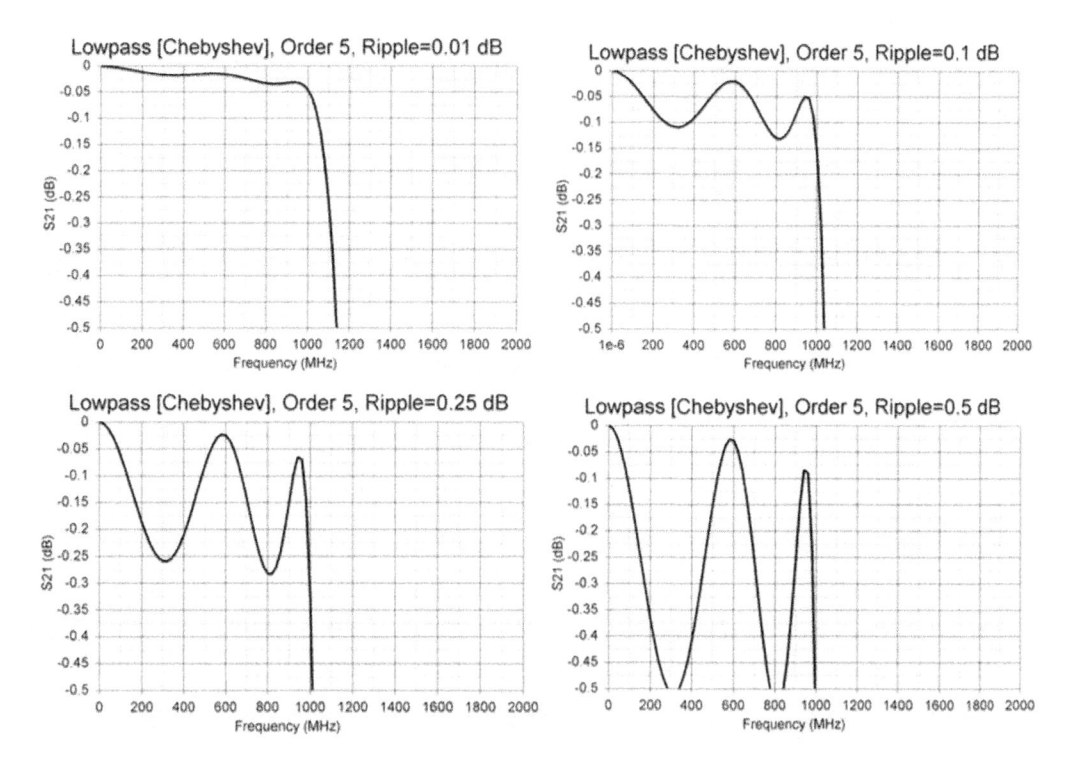

Figure 4-27 Passband ripple values in lowpass Chebyshev filter

In Chapter 2 the relationship for mismatch loss between a source and load was presented. For the Chebyshev and Cauer filters this mismatch loss is the passband ripple. Figure 4-28 shows a solution of Equation (2-52) for selected values of VSWR.

```
1    vswr=[1.1;1.239;1.3;1.355;1.405;1.452;1.538;1.62;1.984]
2    Gamma=(vswr-1)/(vswr+1)
3    ReturnLoss_dB=20*log(Gamma)
4    MismatchLoss=-10*log(1-Gamma^2)
5    Ripple_dB=MismatchLoss
```

vswr ( )	ReturnLoss_dB ( )	Ripple_dB ( )
1.1	26.444	0.01
1.239	19.433	0.05
1.3	17.692	0.075
1.355	16.435	0.1
1.405	15.473	0.125
1.452	14.688	0.15
1.538	13.474	0.2
1.62	12.518	0.25
1.984	9.636	0.5

Figure 4-28 Equation Editor Calculation of filter ripple versus VSWR

Figure 4-29 shows the same filters with the return loss plotted along with the insertion loss, S21. We can see that for a given filter order, there is a tradeoff between filter rejection and the amount of ripple, or return loss, that can be tolerated in the passband. In most RF and microwave filter designs the 0.01 and 0.1 dB ripple values tend to be more popular. This is due to the tradeoff between good impedance match and reasonable filter skirt slope. Figure 4-29 shows good correlation of the worst case return loss with that which is calculated in Figure 4-28. When tuning filters using modern network analyzers it is sometimes easier to see the larger changes in the return loss as opposed to the fine grain ripple as shown in Figure 4-27. For this reason it is common to tune the forward transmission of the filter by observing the level and response of the filter's return loss. Return loss is a very sensitive indicator of the filter's alignment and performance.

Resonant Circuits and Filters 183

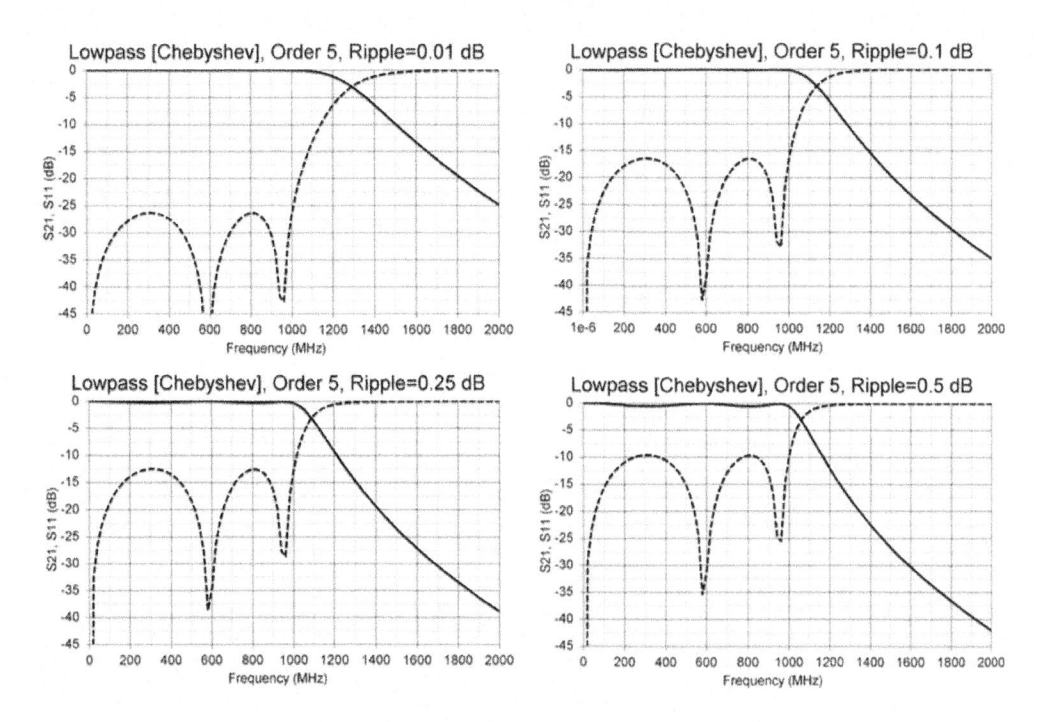

Figure 4-29 Lowpass Chebyshev filter rejection and return loss versus passband ripple

4.8 Lumped Element Filter Design

Classical filter design is based on extracting a prototype frequency-normalized model from a myriad of tables for every filter type and order [1-2]. Fortunately these tables have been built into many filter synthesis software applications that are readily available. In this book we will examine the filter synthesis tool that is built into the Genesys software. We will work through two practical filter examples, one low pass and one high pass using the Genesys filter synthesis tool.

4.8.1 Low Pass Filter Design Example

Example 4.8-1: As a practical filter design, consider a full duplex communication link (simultaneous reception and transmission) through a satellite with the following requirements:

- The uplink signal is around 145 MHz while the downlink is at 435 MHz.
- A 20 W power amplifier is used on the uplink with 25 dB gain.
- It is necessary to provide a low pass filter on the uplink (only pass the 145 MHz uplink signal while rejecting any noise power in the 435 MHz band.
- It is necessary to provide a high pass filter on the downlink so that the 435 MHz downlink signal is received while rejecting any noise power at 145 MHz.

The transmitter and receiver antennas are on the same physical support boom so there is limited isolation between the transmitter and receiver. Even though the signals are at different frequencies, the broadband noise amplified by the power amplifier at 435 MHz will be received by the UHF antenna and sent to the sensitive receiver. Because the receiver is trying to detect very low signal levels, the received noise from the amplifier will interfere or 'de-sense' the received signals. Therefore it is necessary to design a 145 MHz Low Pass filter for this satellite link system. The specifications chosen for the filter design are selected as:

- Select a Chebyshev Response with 0.1 dB pass band ripple.
- Set the passband cutoff frequency (not the -3 dB frequency) at 160 MHz
- The reject requirement is at least -40 dB rejection at 435 MHz.

Solution: Use the Passive Filter synthesis utility to design the Low Pass filter. Select a Low Pass filter of the Chebyshev type. On the Topology tab select a Lowpass filter type with a Chebyshev shape. Select the minimum capacitor subtype. On the Settings tab enter the cutoff frequency of 160 MHz and pass band ripple of 0.1 dB. Also set the cutoff frequency attenuation at 0.1 dB. The Filter Settings Tab is a great place to perform "what-if" analysis. The pass band ripple, filter order, cutoff frequency, and attenuation at cutoff can all be varied while observing their impact on the filter's characteristic. For the design example enter the parameters as shown in Fig. 4-30. The required filter order can be determined by increasing the order until the specification of -40 dBc attenuation at 435 MHz is achieved. As the filter response curve in Fig.4-30 shows, this Low Pass filter must be of fifth order to achieve the required attenuation. Along with the resulting attenuation and return loss the synthesis program creates the filter schematic with the synthesized component values.

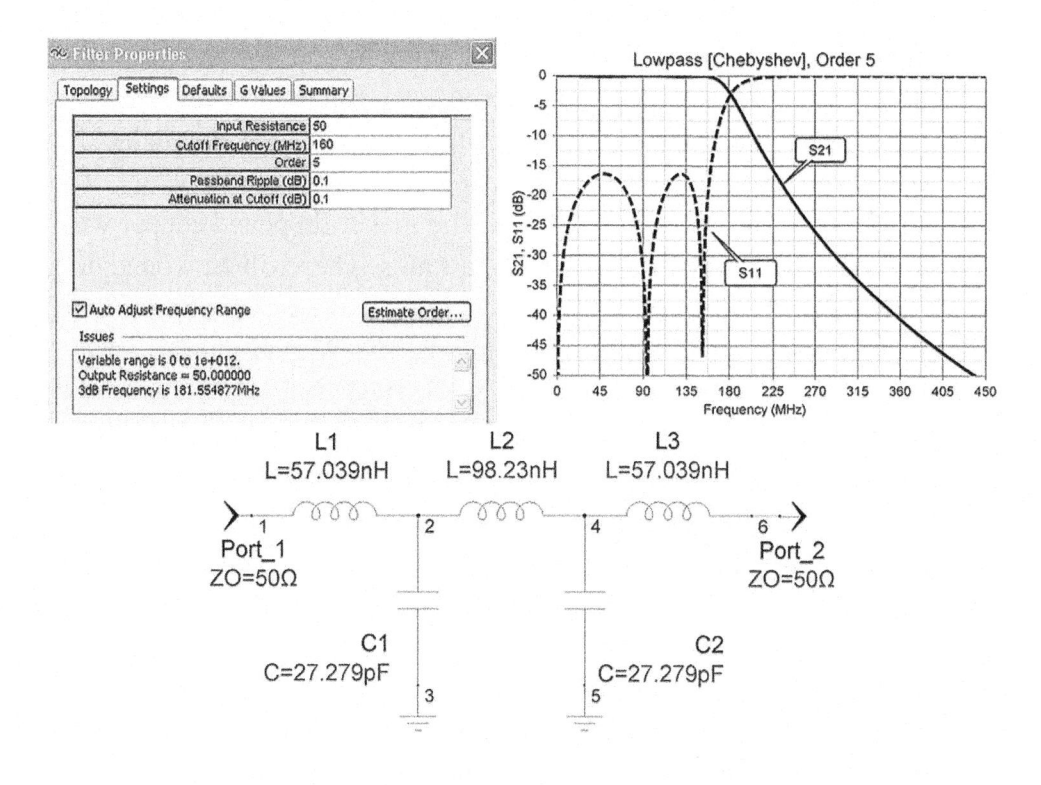

Figure 4-30 Passive filter synthesis utility: topology and settings tab

4.8.2 Physical Model of the Low Pass Filter in Genesys

It is important to realize that the synthesized filter is an ideal design in the sense that ideal (no parasitics and near infinite Q) components have been used. To obtain a good 'real-world' simulation of the filter we need to use component models that have finite Q and parasitics such as multilayer chip capacitors for shunt capacitors. For power handling capability, choose the 700 series chip capacitors from ATC Corporation. We will use measured S parameter files to model the shunt capacitors. The measured S parameters will account for any package parasitic effects and the finite component Q factor. Most microwave chip capacitor manufacturers will supply S parameters for their products. ATC Corporation has a useful application for selection of chip capacitors called ATC Tech Select. This program is available for free download from the ATC web site: www.atceramics.com. Using the Tech Select program the engineer can access complete data sheets and other useful information including current and voltage handling capabilities of the various capacitors. Because the filter is passing relatively high power (20 W), we cannot use small surface mount style chip inductors. Instead we will use air wound coils to realize the series inductors. The inductors will be realized with AWG#16 wire nickel-tin plated copper wire. The wire has a diameter of 0.05 inches or 50 mils. They will be wound on a 0.141 inch diameter form. Use the techniques covered in Chapter one, Section 1.4.1 to design the inductors using the Air Wound inductor model in Genesys. The filter model is then reconstructed using the S parameter files for the shunt capacitors and the physical inductor models for the series inductors. Make sure to model the substrate and the interconnecting printed circuit board traces as microstrip lines. Also model the ground connection of the shunt capacitors as a microstrip via hole. Although these PCB parasitic effects are normally more pronounced at frequencies above 2 GHz, it is often surprising the effect that these parasitics have at lower frequencies. The final filter response and model is shown in Figure 4-32. The response shows that the attenuation specification has been achieved. Because the circuit has physical models replacing the ideal lumped element components, the engineer can be confident that the filter can be assembled and will achieve the designed response. Figure 4-31 is a photo of the prototype low pass filter circuit with SMA coaxial connectors attached to the circuit board.

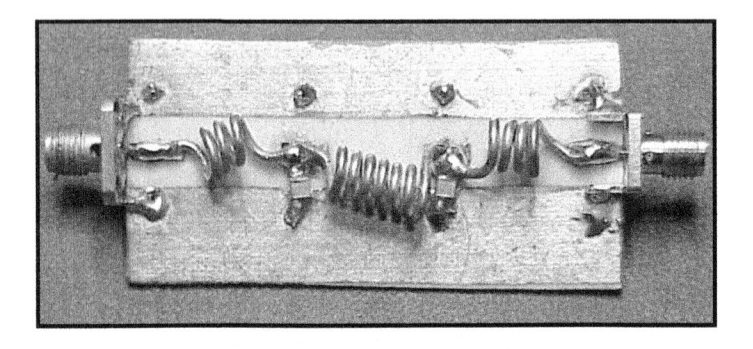

Figure 4-31 Physical prototype of the 146 MHz low pass filter

Resonant Circuits and Filters 187

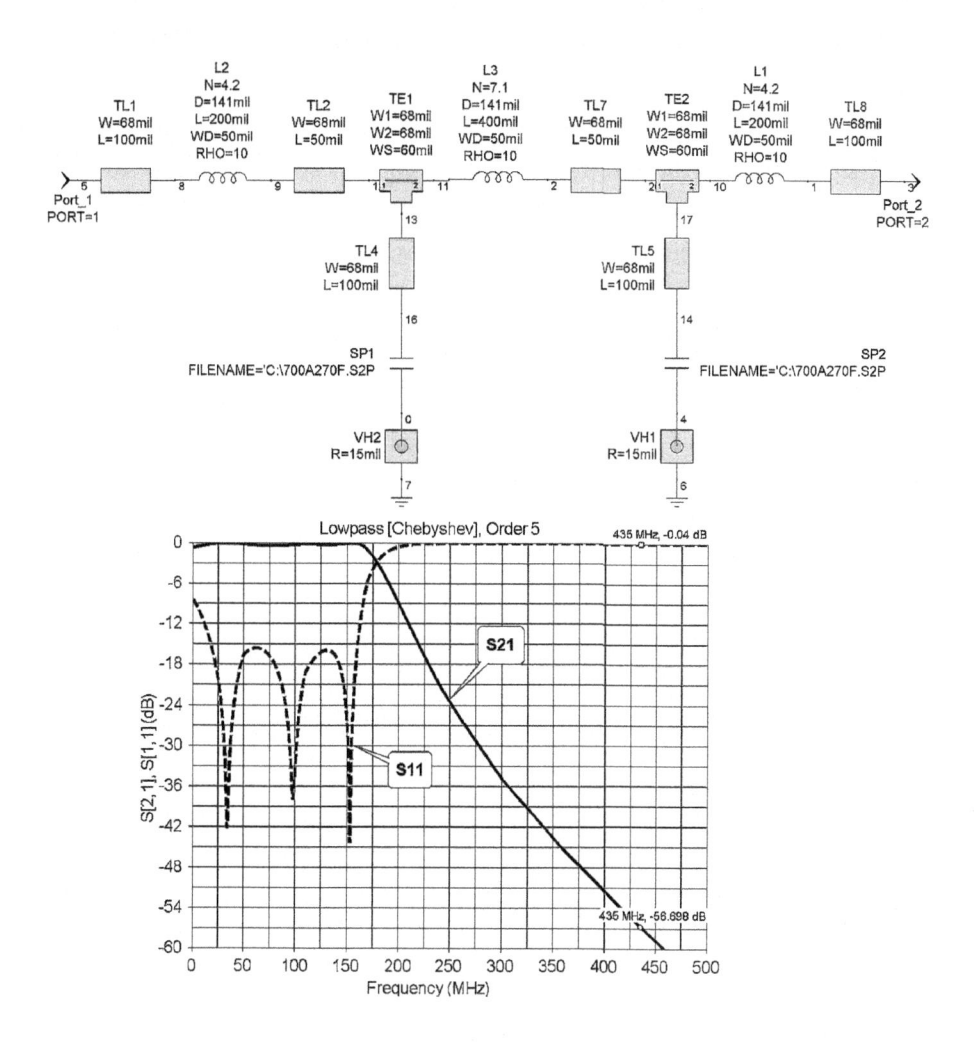

Figure 4-32 Low pass filter model with physical elements

4.8.3 High Pass Filter Design Example

Example 4.8-2: Design a high pass filter that passes frequencies in the 420 MHz to 450 MHz range. This filter could be placed in front of the preamp used in the downlink of the satellite system. This would help to keep out any of the transmit energy or noise power in the 146 MHz transmit frequency range. The High Pass Filter specifications are:

- The pass band cutoff frequency (not the -3 dB frequency) is 420 MHz.
- The filter has a Chebyshev response with 0.1 dB pass band ripple.

- The reject requirement is at least -60 dB rejection at 146 MHz.

Solution: Using the Passive Filter Synthesis tool in Genesys, vary the filter order until sufficient attenuation is achieved at 146 MHz. It is always a good practice to design for some additional rejection (margin) that exceeds the minimum requirement. The Filter synthesis session recommends a filter of 7th order.

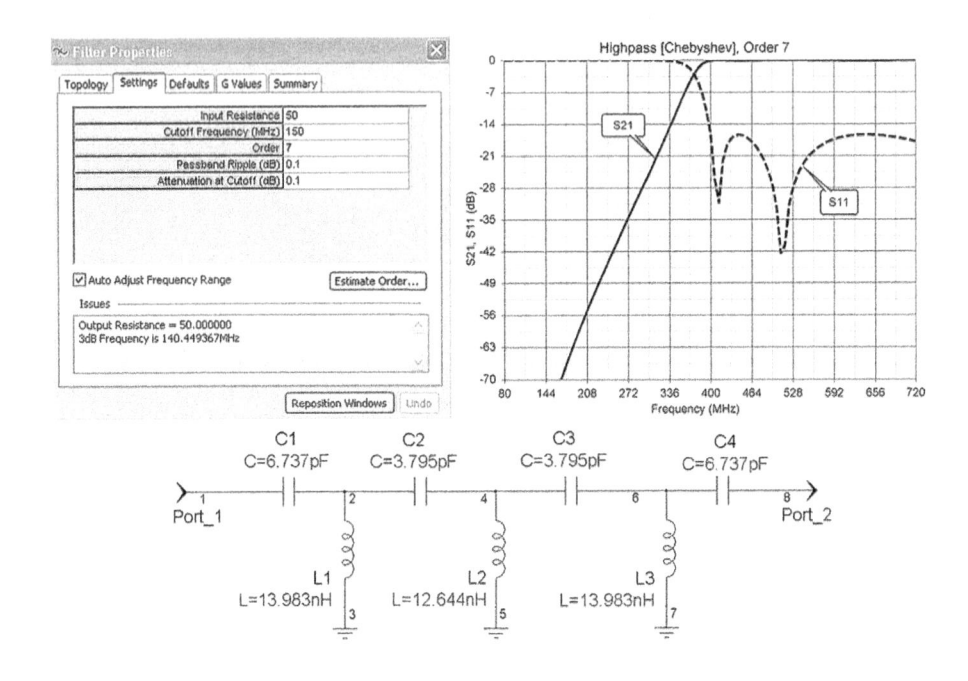

Figure 4-33 Passive filter synthesis model of the high pass filter

4.8.4 Physical Model of the High Pass Filter in Genesys

Using the same techniques as described for the Low Pass Filter design we can proceed with the High Pass Filter realization. The Passive Filter Synthesis application calculated series capacitance values of 6.7 pF and 3.8 pF. Looking through the available ATC 700 series chip capacitors, the nearest values are 6.8 pF and 3.9 pF capacitors. We will select the S parameters files for these chip capacitors to use in the final filter model. The shunt inductors are realized using the Air Wound inductor model. Figure 4-34 shows the final circuit model and filter response. From the response we can see that the required rejection specification of S21 < -60 dBc has been

Resonant Circuits and Filters

maintained. Figure 4-35 shows the completed assembly of the High Pass Filter on a printed circuit board with coaxial SMA connectors.

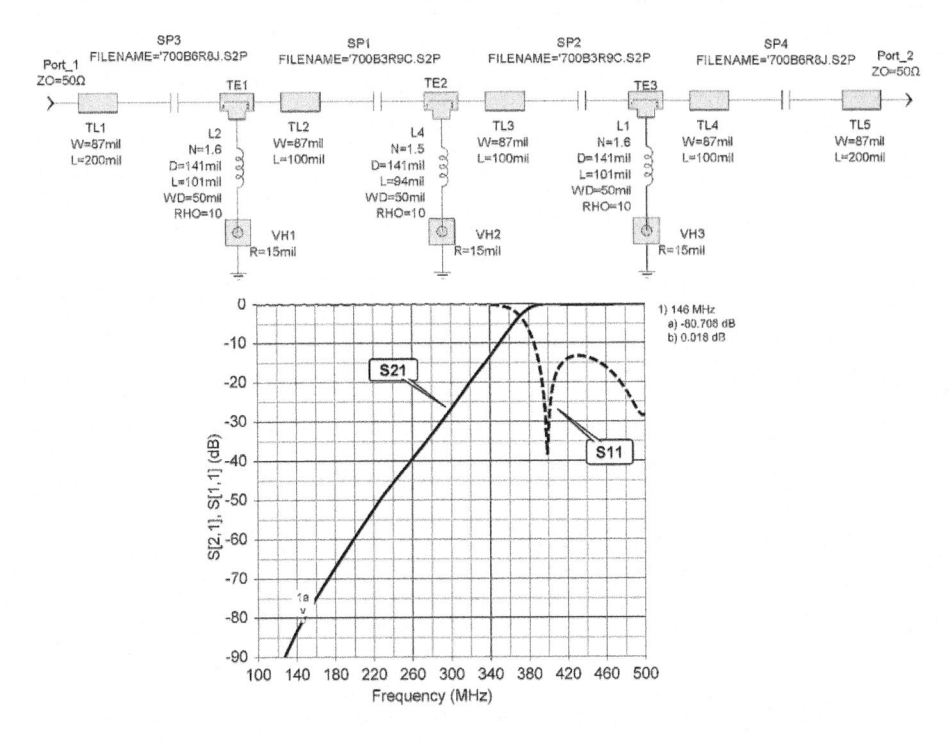

Figure 4-34 High pass filter model with physical elements

Figure 4-35 Physical prototype of the 420 MHz high pass filter

4.8.5 Tuning the High Pass Filter Response

In many cases we would like to tune the component values to change or tweak the filter's characteristic to fit the desired response. The shunt inductors in the high pass filter are very easy to tune because they are modeled with the native inductor model from the Genesys library. To tune the inductors simply check the tune box on the properties tab of the parameter to be tuned. Enable tuning of the inductor length of all three inductors. Using the Tune Window the length of each inductor can be increased or decreased by the step size or percentage selected. Each time the value is changed the analysis is run and the graph is updated. However the capacitors cannot be tuned because their physical model is based on an S parameter file that describes the capacitor's physical model. To change to another capacitor's S parameter file, we must edit the data file and browse to select a new S parameter file. Then we must sweep the circuit to observe the new response. This process involves several steps and we lose the 'real time' sense of tuning the capacitor values and seeing the response change quickly.

4.8.6 S Parameter File Tuning with VBScript

It is helpful to use measured S parameter files of inductors and capacitors to improve the accuracy and manufacturability of filter designs. The designer should be cautious however because a manufacturer's S parameter data file may not contain enough points to be used in a high Q circuit design. There can also be some variation in a component's S parameters based on the type of substrate and the techniques in which the S parameters are measured. In these cases it may be necessary to perform one's own measurement of component S parameters rather than rely on the manufacturer. Another issue in filter design is that the S parameters are typically measured in 50 Ω systems. When combined with other highly reactive S parameter files numerical instabilities may result. One of the inconveniences of S parameter files is that it is difficult to perform real time component tuning. A new S parameter file must be loaded into the schematic and the analysis is re-simulated every time that a new component is selected. One technique around this problem is to create a method of real time tuning of S parameter files using a VBScript application. Genesys has the capability to incorporate Visual Basic Script, VBScript, files to perform various operations on the simulator. We can take advantage of this capability to create an application within the Workspace that performs the required tasks of selecting S

Resonant Circuits and Filters 191

parameter files from disk, performing the linear sweep, and plotting the results on the graph. This allows us to create a real-time tuning tool using individual S parameter files from disk. The VBScript capability within Genesys allows all of the standard text based visual basic programming operations. To understand how to interact with objects within the Genesys Workspace use the VBBrower.exe application that is located in the Scripting folder in the Genesys Examples section. This application will show the proper syntax for accessing and setting the values and methods of the various objects in the Genesys workspace. A new schematic of the High Pass Filter that has been modified for tuning the S parameter files of the capacitors is shown in Figure 4-36.

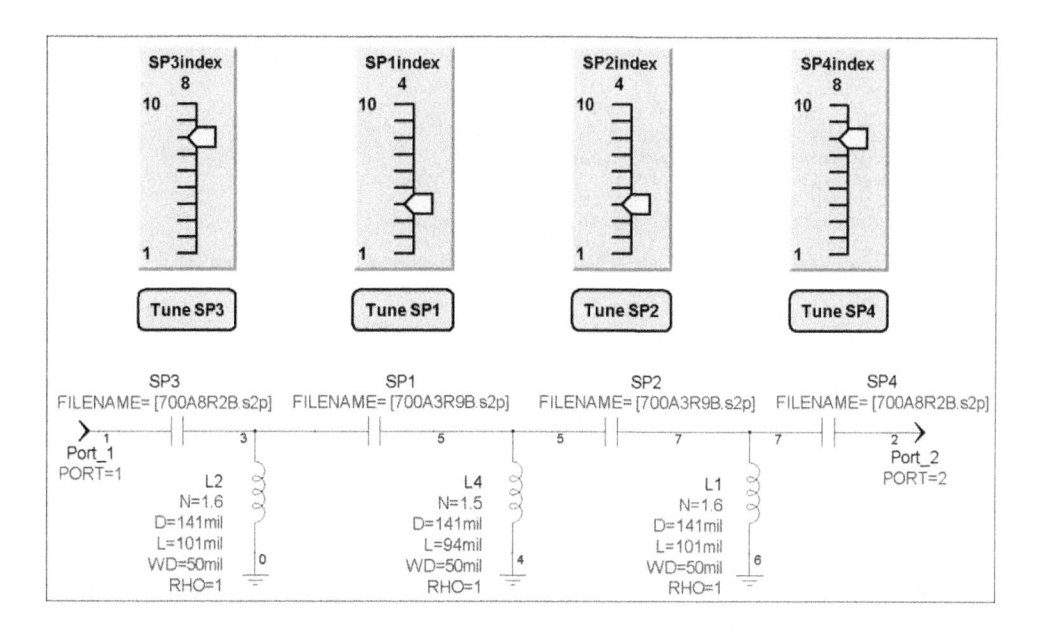

Figure 4-36 High pass filter schematic with slider controls and command buttons for tuning S parameter files

The VBScript code listing for the S parameter Tuner is shown in Figure 4-37. Lines 1 through 3 define the objects and variables used in the program. Lines 5 through 14 contain the S parameter files to be used in the tuning process. You can copy as many S parameter files as you desire to compare in your design. In this example we have chosen ATC700 chip capacitors from 2.2 pF to 12 pF. Make sure to include the full filename and path to the location of your S parameter files. This program listing enables real time tuning by selecting among ten different chip capacitor S parameter files. The index variable, i, is used by the program to select the appropriate file from the list of S parameters.

```
1    Dim WsDoc
2    Dim My_Tuner, My_Analysis
3    Dim name(10), i, number, pos
4    '....Copy S-Parameter Path & Filename here....................
5    name(1)="C:\700A2R2C.s2p"
6    name(2)="C:\700A2R7C.s2p"
7    name(3)="C:\700A3R3B.s2p"
8    name(4)="C:\700A3R9B.s2p"
9    name(5)="C:\700A4R7C.s2p"
10   name(6)="C:\700A5R6C.s2p"
11   name(7)="C:\700A6R8J.s2p"
12   name(8)="C:\700A8R2B.s2p"
13   name(9)="C:\700A100K.s2p"
14   name(10)="C:\700A120F.s2p"
15   '_____
16
17   WsDoc= theApp.GetWorkspaceByIndex(0)    'define the workspace object
18   My_Analysis=WsDoc.Designs.Linear1
19   My_Analysis.ClearModelCache             'clear the memory
20   'Get index value from Equation block
21   i = WsDoc.Equation.VarBlock.SP1index.GetValue()
22   'Load the S-Parameter file from disk
23   My_Tuner=WsDoc.Designs.Sch1.PartList.SP1.ParamSet.FILENAME.SetValue(name(i))
24   My_Tuner=WsDoc.Designs.Sch1.PartList.SP1.ParamSet.FILENAME.Set(name(i))
25   My_Analysis.ClearModelCache             'clear the memory
26   My_Analysis.RunAnalysis                 'simulate the circuit
27
28   Graph=WsDoc.Designs.Graph1              'select the graph
29   Graph.OpenWindow                        'Display Graph
```

Figure 4-37 VBScript code listing for S parameter file tuner

Slider controls are placed on the schematic as a means of selecting from the list of S parameter files. Variables in the equation block can be assigned to the Slider Controls. Equation block variables can also be input to the VBScript code. Therefore the Equation block is useful for exchanging variables between the program code and objects on the schematic. Define an index variable for each of the four S parameter data file elements, as shown in Figure 4-38. Make each variable tunable by placing a {=?} before the numerical value. Enter any integer value for each variable. Line 21 reads the index value from the Equation block and assigns it to the variable, i. Line 24 then uses this variable as an index to select the appropriate S parameter file to load from disk file and assign it to the S parameter data element.

Resonant Circuits and Filters 193

```
1    SP1index=?4
2    SP2index=?4
3    SP3index=?7
4    SP4index=?7
```

Figure 4-38 Index variables for each S parameter file, SP1-SP4

Lines 25 to 29 then execute the linear simulation and graph the response. The VBScript program code can then be copied into a Command Button object so that the code can be executed from the Genesys schematic. After placing a Command Button on the schematic expose the Properties tab and copy and paste the code from the VBScript editor to the command button. Figure 4-36 shows the schematic with a Slider Control and Command button for each S parameter file. Make sure to edit lines 21 and 24 in each Command Button to change the variable name, SP1index, and the file reference, SP1, because these are unique for each S parameter files. Figure 4-39 shows the VBScript code listing inside the Properties Tab of the Command Button.

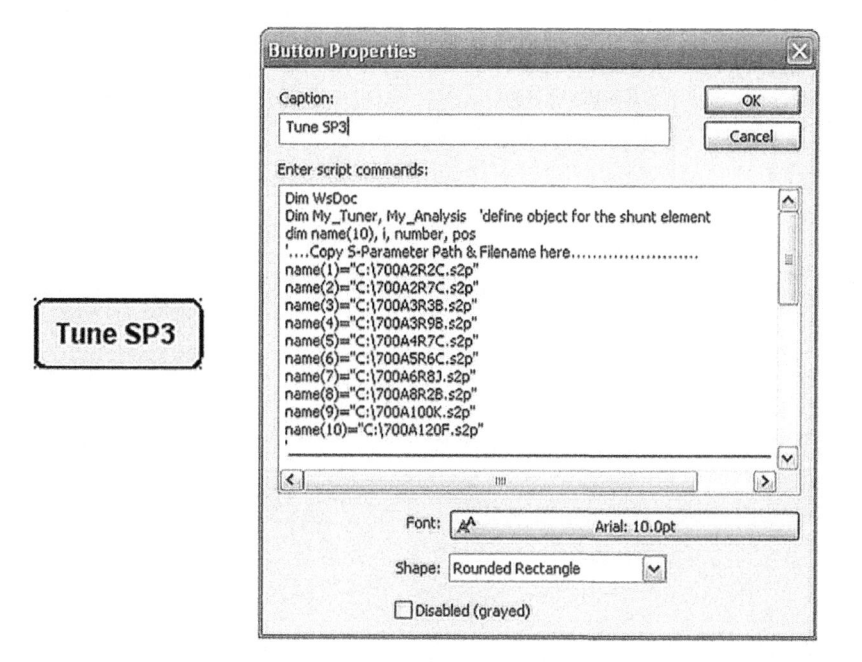

Button Properties

Caption: Tune SP3

Enter script commands:
```
Dim WsDoc
Dim My_Tuner, My_Analysis   'define object for the shunt element
dim name(10), i, number, pos
'....Copy S-Parameter Path & Filename here..........................
name(1)="C:\700A2R2C.s2p"
name(2)="C:\700A2R7C.s2p"
name(3)="C:\700A3R3B.s2p"
name(4)="C:\700A3R9B.s2p"
name(5)="C:\700A4R7C.s2p"
name(6)="C:\700A5R6C.s2p"
name(7)="C:\700A6R8J.s2p"
name(8)="C:\700A8R2B.s2p"
name(9)="C:\700A100K.s2p"
name(10)="C:\700A120F.s2p"
```

Font: Arial: 10.0pt
Shape: Rounded Rectangle
☐ Disabled (grayed)

[Tune SP3]

Figure 4-39 Property tab for the tune button command showing VBScript

The Slider Control Properties Tab is also shown in Figure 4-40. Using the drop-down list, select the Equation variable to assign to the Slider. Set the "Max" limit to the number of S parameter files that you have entered into the VBScript code listing. Uncheck the 'Run simulation' box because we will use the Command Button to run the simulation. Then select 'Snap to Integer values' so that the Slider Control can choose the appropriate S parameter file. Once the Slider Control has been moved, press the Command Button to sweep the circuit and plot the response. Now we can tune the S parameter files of the capacitors just as easily as we can tune the inductor model parameters.

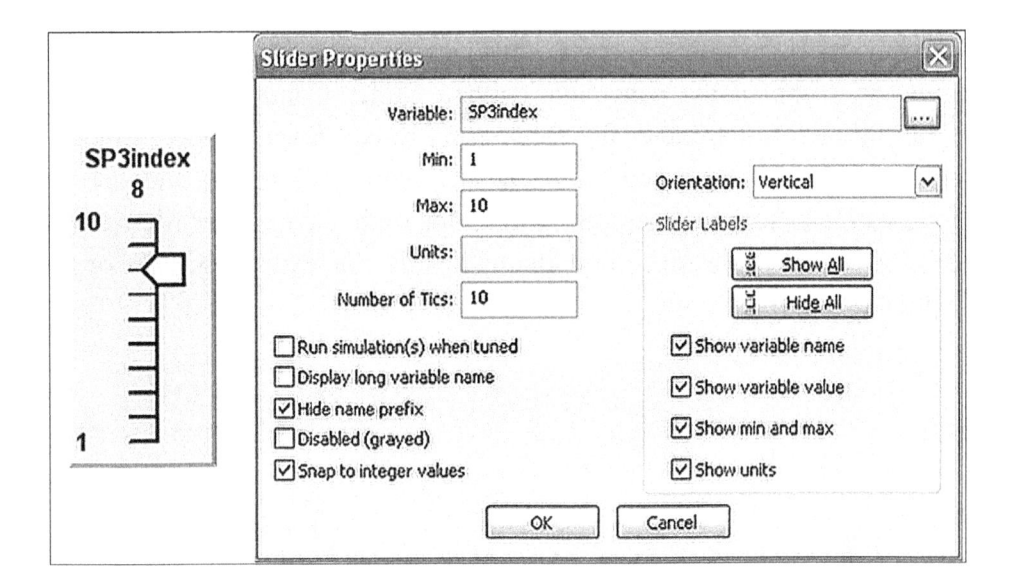

Figure 4-40 Slider control Tab for S parameter file selection

Resonant Circuits and Filters 195

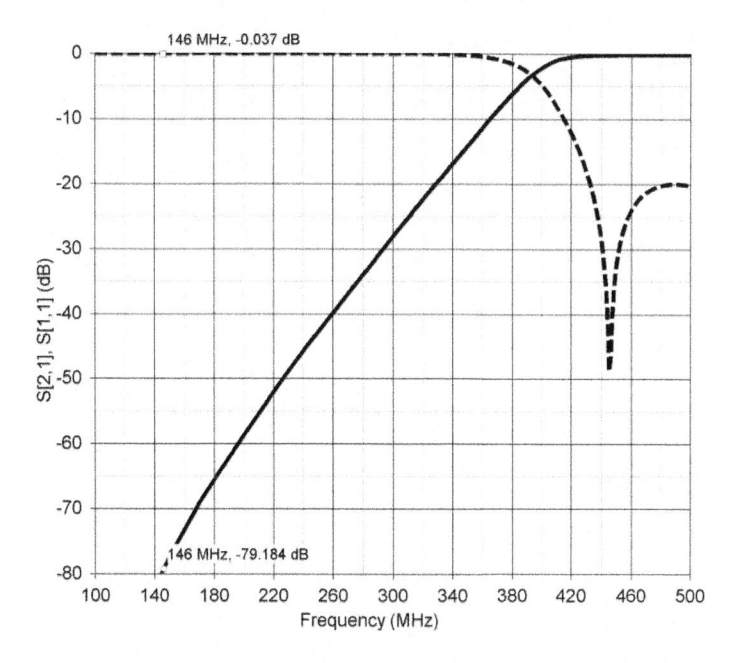

Figure 4-41 High pass filter response with tuned S parameter files

4.9 Distributed Filter Design

4.9.1 Microstrip Stepped Impedance Low Pass Filter Design

In the microwave frequency region filters can be designed using distributed transmission lines. Series inductors and shunt capacitors can be realized with microstrip transmission lines. In the next section we will explore the conversion of a lumped element low pass filter to a design that is realized entirely in microstrip.

4.9.2 Lumped Element to Distributed Element Conversion

Example 4.9-1: Consider the lumped element 2 GHz low pass filter schematic and response shown in Figure 4-42. This low pass filter has a 3 dB bandwidth of approximately 2470 MHz. Use the microstrip equivalent models of series inductance and shunt capacitance to realize the filter in microstrip. The microstrip substrate is Rogers's 6010 material with a 0.025 dielectric thickness.

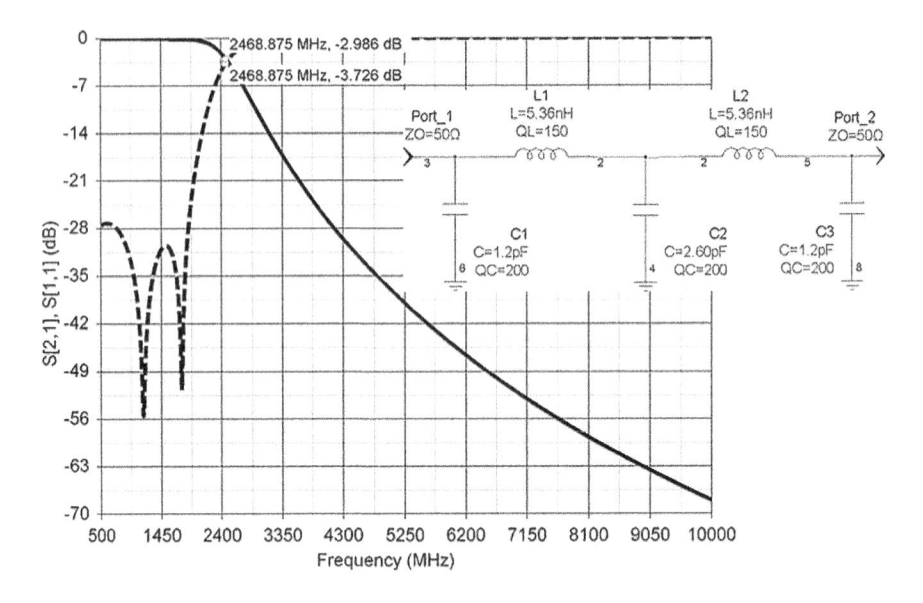

Figure 4-42 Lumped element 2.2 GHz low pass filter and frequency response

Solution: The series inductors will be realized as 80 Ω transmission lines of sufficient length to act as a 5.36 nH inductor. The TLINE Transmission Line synthesis program is used to calculate the microstrip line width for an 80 Ω transmission line on the Rogers 0.025 in. RO3010 material. As Figure 4-43 shows the 80 Ω transmission line has a line width of 6.26 mils with an effective dielectric constant $\varepsilon_r = 6.08$. To realize the required inductance value a specific length of 80 Ω transmission line is required. Equation (2-69) is solved in the Equation Editor of Figure 4-44 to calculate the length of microstrip line that is required to realize the inductance values used in the filter. The length of line required for a 5.36 nH inductor is then found to be 321 mils. The shunt capacitors will be realized as 20 Ω transmission lines. Using TLINE the 20 Ω line width is calculated to be 102.8 mils with an effective dielectric constant $\varepsilon_r = 8.00$. Equation (2-70) is solved in the Equation Editor of Figure 4-45 to find the length of 20 Ω transmission line required to behave as 2.6 pF shunt capacitor is 217 mils. The line length for the 1.2 pF capacitors is then found to be 100 mils.

Resonant Circuits and Filters

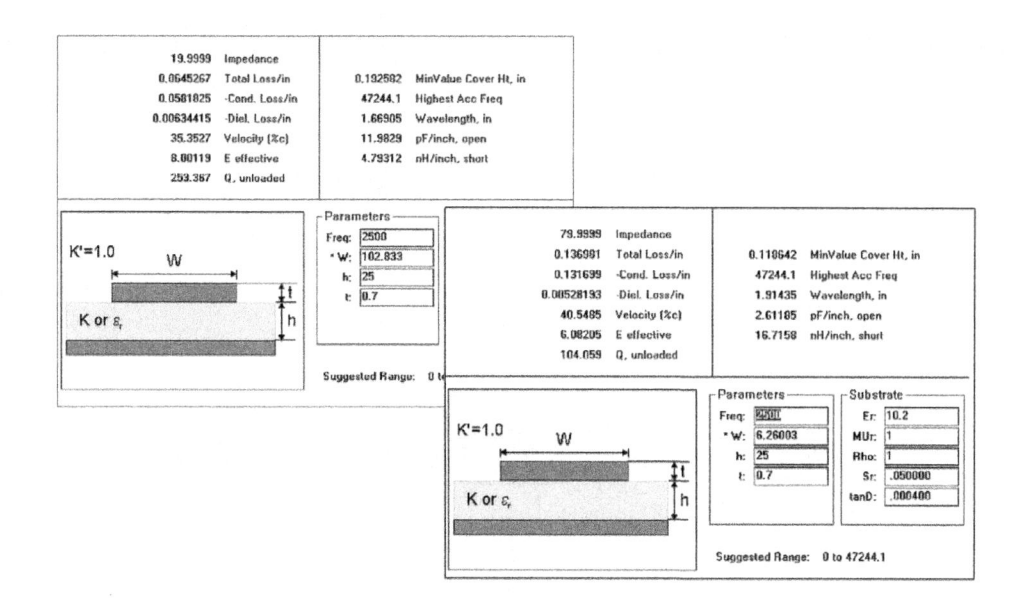

Figure 4-43 TLINE calculations for 20 Ω and 80 Ω microstrip lines

```
1    %Enter design parameters
2    F1=2.5e9
3    Lambda1=3e8/F1              %free space wavelength
4    erEFF1=6.08                 %effective dielectric constant
5    LambdaG1=Lambda1/sqrt(erEFF1) %wavelength in meters
6    LambdaG1=LambdaG1/.0254     %wavelength in inches
7    L=5.36e-9
8    Z1=80
9    %Microstrip Line Length is:
10   Length1=F1*LambdaG1*L/Z1
```

Figure 4-44 Calculating the length of inductive microstrip line

```
1    %Enter frequency, imegdance and the effective dielectric constant
2    F=2.5e9
3    Z=20
4    erEFF=8.00
5    %Enter the desired shunt Capacitance
6    C=2.6e-12
7    %Calculate the following....
8    Lambda=3E8/F    %free space wavelength in meters
9    LambdaG=Lambda/sqrt(erEFF)/.0254 %wavelength in meters
10   %Microstrip Line Length is:
11   Length=LambdaG*F*Z*C %inches
```

Figure 4-45 Calculating the length of capacitive microstrip line

The length of the microstrip lines are 0.321 and 0.217 inches, respectively. Adding a short 50 Ω section to the input and output, the initial low pass filter schematic and response is shown in Figure 4-46.

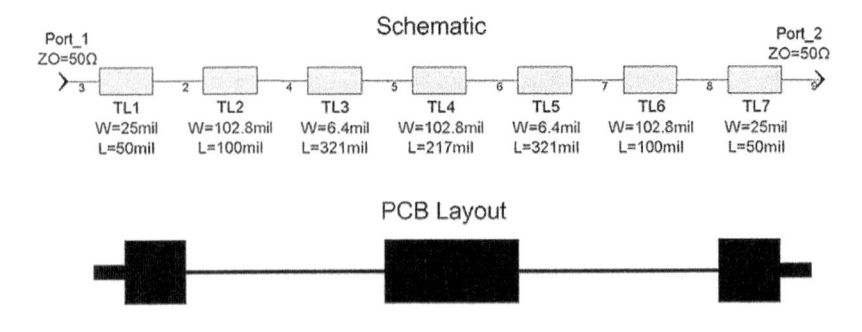

Figure 4-46 Initial schematic and PCB layout of the low pass filter

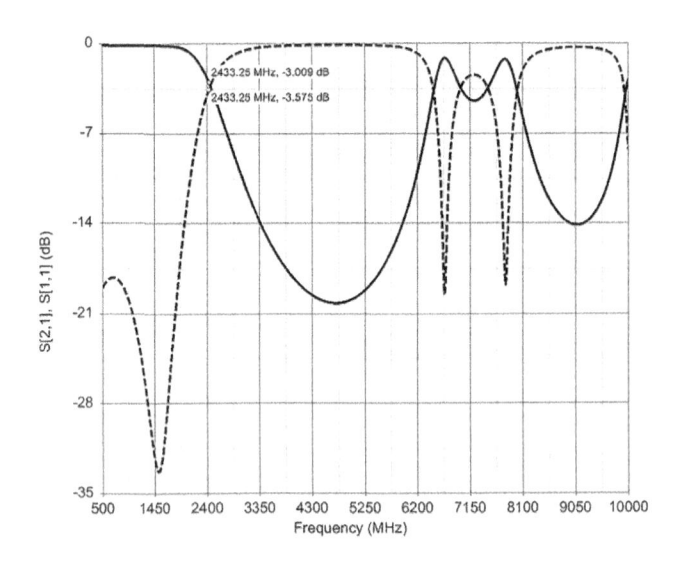

Figure 4-47 Initial Schematic and Response of the Low Pass Filter

Examine the printed circuit board, PCB, layout of the Low Pass Filter of Figure 4-46. Note the change in geometry as the impedance transitions from 50 Ω to 20 Ω and from 20 Ω to 80 Ω. These abrupt changes in geometry are known as discontinuities. Discontinuities in geometry result in fringing capacitance and parasitic inductance that will modify the frequency response of the circuit. At RF and lower microwave frequencies (up to

Resonant Circuits and Filters

about 2 GHz) the effects of discontinuities are minimal and sometimes neglected [4]. As the operation frequency increases, the effects of discontinuities can significantly alter the performance of a microstrip circuit. Genesys has several model elements that can help to account for the effects of discontinuities. These include: T-junctions, cross junctions, open circuit end effects, coupling gaps, and bends. A Microstrip Step element can be placed between series lines of abruptly changing geometry to account for the step discontinuity. Place the Microstrip Step element at each impedance transition in the filter. Make sure that the narrow side and wide side are directed appropriately. The Microstrip Step element will automatically use the adjacent width in its calculation. Figure 4-48 shows the low pass filter schematic with the step elements added between transmission line sections.

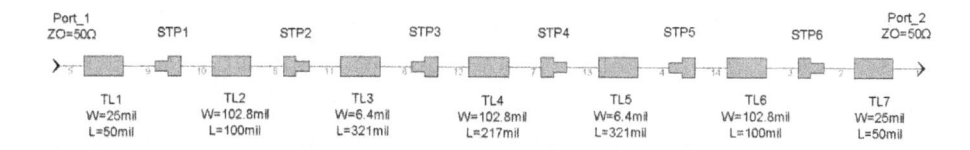

Figure 4-48 Stepped impedance filter with added "step" elements

A comparison of the initial low pass filter model and the modified model is shown in Figure 4-49. As the Figure shows there is a slight difference in the filter insertion loss, S21, particularly as the frequency increases from the cutoff at 2400 MHz.

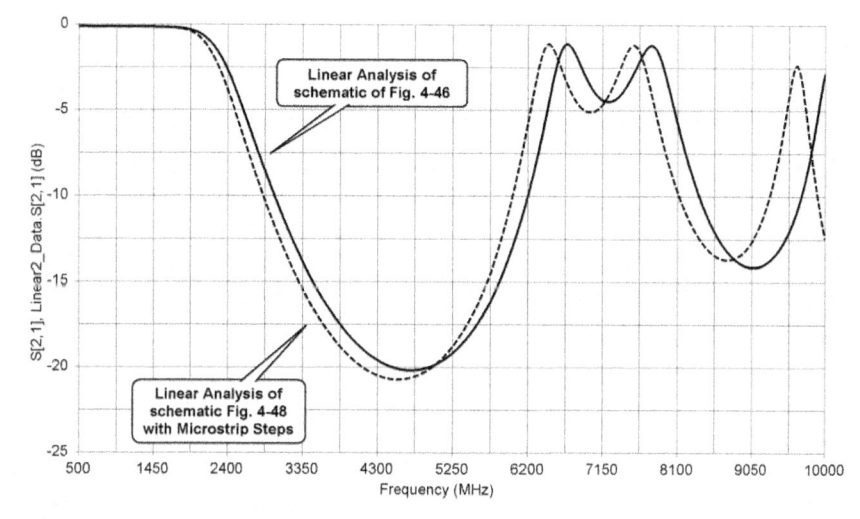

Figure 4-49 Filter cutoff frequency shift due to step discontinuities

4.9.3 Electromagnetic Modeling of the Stepped Impedance Filter

Electromagnetic, EM, modeling is a useful tool in microstrip circuit design as it offers a means of potentially more accurate simulation than linear modeling. The linear microstrip component models used to model the stepped impedance filter are based on closed form expressions developed over many years. For many designs the linear model is quite acceptable. Microstrip circuits that contain several distributed components in a dense printed circuit layout will be affected by cross coupling and enclosure effects. This is because the microstrip circuitry is quasi-TEM with some portion of the EM fields in the free-space above the dielectric material. These effects are very difficult to accurately simulate with linear modeling techniques. The Genesys software suite has a very useful electromagnetic (EM) simulation engine named Momentum. Momentum is based on the method-of-moments (MoM) numerical solution of Maxwell's equations [3]. Unlike some EM simulation software Momentum's solutions are presented in the S parameter format that is familiar to the microwave circuit designer. A dataset is created that can be graphed just like any linear simulation. The Momentum model is created from the circuit layout rather than the schematic. In this section we will create a Momentum model from the layout that was created by the linear schematic. However we could import an arbitrary PCB artwork from any CAD program. Right click on the Layout window to expose the Layout Properties. On the General Tab make sure to specify the correct units (mils) that represent the drawing. Also check the 'show EM Box' check box so that a proper enclosure is modeled for the circuit. The box represents a metal enclosure that will serve as the boundary conditions for the EM simulation. The simulator will identify any box resonances that may occur which could have an adverse effect on the circuit design. The box sides must be lined up perpendicular to the input and output ports. The box size (length and width) can be adjusted by entering the desired dimensions in the Box Width (X) and Box Height (Y) settings as shown in Figure 4-50. The box height is specified in the Layers tab as the air above the metal conductor or 250 mils. On the Layer Tab make sure that the microstrip dielectric material is defined on the Substrate

Resonant Circuits and Filters 201

line. Once all of the Layer settings have been specified add the Momentum Analysis to the Workspace as shown in Figure 4-52.

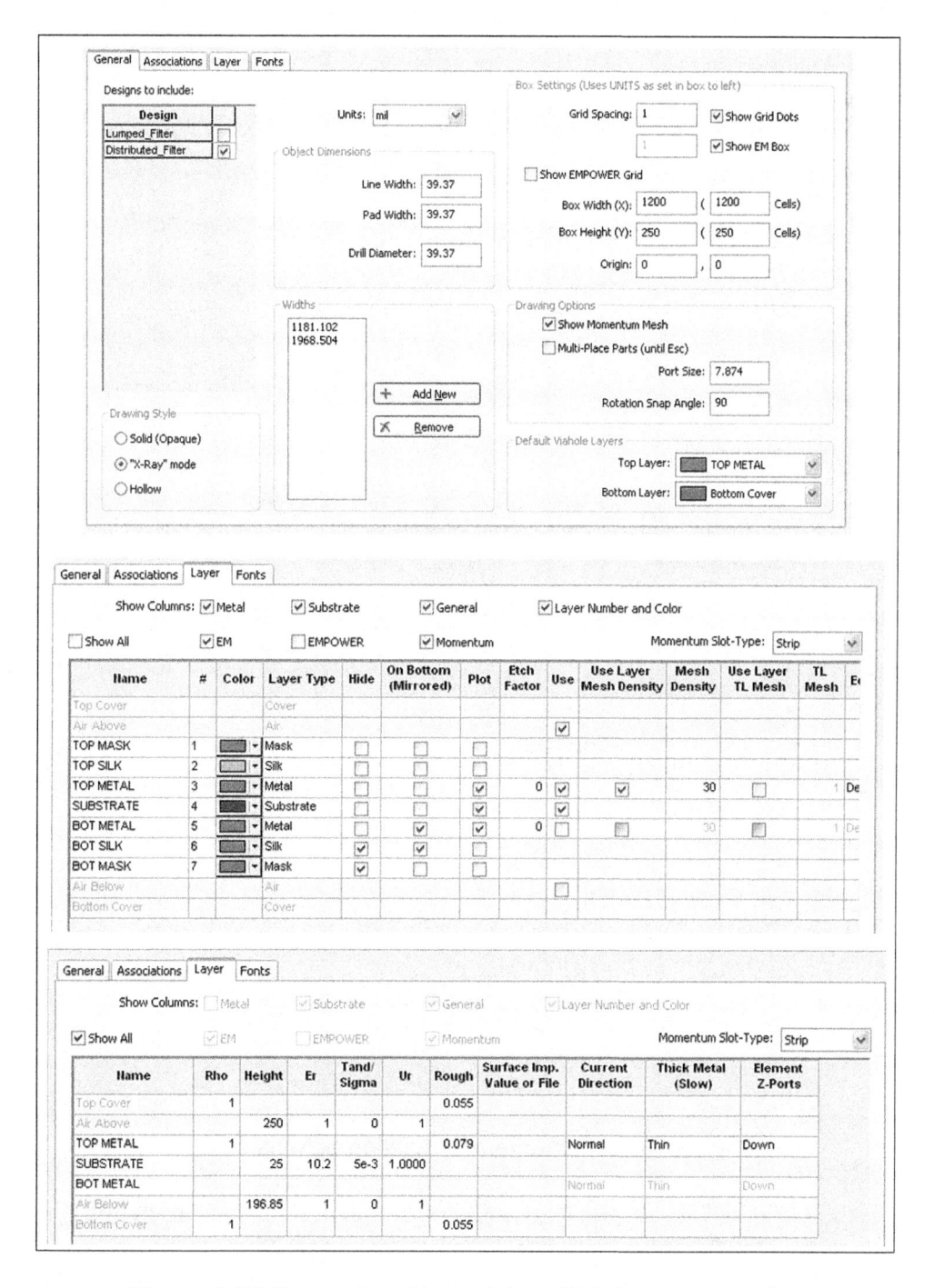

Figure 4-50 General and layer tabs of the layout properties

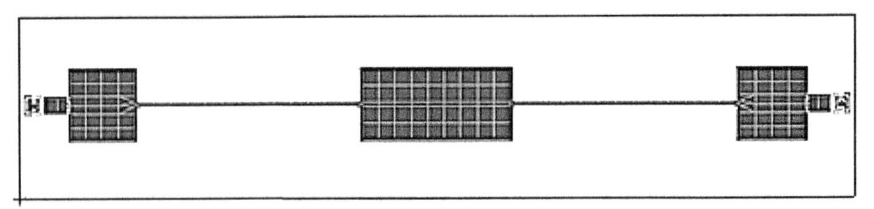

Figure 4-51 Filter layout showing box outline and conductor mesh

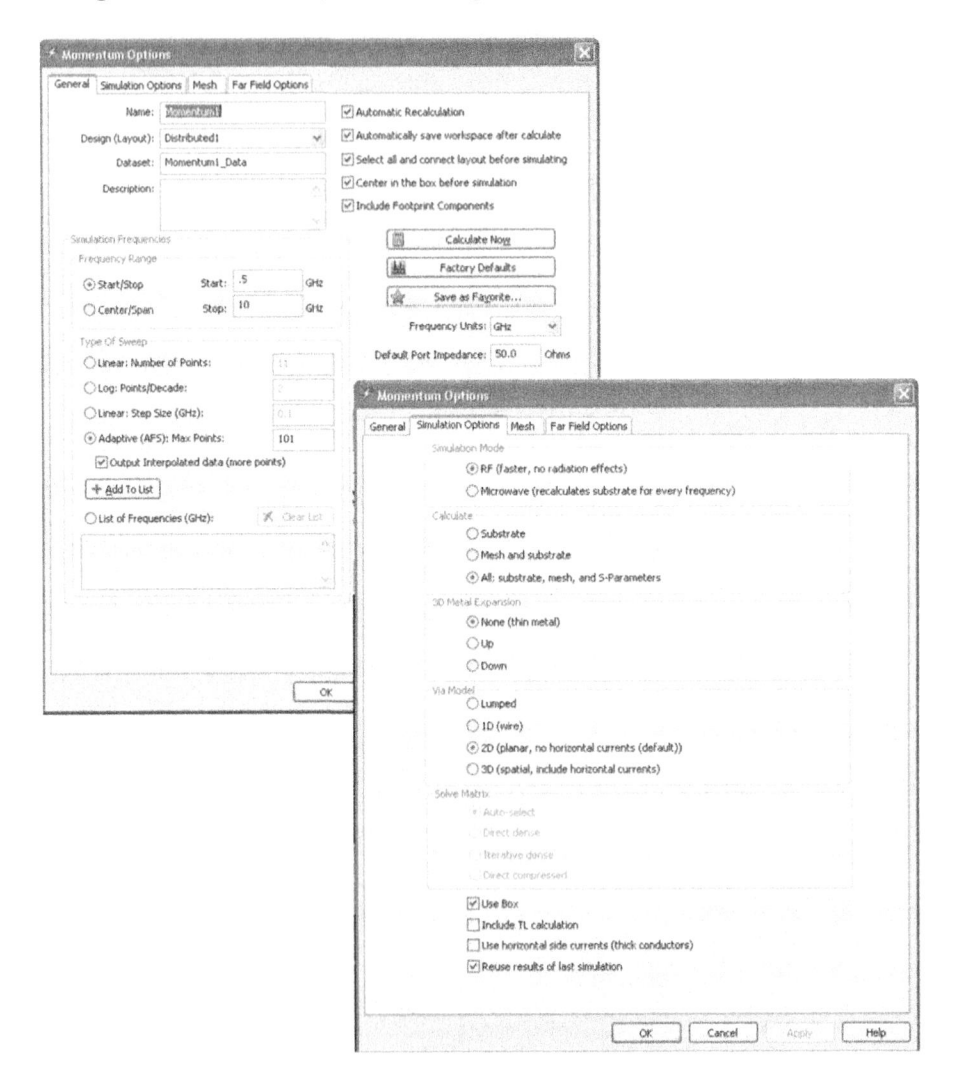

Figure 4-52 Momentum simulation options setup

On the General Tab, set the start and stop frequency for the simulation and select the adaptive sweep type. The adaptive sweep reduces frequency point interpolation error. On the Simulation Options Tab choose the RF

Resonant Circuits and Filters

simulation mode. The RF simulation mode is a much faster EM simulation and is suitable for lower microwave (RF) frequencies where there is not a significant amount of coupling among transmission lines. The Microwave simulation mode is a full wave EM analysis that includes all coupling radiation within the box. Also check the 'Calculate – All' button so that the metal mesh and the substrate are used in the solution.

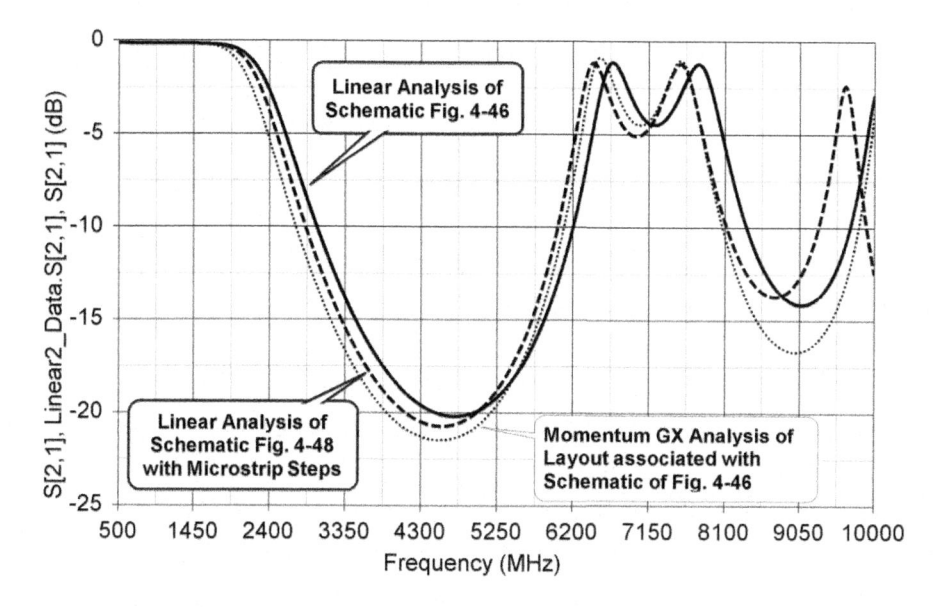

Figure 4-53 Linear simulation and momentum simulation comparison

The comparison between the Linear and Momentum simulations shows that there is some further deviation in the filter rejection as the frequency increases above 2 GHz.

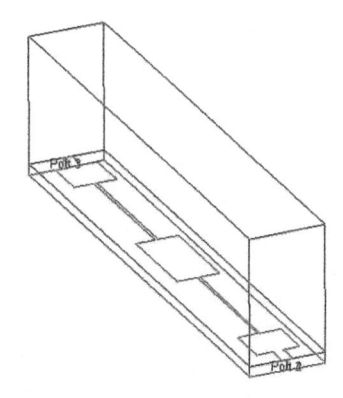

Figure 4-54 3D View of filter and prototype filter printed circuit board

4.9.4 Reentrant Modes

In Chapter 3 section 3.7 it was demonstrated that distributed transmission lines have repetitive impedance characteristics which are dependent on the physical length and frequency. In planar distributed filter design this leads to the creation of reentrant responses in the filter's passband characteristic. Reentrant modes can be seen in the filter response of Figure 4-53 at frequencies near 6667 MHz, 7826 MHz and 10 GHz. Depending on the filter design goals these modes may be harmless. If the distributed low pass filter's goal is to reject frequencies near 5 GHz only, then the circuit can be used as designed and the reentrant modes are of no consequence. If however the filter is required to have > 20 dB rejection of all frequencies from 5 GHz through 10 GHz then the reentrant modes present a problem as these frequencies will be passed by the filter.

Example 4.9-2: One technique to remove reentrant modes from a filter response is to cascade the filter with a 'clean-up' filter. For the filter design of Figure 4-53 the 'clean-up' filter is a stepped impedance low pass filter with a higher cutoff frequency. Check the reentrant response of the filter.

Solution: Figure 4-55 shows a low pass filter with a 4 GHz cutoff frequency in cascade with the 2 GHz filter.

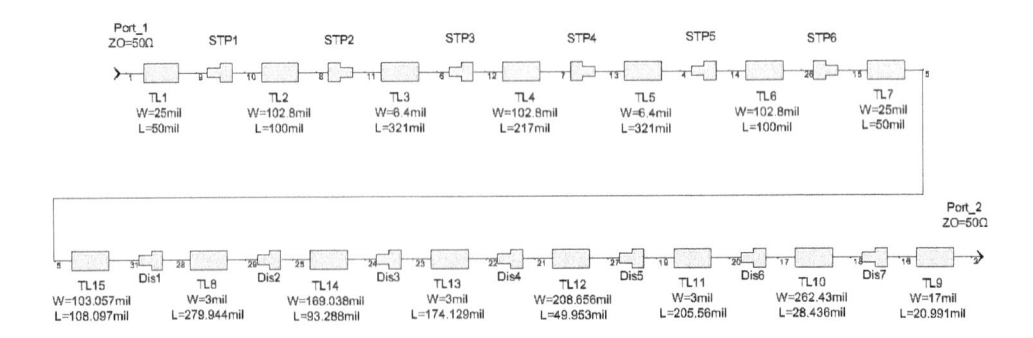

Figure 4-55 Cascaded stepped impedance low pass filter schematic

The response of the cascaded filters is shown in Figure 4-56. Compared to the response of Figure 4-53, the reentrant responses have been reduced greater than 20 dB. The filter now has a very nice ultimate band rejection characteristic through 10 GHz.

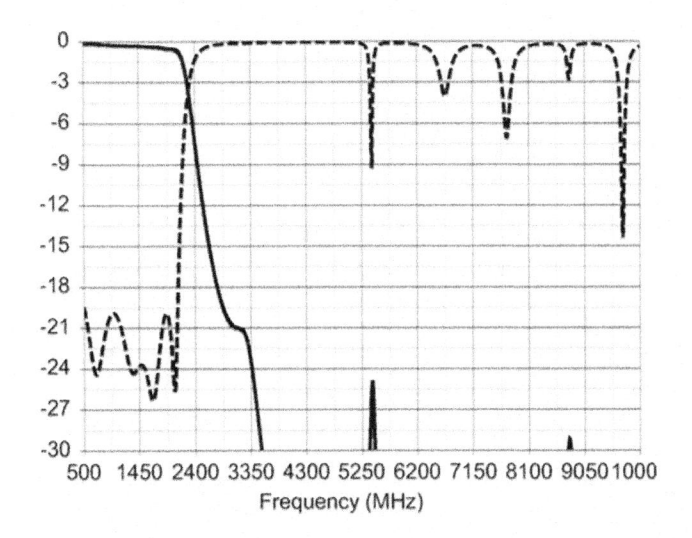

Figure 4-56 Cascaded stepped impedance low pass filter response

4.9.5 Microstrip Coupled Line Filter Design

The edge coupled microstrip line is very popular in the design of band pass filters. A cascade of half-wave resonators in which quarter wave sections are parallel edge coupled lines, are very useful for realizing narrow band, band pass filters. This type of filter can typically achieve ≤ 15% fractional bandwidths [4].

Example 4.9-3: Design a band pass filter at 10.5 GHz. The filter is designed on RO3010 substrate (ε_r = 10.2) with a dielectric thickness of 0.025 inches. The filter should have a pass band of 9.98 – 11.03 GHz. As a design goal the filter should achieve at least 20 dB rejection at 9.65 GHz. In other words the filter is required to have > 20 dB rejection at 330 MHz below the lower passband frequency.

Solution: The Microwave Filter synthesis utility in Genesys is used to design the filter network. Figure 4-57 shows the entry of the design parameters into the Topology, Settings, and Options tabs.

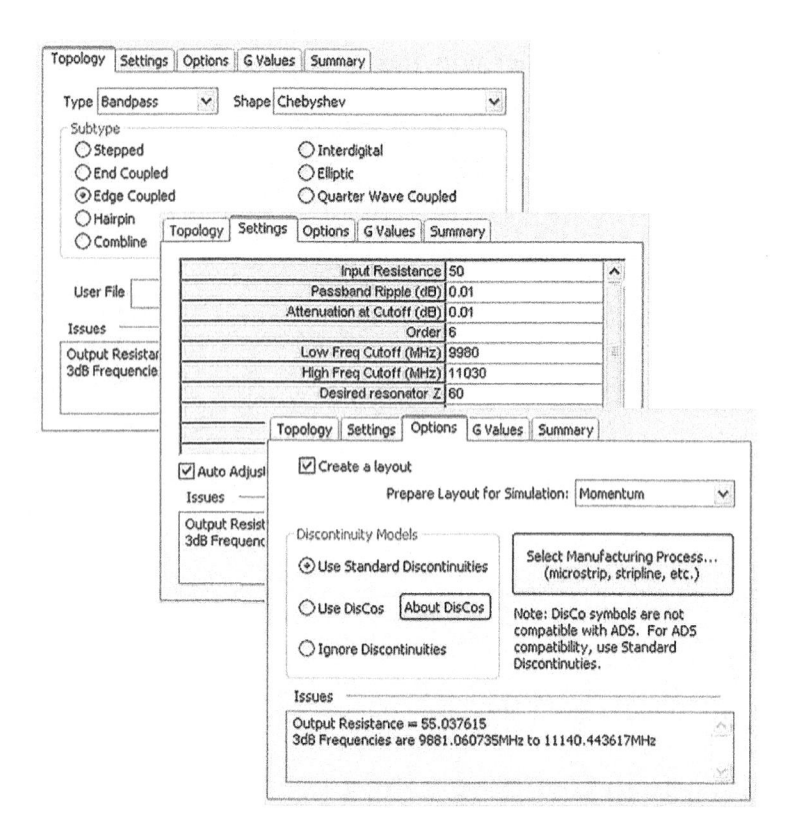

Figure 4-57 Bandpass filter synthesis selections

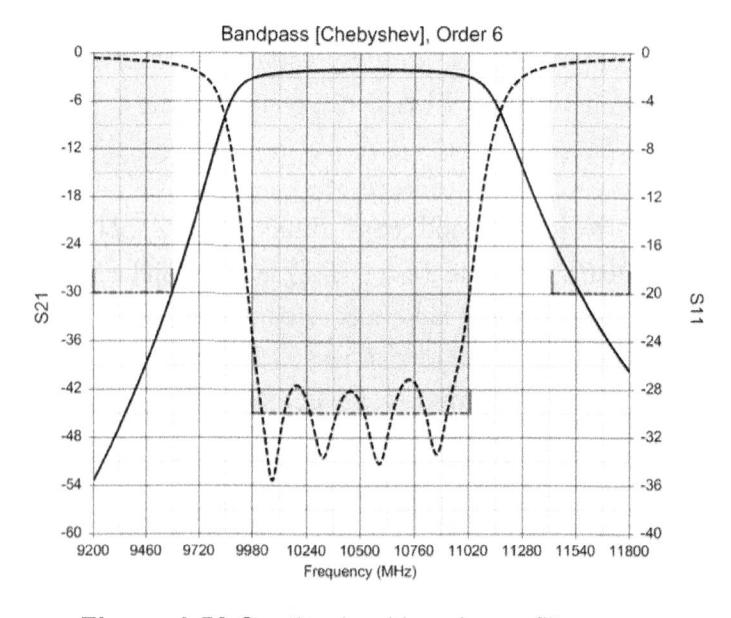

Figure 4-58 Synthesized bandpass filter response

Resonant Circuits and Filters 207

On the Topology tab set the design for an edge coupled band pass filter with a Chebyshev shape. On the Settings tab, enter the pass band frequencies per the filter specification. Vary the order until the desired filter rejection is achieved. The Microwave Filter synthesis program shows that a 6th order filter should meet the rejection specifications. On the Options tab, select standard discontinuities and check the 'Create a Layout'. The synthesized filter schematic is shown in Figure 4-59.

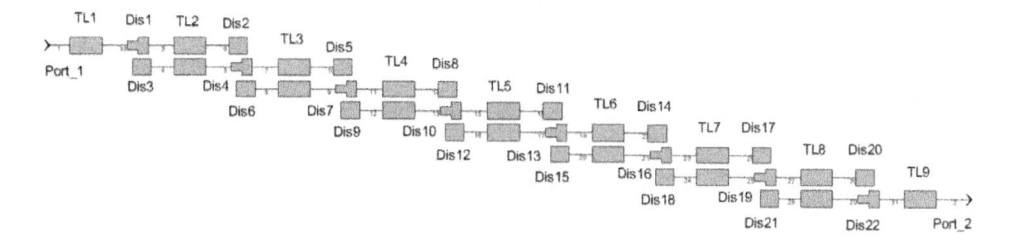

Figure 4-59 Parallel line edge coupled microstrip bandpass filter schematic

4.9.6 Electromagnetic Analysis of the Edge Coupled Filter

The designer must be cautious when using Linear Analysis techniques to design circuits with multiple edge coupled microstrip lines. We know that there is a large percentage of the parallel line coupling that occurs in the free space above the microstrip substrate and the conductors. This can lead to considerable error when relying on the linear simulation results. The coupled line models used by the linear simulator are based on closed form expressions derived from coupling measurements on parallel lines with loosely defined boundary conditions.

Example 4.9-4: Use the microwave mode in the Momentum software to simulate the edge coupled filter.

Solution: Using the circuit model created in the last section examine the layout of the filter. On the Layout-Properties window, make sure that the "Show EM Box" is selected. Then adjust the size of the box to fit the filter. Set the filter box width (Y) to 400 mils and the box height (X dimension) to 760 mils. The EM box will show up as a red rectangle on the layout window. Next edit the Layer tab. Check the "Show All" box. Scroll across to make sure that the substrate thickness, dielectric constant, and loss tangent parameters are correct. Check "Use" boxes as appropriate. Set the initial box height (Z dimension) at 300 mils.

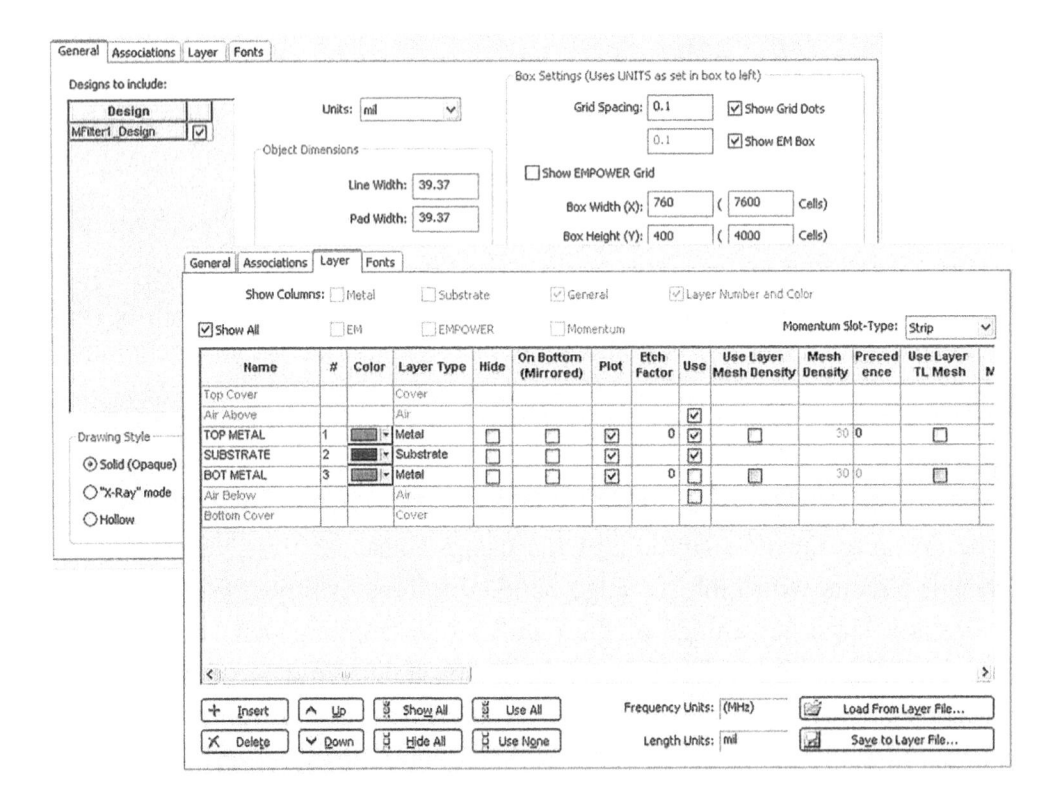

Figure 4-60 Layer setup for momentum analysis of edge coupled filter

Add a Momentum GX Analysis to the workspace and set the parameters as shown in Figure 4-61. Set the analysis frequency range of 9 to 12 GHz with an Adaptive sweep type. On the Simulation Options tab, make sure that the "use box" check box has been selected. This forces a 3D mesh of the entire filter and box. This time we must select the microwave simulation mode so that the box radiation effects are properly modeled.

Resonant Circuits and Filters 209

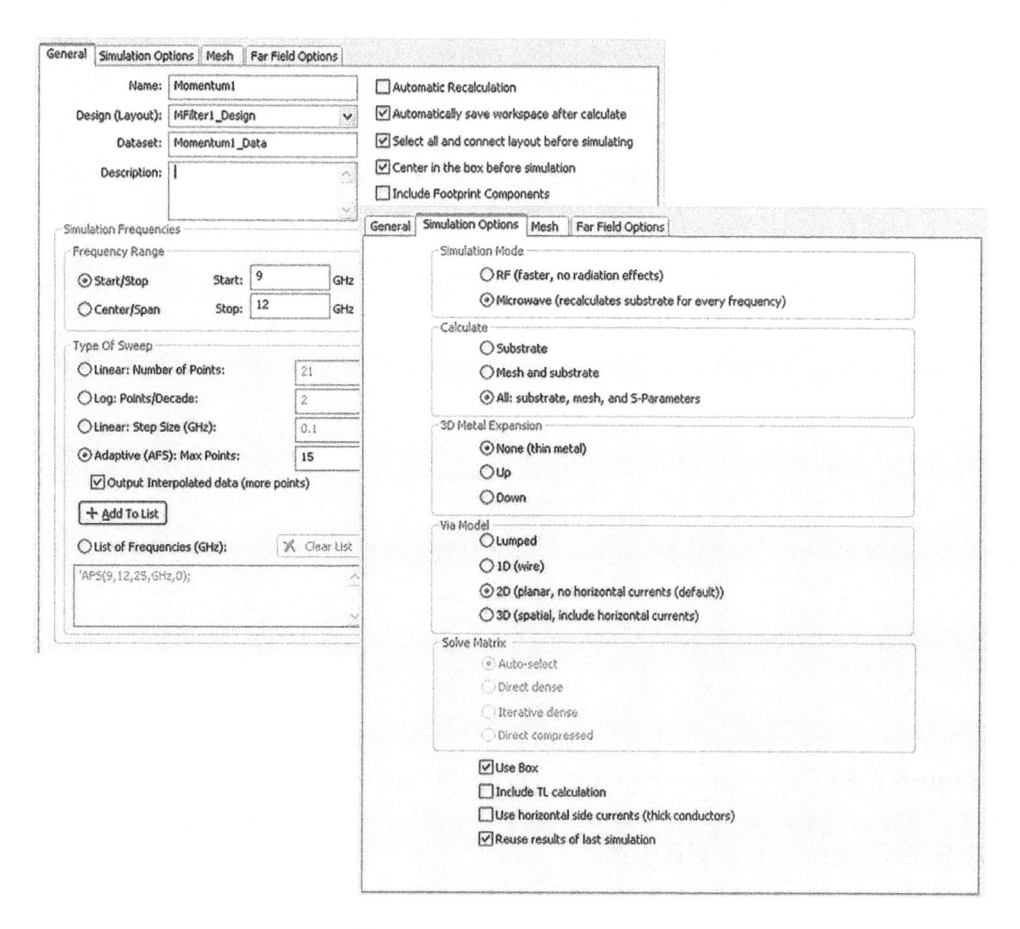

Figure 4-61 Momentum GX general setup

Perform a Momentum simulation on the filter and observe the response. The edge-coupled filter is particularly sensitive to the sidewalls providing the proper boundary conditions for the coupled energy between the parallel-coupled lines. Begin the tuning process by first reducing the box height to 250 mils. Keep the cover height at 250 mils. A 3D picture of the filter is shown in Figure 4-64 to have a good appreciation of the physical geometry of the filter. Figure 4-62 shows the comparison between the Linear model and the Momentum EM model. We see that the actual pass band is shifted up in frequency slightly while the bandwidth as been reduced. Varying the box dimensions shows that the filter skirts are heavily dependent on the box around the filter.

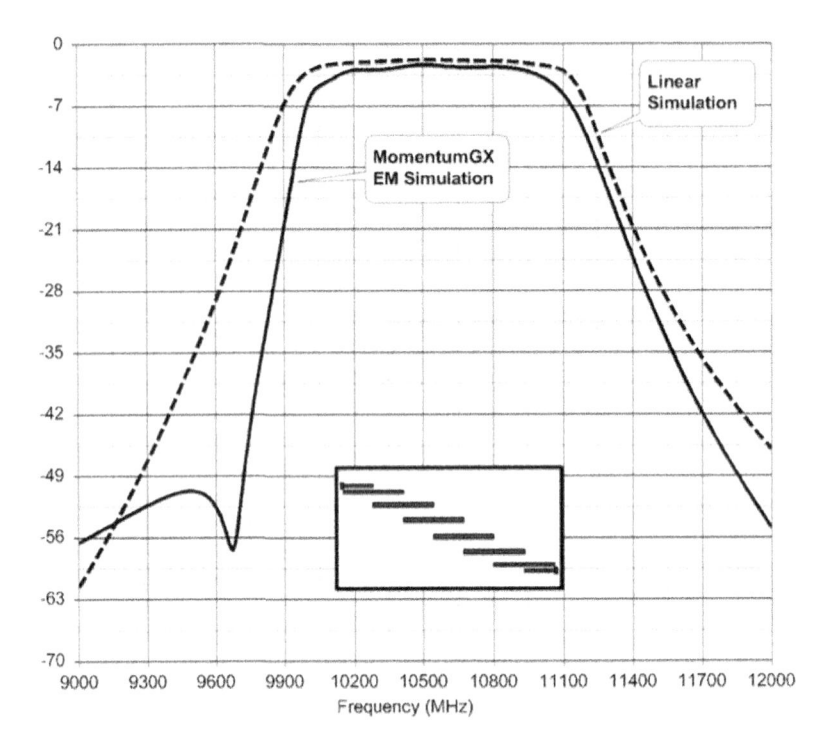

Figure 4-62 Comparison between the linear and the Momentum EM model

4.9.7 Enclosure Effects

Another issue with the box or enclosure design is the possibility of cavity-like resonances that can occur in the physical surroundings of a microwave circuit. Resonances within the metallic enclosure can allow a parallel path for microwave radiation that could bypass the resonators of the filter. This can cause the filter skirts to have significantly less attenuation than the design predicts. In active circuits such as amplifiers the box resonance can result in problems with oscillation.

Example 4.9-5: Analyze the effects of placing the edge coupled filter in a wider 900 mil box as shown in Figure 4-63.

Solution: When the initial EM model is setup the Momentum simulation status window will issue a warning about any box resonances that may exist. As Figure 4-63 shows this box has a dominant resonance at 10.28 GHz. As the resulting filter response of Figure 4-63 shows the box resonance has an extremely detrimental effect on the low side pass band response. Very little filter rejection is achieved on the low side of the filter pass band. EM simulation is a powerful tool that the engineer should consider for accurate microwave circuit design. By modeling radiation effects accurately a better representation of resonator Q factor is realized. This is particularly important in the design of microwave filter networks.

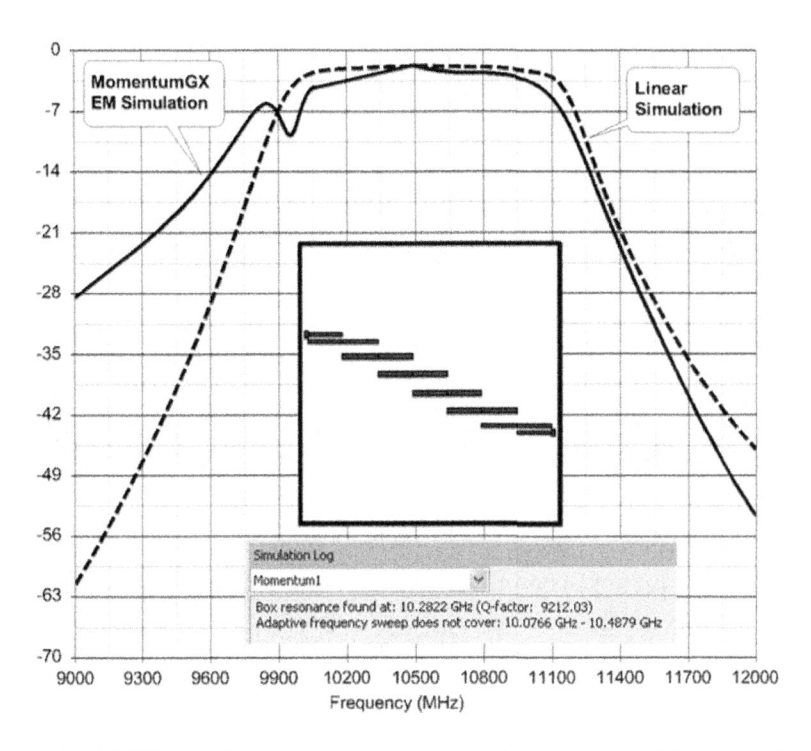

Figure 4-63 Effect of box resonance on edge coupled filter passband

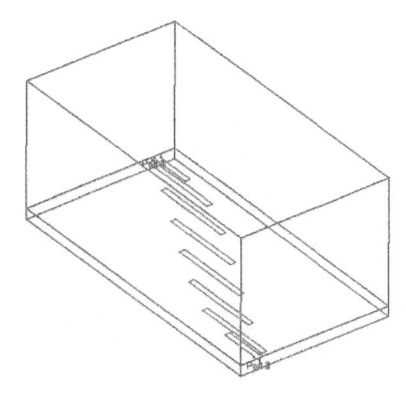

Figure 4-64 3D View of the edge coupled filter showing layer stack

References and Further Reading

[1] Ali A. Behagi and Stephen D. Turner, *Microwave and RF Engineering*, BT Microwave LLC, State College, PA, 2011

[2] RF Circuit Design, Second Edition, Christopher Bowick, Elsevier 2008

[3] Keysight Technologies, *Genesys 2014.03, Users Guide*, www.keysight.com

[4] Foundations for Microstrip Circuit Design, T.C. Edwards, John Wiley & Sons, New York, 1981

[5] Q Factor, Darko Kajfez, Vector Fields, Oxford Mississippi, 1994

[6] Q Factor Measurement with Network Analyzer, Darko Kajfez and Eugene Hwan, IEEE Transactions on Microwave Theory and Techniques, Vol. MTT-32, No. 7, July 1984.

[7] High Frequency Techniques, Joseph F. White, John Wiley & Sons, Inc., 2005

[8] Soft Substrates Conquer Hard Designs, James D. Woermbke, Microwaves, January 1982.

[9] Principles of Microstrip Design, Alam Tam, RF Design, June 1988.

[10] David M. Pozar, *Microwave Engineering*, Fourth Edition, John Wiley & Sons, New York, 2012

Problems

4-1. Consider the one port resonator that is represented as a series RLC circuit as shown. Analyze the circuit, with R = 5 Ω, L = 5 nH, and C = 5 pF. Plot the magnitude of the resonator input impedance and measure the resonance frequency.

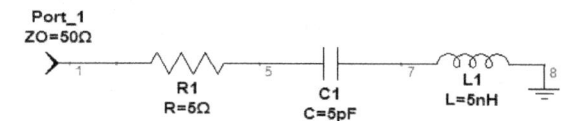

4-2. Consider the one port resonator that is represented as a parallel RLC circuit as shown. Analyze the circuit, with R = 500 Ω, L = 50 nH, and C = 50 pF. Plot the magnitude of the resonator input impedance and measure the resonance frequency.

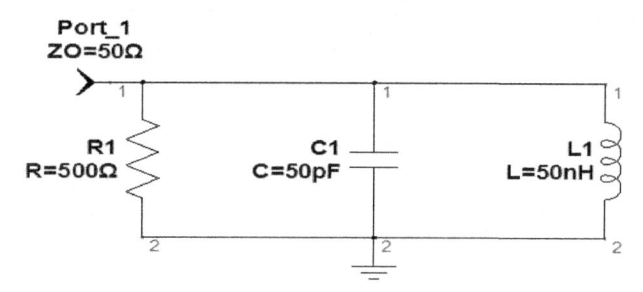

4-3. Design a Butterworth lowpass filter having a passband of 2 GHz with an attenuation 20 dB at 4 GHz. Plot the insertion loss versus frequency from 0 to 5 GHz. The system impedance is 50 Ω.

4-4. Design a 5th order Chebyshev highpass filter having 0.2 dB equal ripples in the passband and cutoff frequency of 2 GHz. The system impedance is 75 Ω. Plot the insertion loss versus frequency from 0 to 5 GHz.

4-5. In a full duplex communication link, the uplink signal is around 200 MHz while the downlink is at 500 MHz. A 25 Watt power amplifier is used on the uplink with 20 dB gain.

(a) Design a low pass filter on the uplink to pass the 200 MHz uplink signal while rejecting any noise power in the 500 MHz band. Design the passband cutoff frequency is at 220 MHz, therefore, the filter should have a Chebyshev Response with 0.1 dB pass band ripple. The reject requirement is at least -40 dB rejections at 500 MHz.

(b) Design a High Pass Filter that passes frequencies in the 480 MHz to 520 MHz range. The High Pass Filter specifications are: The passband cutoff frequency is 480 MHz, therefore, the filter should have a Chebyshev response with 0.1 dB pass band ripple. The reject requirement is at least -60 dB losses at 190 MHz.

4-6. Design a 75 Ω transmission line of sufficient length to act as a 10 nH inductor. Use the TLINE Transmission Line Synthesis Program to calculate the microstrip line width on the Rogers 0.025 inch RO3010 material.

4-7. Using the microwave filter synthesis tool, design a stepped impedance low pass filter on RO3003 material that is 0.010 inches thick. Use a Chebyshev response with a 0.01dB ripple and a cutoff frequency of 4 GHz. Determine the worst case in band return loss and the rejection at 6 GHz.

4-8. For the filter design of Problem 4-7 create an EM simulation using Momentum. Compare the EM simulation to a linear simulation. Comment on the rejection comparison at 6 GHz.

4-9. For the filter design of Problem 4-7 determine the frequencies at which reentrant modes exist up through 20 GHz.

4-10. Design a half wave microstrip resonator at 10 GHz using RO3003 substrate that is 0.020 thick. Initially design the resonator with a 50 W line impedance. Select a coupling capacitor to critically couple the resonator to the 50 Ω source. Then determine the resonator line impedance that results in the highest unloaded Q_o.

4-11. Use the microwave filter synthesis tool to design a parallel edge coupled filter on RO3003 substrate that is 0.010 thick. Use a Chebyshev characteristic with 0.10 dB ripple. Design the passband to cover 10.7 GHz to 12.2 GHz. Determine the filter order required to achieve 30 dB rejection at 9 GHz.

4-12. For the filter design of Problem 4-10, create an EM simulation using Momentum. Compare the linear and EM simulations. Determine the minimum box width (EM Box height, Y) that creates a box resonance frequency. What is the frequency of the box resonance?

Chapter 5

Power Transfer and Impedance Matching

5.1 Introduction

Impedance matching is an integral part of RF and microwave circuit and system design. It is necessary for the efficient transfer of power from a source to the load. For example, in microwave amplifier design, the need for impedance matching arises when the amplifier must be properly terminated at both terminals in order to deliver maximum power from the source to the load. In narrowband applications impedance matching can be achieved, at a single frequency, with a lossless two-element network, known as L-network or L-section. In this chapter the basics of power transfer and the conditions for maximum power transfer are presented. The mathematical equations for the design of discrete L-networks to match arbitrary source and load impedances are derived. Matlab compatible expressions in the Genesys Equation Editor are used to solve the mathematical expressions. This builds a solid foundation for the engineer to handle more complicated matching problems including broadband matching applications. In this chapter both analytical and graphical techniques are used to design narrowband and broadband matching networks. An introduction to the impedance matching synthesis program that is integrated into the Genesys software is also presented. The VBScript capability of Genesys is also used to create a simple L-network matching utility that solves the impedance matching equations for arbitrary source and load impedances. Engineers can use VBScript programming techniques to generate their own synthesis applications within the Genesys software.

5.2 Power Transfer Basics

At low frequencies, where the electrical wavelengths of the signals are much longer than the physical dimensions of the wires and lumped components, the phase of the voltage and current waveforms do not change significantly along the length of wires or components. Therefore, power is transmitted from the source to the load with little loss. At RF and microwave frequencies, where the signal wavelengths are equal to or shorter than the physical dimension of the network components, the amplitude and phase of the voltage and current waveforms change significantly as they travel from source to the load. In this case power is reflected at discontinuities and the maximum power will not reach the load. To achieve maximum power transfer we need to eliminate reflections at discontinuities by inserting proper impedance matching networks.

5.2.1 Maximum Power Transfer Conditions

For the network of Figure 5-1 a voltage source, V_S, and the series impedance $Z_S = R_S + jX_S$ are connected to a network, having the input impedance $Z_{IN} = R_{IN} + jX_{IN}$, the power transferred to the network is given by Equation (5-1).

$$P_{Network} = \frac{\text{Re}[V_{IN} I_{IN}^*]}{2} \quad (5\text{-}1)$$

In Equation (5-1) Re denotes the real part and the symbol * denotes the conjugate value.

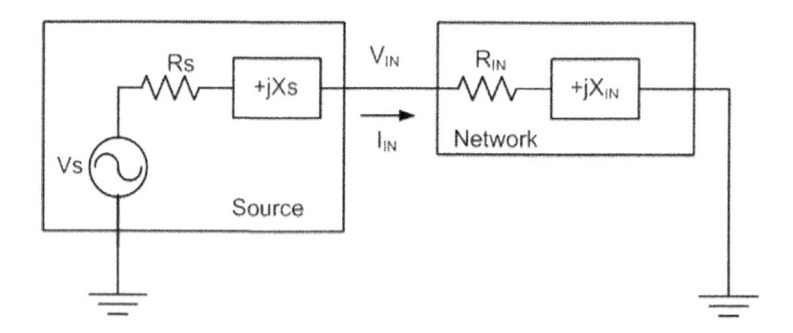

Figure 5-1 Voltage source connected to complex load impedance

In Figure 5-1, we assume V_S is sinusoidal steady state voltage source, R_S and R_{IN} are positive, and X_S and X_{IN} are real numbers. The input voltage and current to the network can be related to the source voltage as:

$$V_{IN} = \frac{V_S(R_{IN} + jX_{IN})}{(R_{IN} + R_S) + j(X_{IN} + X_S)}$$

and,

$$I_{IN}^* = \frac{V_S}{(R_{IN}+R_S) - j(X_{IN}+X_S)}$$

Therefore, multiplying V_{IN} by I_{IN}^*, Equation (5-1) can be written as:

$$P = \frac{\left(\frac{V_S^2 R_{IN}}{2}\right)}{(R_{IN}+R_S)^2 + (X_{IN}+X_S)^2} \qquad (5\text{-}2)$$

To maximize the power transfer, we differentiate Equation (5-2) with respect to R_{IN} and X_{IN} and set them equal to zero,

$$\frac{V_S^2}{2}\left[(R_{IN}+R_S)^2 + (X_{IN}+X_S)^2\right] - \frac{V_S^2 R_{IN}}{2}[2R_{IN}+2R_S] = 0$$

$$-\frac{V_S^2 R_{IN}}{2}(2X_{IN}+2X_S) = 0$$

A simultaneous solution of these two equations for R_{IN} and X_{IN}, leads to:

$$R_{IN} = R_S \qquad (5\text{-}3)$$

and

$$X_{IN} = -X_S \qquad (5\text{-}4)$$

Therefore, the maximum power transfer condition is that the load impedance be equal to the conjugate of the source impedance given in Equation (5-5).

$$Z_{IN} = Z_S^* \qquad (5\text{-}5)$$

This maximum power transfer condition between the source and load impedance can be divided into three cases.

Case 1:

If the source impedance is purely resistive the load impedance must also be purely resistive and equal to source resistance.

$$Z_{IN} = R_{IN} = R_S$$

Case 2:

If the source impedance is a resistor in series with a capacitor, $Z_S = R_S - jX_S$, the load impedance must be a resistor in series with an inductor such that:

$$Z_{IN} = R_S + jX_S$$

Case 3:

If the source impedance is a resistor in series with an inductor, $Z_S = R_S + jX_S$, the load impedance must be a resistor in series with a capacitor such that:

$$Z_{IN} = R_S - jX_S$$

5.2.2 Maximum Power Transfer with Purely Resistive Source and Load Impedance

This section explores in detail the three cases for maximum power transfer along with solutions using the Genesys software.

Example 5.2-1: Prove the maximum power transfer condition where the source and load impedance is purely resistive as shown in Figure 5-2.

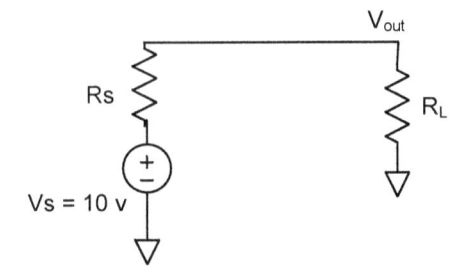

Figure 5-2 Network with purely resistive source and load impedance

Power Transfer and Impedance Matching

Solution: When the source impedance is purely resistive and the load resistance is R_L, the maximum power is transferred to the load when $R_L = R_S$. Using the resistor voltage-divider principle we can determine the output voltage, V_{out} for the following three possible cases:

$$
\begin{array}{ll}
\text{Case I:} & R_L = R_S \\
\text{Case II:} & R_L < R_S \\
\text{Case III:} & R_L > R_S
\end{array}
$$

Case I: If the input voltage is 10 VDC and $R_L = R_S = 50\ \Omega$, the output voltage is 5 Volts and the output power is 0.5 Watts. This is the maximum power that can be transferred.

$$V_{out} = V_S \frac{R_L}{(R_S + R_L)} = 10 \frac{50}{(50+50)} = 5\ volts$$

$$P_L = \frac{V_{out}^2}{R_L} = \frac{5^2}{50} = 0.5\ Watts$$

Case II: If $R_L = 25\ \Omega$ and $R_S = 50\ \Omega$, the output voltage is 3.333 Volts and the output power is 0.444 Watts.

$$V_{out} = V_S \frac{R_L}{(R_S + R_L)} = 10 \frac{25}{(50+25)} = 3.333\ volts$$

$$P_L = \frac{V_{out}^2}{R_L} = \frac{3.333^2}{25} = 0.444\ Watts$$

Case III: If $R_L = 100\ \Omega$ and $R_S = 50\ \Omega$, the output voltage is 6.666 Volts and the output power is 0.444 Watts.

$$V_{out} = V_S \frac{R_L}{(R_S + R_L)} = 10 \frac{100}{(50+100)} = 6.666\ volts$$

$$P_L = \frac{V_{out}^2}{R_L} = \frac{6.666^2}{100} = 0.444 \ Watts$$

Notice that the power output in Case I is greater than either Cases II or III. Therefore, the maximum power transfer is achieved only when $R_L = R_S$.

5.2.3 Maximum Power Transfer Validation in Genesys

The electronic design automation techniques, EDA, used in this book are based solely on the linear simulation capabilities of the Genesys software which in turn is based on the matrix manipulation of S parameters.

Example 5.2-2: Use the Genesys linear simulation techniques to demonstrate the maximum power transfer condition in a purely resistive system.

Solution: This is a very simple Genesys schematic with only input and output ports, as shown in Figure 5-3. The input port represents an RF signal source with 50 Ω source impedance and the output port is simply a 50 Ω resistive load. Use a Linear Analysis to sweep the schematic from 450 to 550 MHz and add a rectangular graph with the insertion loss, S21, in dB. Figure 5-3 shows the schematic and plot indicating that S21 = 0 dB. Because there is zero insertion loss between the source and load, there is maximum power transfer.

Power Transfer and Impedance Matching

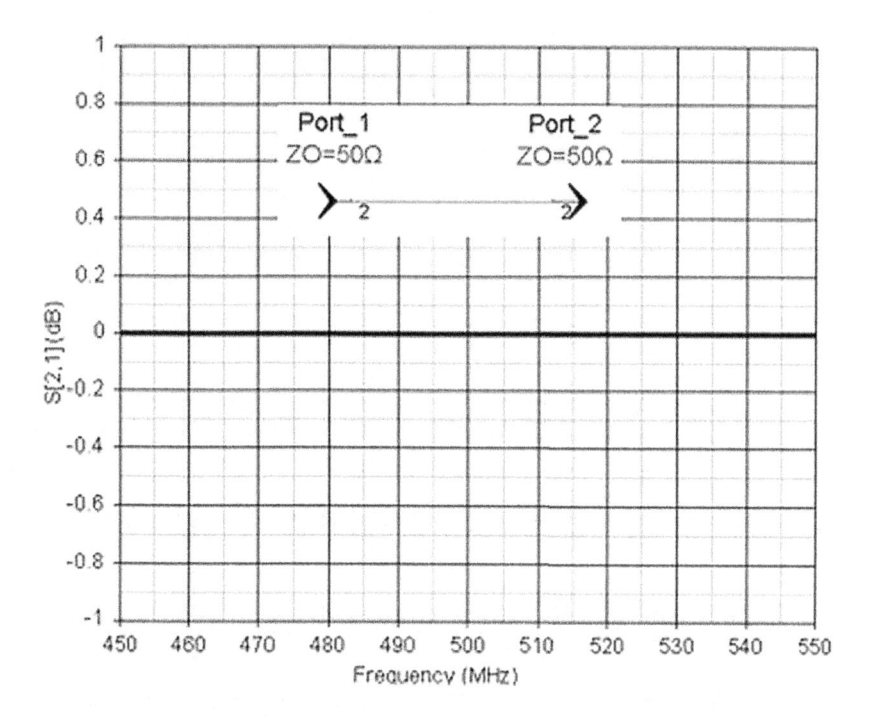

Figure 5-3 Case I power transfer with $R_L = R_S$

Change the load impedance value to 25 Ω to verify the power transfer of the Case II example. The plot of S21 and revised schematic are shown in Figure 5-4. There is now 0.512 dB loss in power from the source to the load. We know from Case I that the maximum power transfer is 0.5 Watts. To determine the amount of 0.512 dB power loss we first convert the 0.5 Watts, maximum output power, to dBm which is the power in dB relative to 1 mW. Utilizing the conversion equation,

$$10 \log (\text{power in mW}) = \text{power in dBm}.$$

$$10 \log (500 mW) = 26.98 \; dBm$$

The loss of 0.512 dB is subtracted from the 26.98 dBm to result in (26.98 dBm − 0.512 dB) = 26.47 dBm. Then converting from dBm back to mW we get 444 mW as calculated in Case II.

$$10^{\left(\frac{26.47}{10}\right)} = 444 \; mW \; \text{ or } \; 0.444 \; Watts$$

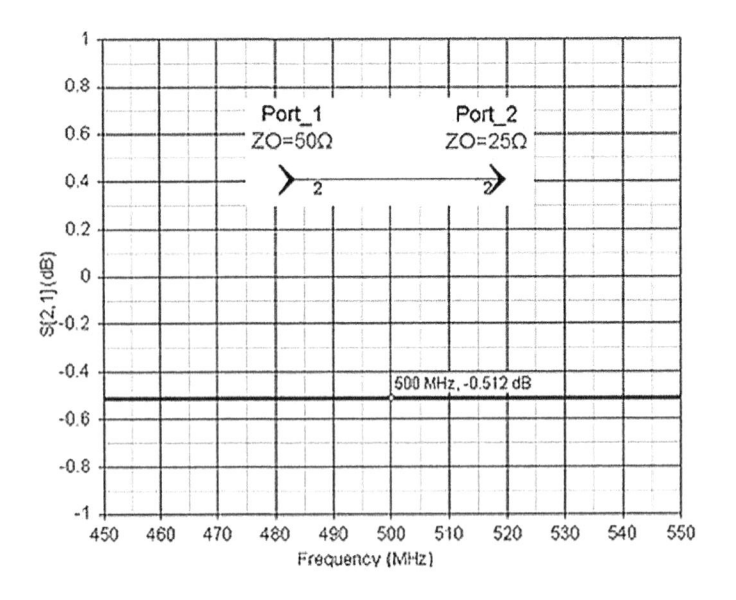

Figure 5-4 Case II power transfer with $R_L < R_S$

5.2.4 Maximum Power Transfer with Complex Load Impedance

According to Equation 5-5 maximum power transfer occurs when $Z_S = Z_L^*$. Therefore, if $Z_L = R_L - jX_L$, then for maximum power transfer we must have $Z_S = R_L + jX_L$.

Example 5.2-3: If the load is 50 Ω in series with a 15 pF series capacitance, find the source impedance to have maximum power transfer at 500 MHz.

Solution: We can find a source inductance that cancels the reactance of the load at this frequency. The reactance of the 15 pF capacitor is:

$$X_C = \frac{1}{2\pi f C} = 21.231 \ \Omega$$

therefore,

$$Z_L = 50 - j21.231 \ \Omega$$

For maximum power transfer the source impedance must be:

$$X_S = 50 + j21.231 \ \Omega$$

Power Transfer and Impedance Matching

At 500 MHz the value of the source series inductor is:

$$L = \frac{21.231}{2\pi f} = 6.76 \quad nH$$

To demonstrate the maximum power transfer, create a schematic in Genesys as shown in Figure 5-5. Place the specified 15 pF capacitance in series with the load and the 6.76 nH inductance in series with the source. Plot the insertion loss and VSWR on a rectangular graph as shown in Figure 5-5. Note that unlike the purely resistive source and load case, a complex conjugate match occurs at a single frequency. A perfect 1:1 VSWR is achieved at the conjugate match frequency of 500 MHz.

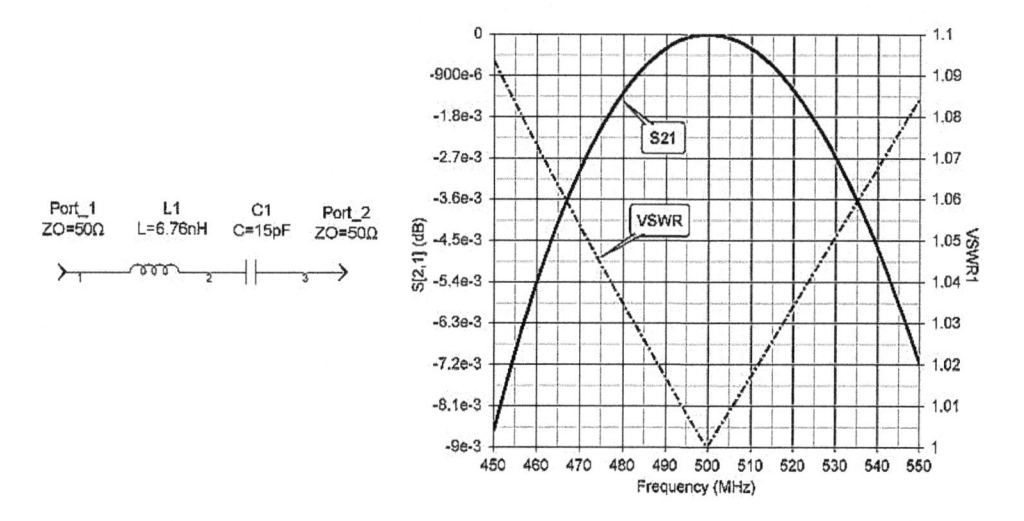

Figure 5-5 Maximum power transfer with complex source and load impedance

5.3 Analytical Design of Impedance Matching Networks

One of the important tasks in RF and microwave engineering is the determination of how an arbitrary complex load impedance, $Z_L = R_L + jX_L$, is analytically matched to any complex source impedance, $Z_S = R_S + jX_S$, as shown in Figure 5-6. This problem arises mainly in the design of inter-stage matching networks between active devices or between an antenna and a transmitter.

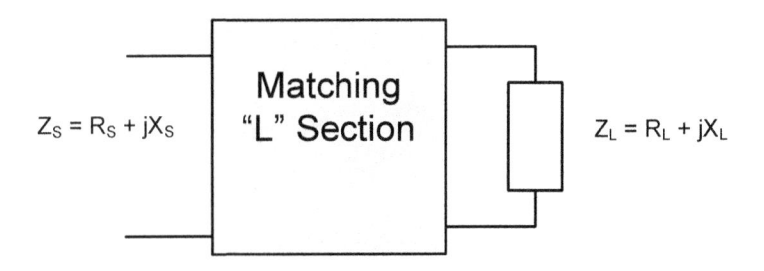

Figure 5-6 General impedance matching with an L-network

In Figure 5-6 the complex load impedance, $Z_L = R_L + j X_L$, is to be matched to the source impedance $Z_S = R_S + jX_S$. The only condition for impedance matching is that both R_S and R_L must be nonnegative while X_S and X_L could take any real value. In section 5.2 it was shown that maximum power would be transferred from the source to the load when the load impedance is the conjugate of the source impedance. For L-network matching there are two configurations that can match arbitrary load impedances to arbitrary source impedances. In the first configuration the first element adjacent to the load is a series element as shown in Figure 5-7.

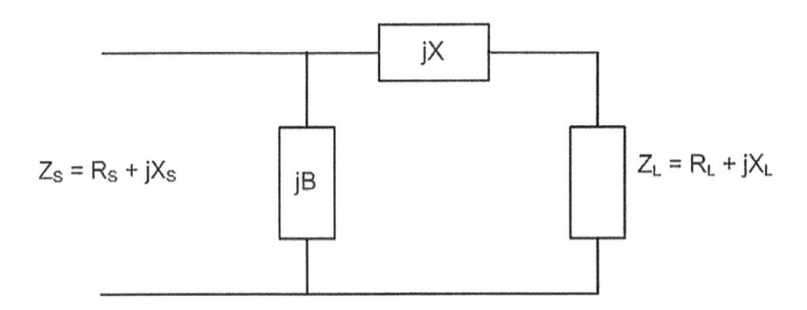

Figure 5-7 First impedance matching L-network with series element adjacent to load impedance

In the second configuration the first element adjacent to the load is a shunt element as shown in Figure 5-8.

Power Transfer and Impedance Matching 227

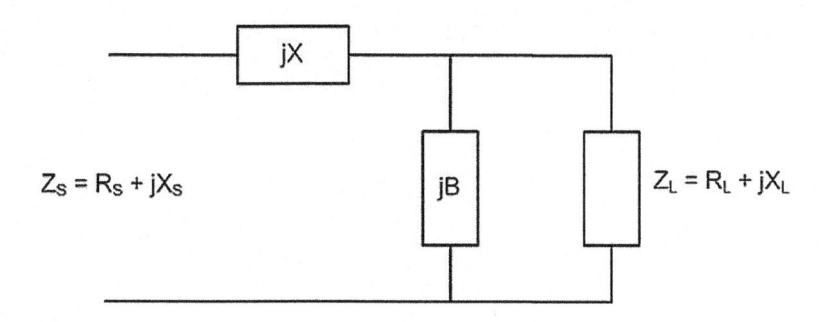

Figure 5-8 Second impedance matching L-network with shunt element adjacent to load impedance

5.3.1 Matching a Complex Load to Complex Source Impedance

To match any complex load to any complex source impedance with a single L-network, either the first or second configuration may be used. The choice of configurations depends on the conditions that source and load impedances dictate. In this section we first calculate the values for B and X and then, based on positive or negative values of B and X, determine the element values of the matching networks. The calculation of B and X is based on the requirement for maximum power transfer as stated in Equation (5-5). Applying Equation $Z_S^* = Z_{IN}$ to the first matching configuration in Figure 5-7 we get:

$$R_S - jX_S = \frac{1}{jB + \left(\dfrac{1}{jX + R_L + jX_L}\right)} \quad (5\text{-}6)$$

By separating the real and imaginary parts of Equation (5-6) we obtain two solutions for B and X as follows:

$$B_1 = \frac{R_L X_S + \sqrt{R_L R_S (R_S^2 + X_S^2 - R_L R_S)}}{R_L (R_S^2 + X_S^2)} \quad (5\text{-}7)$$

$$X_1 = \frac{R_L X_S - R_S X_L}{R_S} + \frac{R_S - R_L}{B_1 R_S} \tag{5-8}$$

and

$$B_2 = \frac{R_L X_S - \sqrt{R_L R_S (R_S^2 + X_S^2 - R_L R_S)}}{R_L (R_S^2 + X_S^2)} \tag{5-9}$$

$$X_2 = \frac{R_L X_S - R_S X_L}{R_S} + \frac{R_S - R_L}{B_2 R_S} \tag{5-10}$$

Similarly, applying Equation $Z_S^* = Z_{IN}$ to the second configuration in Figure 5-8 we have:

$$R_S - jX_S = jX + \cfrac{1}{jB + \left(\cfrac{1}{R_L + jX_L}\right)} \tag{5-11}$$

Separating the real and imaginary parts of Equation (5-11), we also get two sets of solutions for B and X:

$$B_3 = \frac{R_S X_L + \sqrt{R_L R_S (R_L^2 + X_L^2 - R_L R_S)}}{R_S (R_L^2 + X_L^2)} \tag{5-12}$$

$$X_3 = \frac{R_S X_L - R_L X_S}{R_L} + \frac{R_L - R_S}{B_3 R_L} \tag{5-13}$$

and

$$B_4 = \frac{R_S X_L - \sqrt{R_L R_S (R_L^2 + X_L^2 - R_L R_S)}}{R_S (R_L^2 + X_L^2)} \tag{5-14}$$

$$X_4 = \frac{R_S X_L - R_L X_S}{R_L} + \frac{R_L - R_S}{B_4 R_L} \tag{5-15}$$

Power Transfer and Impedance Matching 229

Conditions for the validity of solutions are that the arguments of the square roots in Equations (5-7), (5-9), (5-12) and (5-14) be positive or zero.

- If $R_S^2 + X_S^2 - R_L R_S > 0$ and $R_L^2 + X_L^2 - R_L R_S < 0$, the two solutions obtained from Equations (5-7) through (5-10) are the only valid solutions.

- If $R_S^2 + X_S^2 - R_L R_S < 0$ and $R_L^2 + X_L^2 - R_L R_S > 0$, the two solutions obtained from Equations (5-12) through (5-15) are the only valid solutions.

- If $R_S^2 + X_S^2 - R_L R_S > 0$ and $R_L^2 + X_L^2 - R_L R_S > 0$, all four solutions obtained from Equations (5-7) through (5-10) and Equations (5-12) through (5-15) are valid.

Once the real values for B and X are calculated, the values of the matching elements are obtained from the following equations:

If B is positive, the matching element is a capacitor given by:

$$C = \frac{B}{2\pi f} \qquad (5\text{-}16)$$

If B is negative, the matching element is an inductor given by:

$$L = -\frac{1}{2\pi f B} \qquad (5\text{-}17)$$

If X is positive, the matching element is an inductor given by:

$$L = \frac{X}{2\pi f} \qquad (5\text{-}18)$$

If X is negative, the matching element is a capacitor given by:

$$C = -\frac{1}{2\pi f X} \tag{5-19}$$

In the above equations, if the frequency is in Hz, capacitors and inductors are in Farad and Henry, respectively. Now by utilizing Equations (5-7) through (5-10) and Equations (5-12) through (5-15) we can analytically design L-networks to match any complex load to any complex source impedance. Example 5.3-1 is a case with four different solutions. Each solution has a different matching bandwidth. We use the bandwidth at 20 dB return loss as the matching bandwidth.

Example 5.3-1: Analytically design L-networks matching a complex load impedance, $Z_L = 10 - j15$ Ω, to a complex source impedance, $Z_S = 15 - j20$ Ω, at a frequency of 2 GHz. Since $R_S^2 + X_S^2 - R_L R_S = 475 > 0$, and $R_L^2 + X_L^2 - R_L R_S = 175 > 0$ the problem has four solutions.

First Solution: Use Equations (5-7) and (5-8) in an Equation Editor to calculate the matching element values.

```
1   %Enter Design Parameters
2   RS=15
3   XS=-20
4   RL=10
5   XL=-15
6   f=2e9
7   %Calculate B1, X1, and Matching Element Values
8   B1=((RL*XS)+sqrt(RS*RL*(RS^2+XS^2-RS*RL)))/(RL*(RS^2+XS^2))
9   X1=(RL*XS-RS*XL)/RS+(RS-RL)/(B1*RS)
10  L1=X1/(2*pi*f)
11  C2=B1/(2*pi*f)
```

Figure 5-9 Calculation of the matching element values

Viewing Workspace Variables show that $B1 = 0.011$ and $X1 = 32.795$, therefore, the series element is an inductor, L1 = 2.61 nH, and the shunt element is a capacitor, $C2 = 0.852$ pF. The schematic and swept response is shown in Figure 5-10. The input return loss, S11, shows that the network is perfectly matched at 2 GHz and the measured fractional bandwidth at 20 dB return loss is about 11 %, indicating a narrow matching bandwidth. Fractional bandwidths less than 20% are generally considered narrow band. The schematic and response of the first solution is shown in Figure 5-10.

Power Transfer and Impedance Matching 231

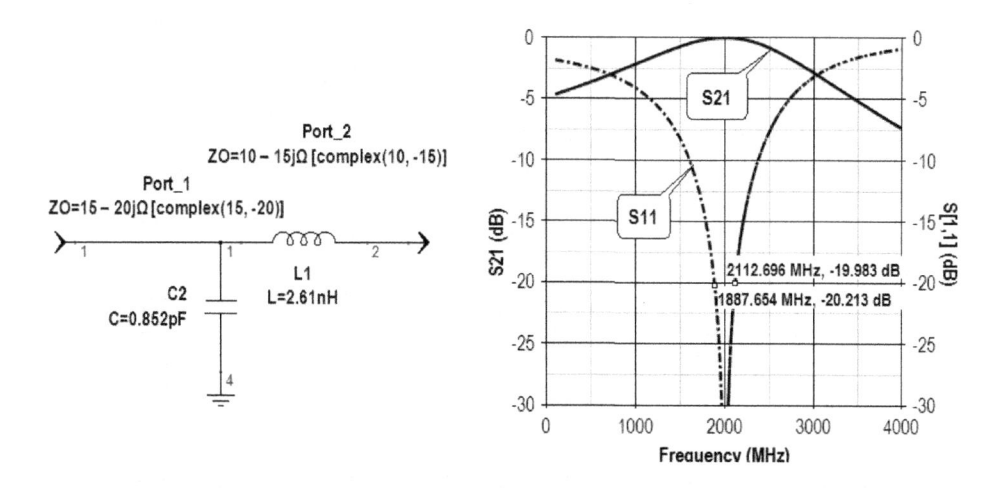

Figure 5-10 Schematic and response of the matching L-network

Second Solution: Use Equations (5-9) and (5-10) in an Equation Editor to calculate the matching element values.

```
1    %Enter Design Parameters
2    RS=15
3    XS=-20
4    RL=10
5    XL=-15
6    f=2e9
7    %Calculate B2, X2 and Matching Element Values
8    B2=((RL*XS)-sqrt(RS*RL*(RS^2+XS^2-RS*RL)))/(RL*(RS^2+XS^2))
9    X2=(RL*XS-RS*XL)/RS+(RS-RL)/(B2*RS)
10   C1=-1/(2*pi*f*X2)
11   L2=-1/(2*pi*f*B2)
```

Figure 5-11 Calculation of the matching element values

Viewing Workspace Variables show that $B2 = -0.075$ and $X2 = -2.795$, therefore, the series element is a capacitor, $C1 = 28.47$ pF and the shunt element is an inductor, $L2 = 1.065$ nH. To plot the response of the matching network, create a new Design with Schematic and use the given element values. Simulate the schematic and display the return loss, S11, and insertion loss, S21, in dB. Notice that the input return loss, S11, in Figure 5-12 shows that the network is matched at 2 GHz and the fractional bandwidth at 20 dB return loss is about 12 % of the center frequency. The schematic and the swept response are shown in Figure 5-12.

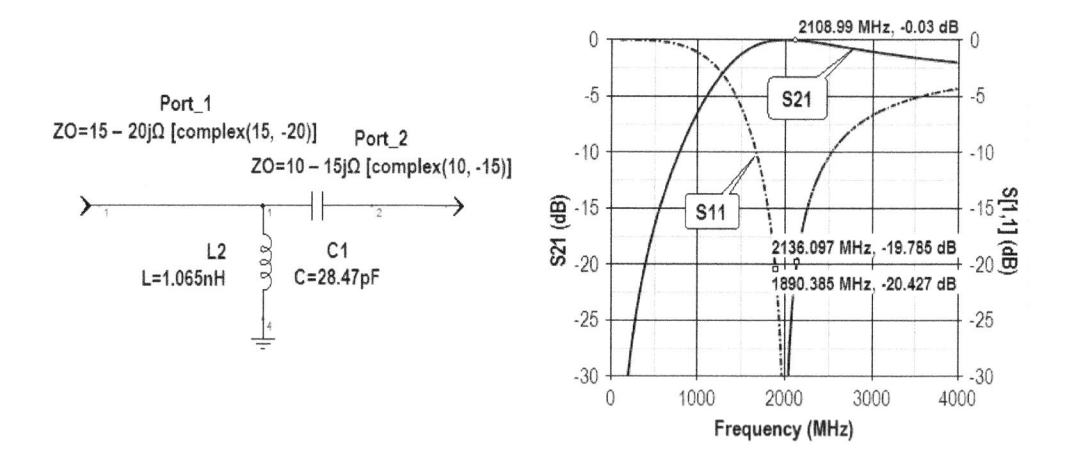

Figure 5-12 Schematic and response of the matching L-network

Third Solution: Use Equations (5-12) and (5-13) in an Equation Editor to calculate the matching element values.

```
1    %Enter Design Parameters
2    RS=15
3    XS=-20
4    RL=10
5    XL=-15
6    f=2e9
7    %Calculate B3, X3, and Matching Element Values
8    B3=((RS*XL)+sqrt(RS*RL*(RL^2+XL^2-RS*RL)))/(RS*(RL^2+XL^2))
9    X3=(RS*XL-RL*XS)/RL+(RL-RS)/(B3*RL)
10   L1=-1/(2*pi*f*B3)
11   L2=X3/(2*pi*f)
```

Figure 5-13 Calculating the matching element values

Viewing the Workspace Variables show that $B3$ = -0.013 and $X3$ = 36.202, therefore, the shunt element is an inductor, $L1$ = 6.16 nH and the series element is also an inductor, $L2$ = 2.881 nH. To plot the response of matching network set up a new Design with Schematic using the given element values. Simulate the schematic and display the input return loss, S11, and insertion loss, S21, in dB. The matching network and the swept responses are shown in Figure 5-14. Notice that the fractional bandwidth at 20 dB return loss is about 16 % of the center frequency indicating a narrowband matching network. The schematic and the swept response are shown in Figure 5-14.

Power Transfer and Impedance Matching

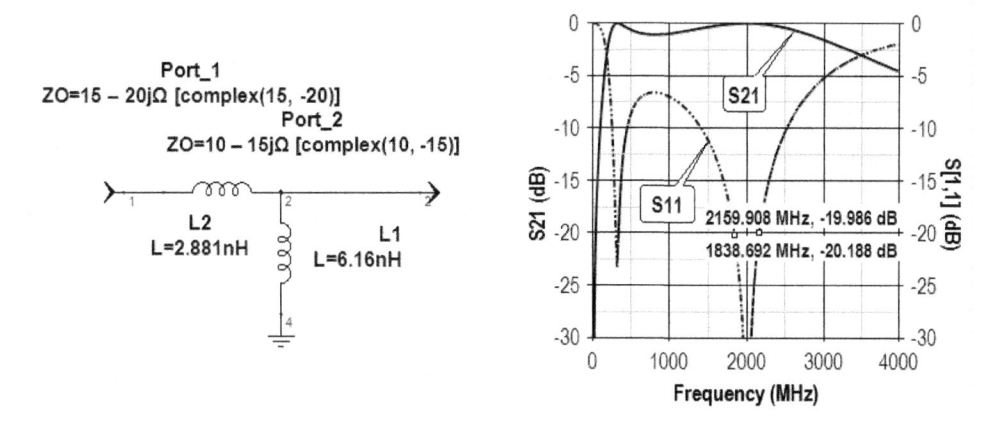

Figure 5-14 Solution and response of the matching network for the third solution

Fourth Solution: Use Equations (5-14) and (5-15) in an Equation Editor to calculate the matching element values.

```
1    %Enter Design Parameters
2    RS=15
3    XS=-20
4    RL=10
5    XL=-15
6    f=2e9
7    %Calculate B4, X4, and Matching Element Values
8    B4=((RS*XL)-sqrt(RS*RL*(RL^2+XL^2-RS*RL)))/(RS*(RL^2+XL^2))
9    X4=(RS*XL-RL*XS)/RL+(RL-RS)/(B4*RL)
10   L1=-1/(2*pi*f*B4)
11   L2=X4/(2*pi*f)
```

Figure 5-15 Calculating the matching element values

Viewing Workspace Variables show that $B4 = -0.079$ and $X4 = 3.798$, therefore, the shunt element is a capacitor, $L1 = 1.002$ nH and the series element is also an inductor, $L2 = 0.302$ nH. To plot the response of the fourth solution, set up a new Design with Schematic in Genesys using the given element values. Simulate the schematic and display the Return Loss, S11, and insertion loss, S21, in dB. Notice that the input return loss, S11 in Figure 5-16 shows that the network is matched at 2 GHz and the measured fractional bandwidth at 20 dB return loss is about 15 % of the center frequency. The schematic and the swept response are shown in Figure 5-16.

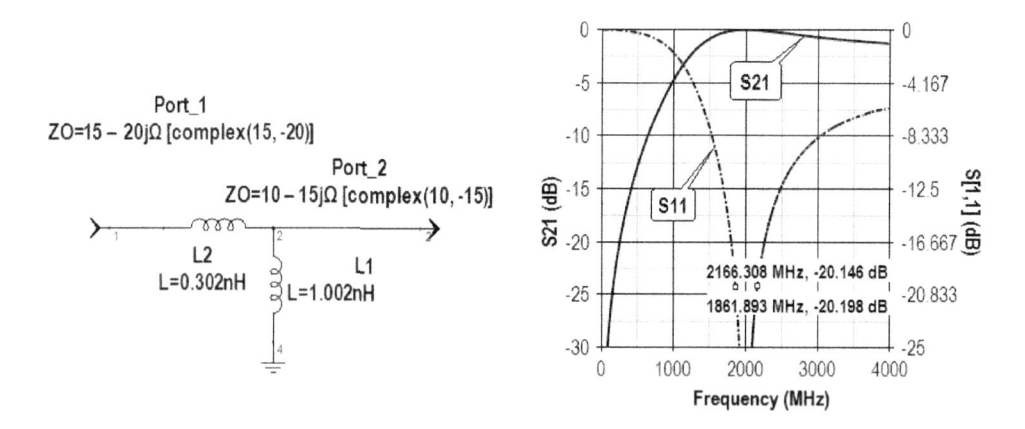

Figure 5-16 The matching network and response of the fourth solution

5.3.2 Matching a Complex Load to a Real Source Impedance

In amplifier design a common matching problem is the matching of a complex impedance, $Z_L = R_L + jX_L$ to a real impedance, $Z_S = R_S$. The complex impedance is usually the load and the real impedance is the characteristic impedance of the transmission line connected to the source. To design the impedance matching networks both configurations of Figures 5-7 and 5-8 are used. For the case of real source impedance the configuration of Figure 5-6 is redrawn in Figure 5-17.

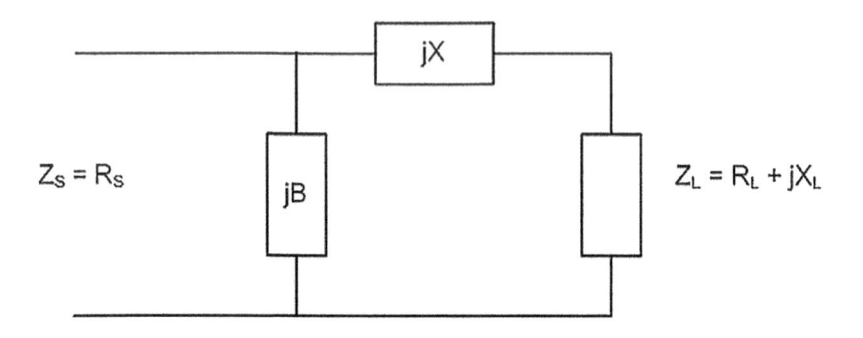

Figure 5-17 Matching complex load to resistive source ($R_L < R_S$)

To derive the analytical expressions for B and X we utilize the maximum power transfer condition and set the conjugate of the source impedance equal to the input impedance of the matching network connected to the load impedance, as given in Equation (5-20).

$$R_S = \frac{1}{jB + \left(\dfrac{1}{jX + R_L + jX_L}\right)} \qquad (5\text{-}20)$$

The two solutions for B and X are easily obtained by substituting $X_S = 0$ in Equations (5-7) through (5-10).

$$B_1 = \frac{+\sqrt{R_S - R_L}}{R_S \sqrt{R_L}} \qquad (5\text{-}21)$$

$$X_1 = +\sqrt{R_L(R_S - R_L)} - X_L \qquad (5\text{-}22)$$

and,

$$B_2 = \frac{-\sqrt{R_S - R_L}}{R_S \sqrt{R_L}} \qquad (5\text{-}23)$$

$$X_2 = -\sqrt{R_L(R_S - R_L)} - X_L \qquad (5\text{-}24)$$

Note that the solutions given in Equations (5-21) through (5-24) are only valid if $R_L < R_S$. To calculate B and X, when $R_L > R_S$, use the second matching configuration given in Figure 5-18,

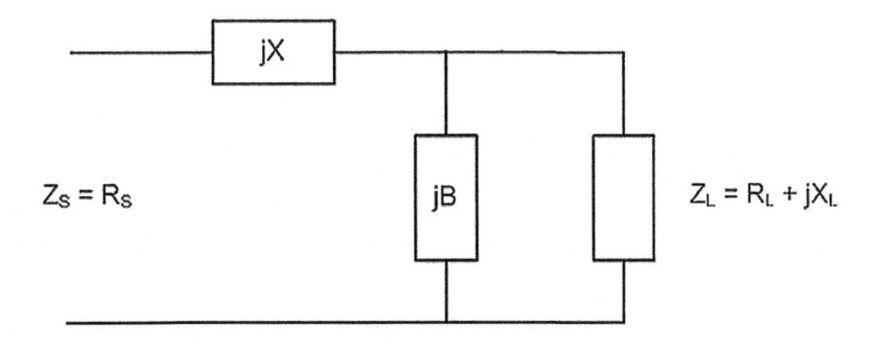

Figure 5-18 Matching complex load to resistive source ($R_L > R_S$)

Applying the maximum power condition to the matching network in Figure 5-18, we have:

$$R_S = jX + \cfrac{1}{jB + \left(\cfrac{1}{R_L + jX_L}\right)} \tag{5-25}$$

The solutions for B and X in Equation (5-25) can be obtained by reusing Equations (5-12) and (5-14) and substituting $X_S = 0$ in Equations (5-13) and (5-15).

$$B_3 = \frac{R_S X_L + \sqrt{R_L R_S (R_L^2 + X_L^2 - R_L R_S)}}{R_S (R_L^2 + X_L^2)} \tag{5-26}$$

$$X_3 = \frac{R_S X_L}{R_L} + \frac{R_L - R_S}{B_3 R_L} \tag{5-27}$$

and

$$B_4 = \frac{R_S X_L - \sqrt{R_L R_S (R_L^2 + X_L^2 - R_L R_S)}}{R_S (R_L^2 + X_L^2)} \tag{5-28}$$

$$X_4 = \frac{R_S X_L}{R_L} + \frac{R_L - R_S}{B_4 R_L} \tag{5-29}$$

The conditions for the valid solutions are that the arguments of the square roots in Equations (5-26) through (5-29) be non-negative. Therefore, the two solutions obtained from Equations (5-26) through (5-29) are valid only if $R_L > R_S$. Combined conditions for valid solutions are summarized in Table 5-1.

Case #	First Condition	Second Condition	# of Solutions	Equations Used
1	$R_L < R_S$	$R_L^2 + X_L^2 - R_L R_S > 0$	4	(5-21) to (5-24) (5-26) to (5-29)
2	$R_L < R_S$	$R_L^2 + X_L^2 - R_L R_S < 0$	2	(5-21) to (5-24)
3	$R_L > R_S$	N/A	2	(5-26) to (5-29)

Table 5-1 Impedance matching conditions and the number of solutions

Power Transfer and Impedance Matching

The solutions in Equations (5-26) through (5-29) can be simplified by normalizing the load impedance with respect to the source resistor. Therefore, if we let the source resistor be equal to Z_0, the normalized load resistance and reactance become,

$$r = \frac{R_L}{Z_0} \tag{5-30}$$

and

$$x = \frac{X_L}{Z_0} \tag{5-31}$$

The simplified equations are:

$$B_1 = \frac{\sqrt{\frac{(1-r)}{r}}}{Z_0} \tag{5-32}$$

$$X_1 = Z_0 \left[\sqrt{r(1-r)} - x \right] \tag{5-33}$$

$$B_2 = -\frac{\sqrt{(1-r)}}{Z_0} \tag{5-34}$$

$$X_2 = -Z_0 \left[\sqrt{r(1-r)} + x \right] \tag{5-35}$$

$$B_3 = \frac{x + \sqrt{r(r^2 + x^2 - r)}}{Z_0 (r^2 + x^2)} \tag{5-36}$$

$$X_3 = Z_0 \sqrt{\frac{(r^2 + x^2 - r)}{r}} \tag{5-37}$$

$$B_4 = \frac{x - \sqrt{r(r^2 + x^2 - r)}}{Z_0(r^2 + x^2)} \qquad (5\text{-}38)$$

$$X_4 = -Z_0 \sqrt{\frac{(r^2 + x^2 - r)}{r}} \qquad (5\text{-}39)$$

Example 5.3-2: Design a single L-network that will match a real source impedance $Z_0 = 50\ \Omega$ to a complex load impedance, $Z_L = 7 - j22\ \Omega$, at a frequency of 1 GHz. Notice that for this example $R_L < Z_0$ and $R_L^2 + X_L^2 - R_L Z_0 = 183 > 0$, therefore, the conditions for Case 1 in Table 5-1 are met and the matching network has four solutions.

First Solution: Use Equations (5-32) and (5-33) in an Equation Editor to calculate the matching element values.

```
1    %Enter Design Parameters
2    Z0=50
3    RL=7
4    XL=-22
5    f=1000e6
6    r=RL/Z0
7    x=XL/Z0
8    % Calculate Element Values
9    B1=sqrt((1-r)/r)/Z0
10   X1=Z0*sqrt(r*(1-r))-x*Z0
11   LS1=X1/(2*pi*f)
12   CP2=B1/(2*pi*f)
```

Figure 5-19 Calculating the matching element values

Viewing Workspace Variables show that the series element is an inductor, $LS1 = 6.263$ nH, and the shunt element is a capacitor, $CP2 = 7.889$ pF. To display the frequency response, setup a new Design with Schematic in Genesys and use the given element values. Simulate the schematic and display the input return loss, S11, and insertion loss, S21, in dB.

Power Transfer and Impedance Matching   239

The schematic and response of the first solution is shown in Figure 5-20. The simulated response in Figure 5-20 shows that the load is perfectly matched to the source at 1 GHz and the measured fractional bandwidth at 20 dB return loss is about 5 % indicating a narrowband matching network.

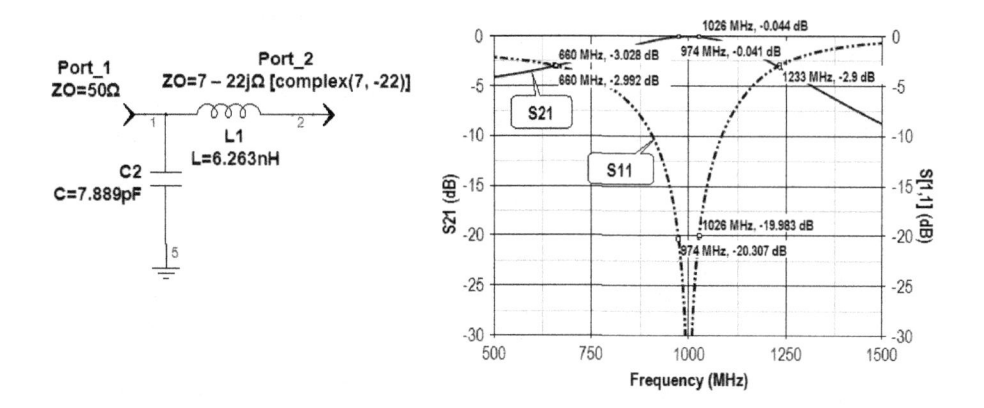

Figure 5-20 Schematic and response of the first solution

Second Solution: Use Equations (5-34) and (5-35) in an Equation Editor to calculate the matching element values.

```
1    %Enter Design Parameters
2    Z0=50
3    RL=7
4    XL=-22
5    f=1000e6
6    %Normalize Load Impedance
7    r=RL/Z0
8    x=XL/Z0
9    % Calculate Element Values
10   B2=-sqrt((1-r)/r)/Z0
11   X2=-Z0*sqrt(r*(1-r))-x*Z0
12   L1=X2/(2*pi*f)
13   L2=-1/(2*pi*f*B2)
```

Figure 5-21 Calculating the matching element values

Viewing Workspace Variables show that the series element is an inductor, $L1 = 0.740$ nH and the shunt element is another inductor $L2 = 3.211$ nH. To display the frequency response, set up a new Design with Schematic in Genesys and use the given element values. Simulate the schematic from 500 to 1500 MHz and display S11, and S21, in dB.

The schematic and simulated response in Figure 5-22 shows that the impedance matching network perfectly matches the complex load to a 50 Ω source resistor at 1 GHz. The measured fractional bandwidth at 20 dB return loss is about 13 % indicating a narrowband matching network.

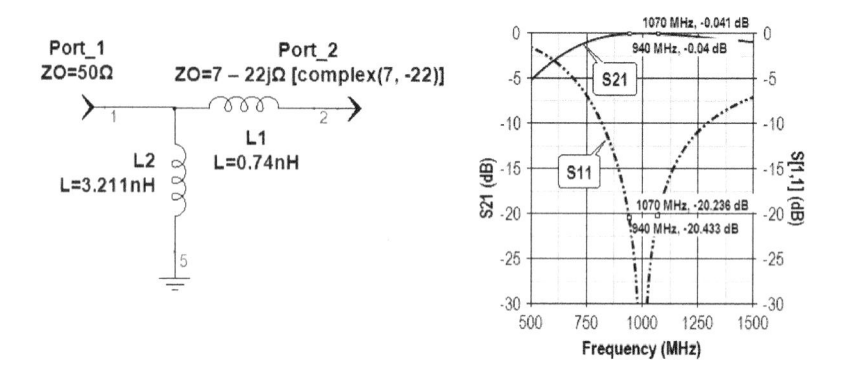

Figure 5-22 Schematic and response of the second solution

Third Solution: Use Equations (5-36) and (5-37) in an Equation Editor to calculate the matching element values.

```
1    %Enter Design Parameters
2    Z0=50
3    RL=7
4    XL=-22
5    f=1000e6
6    r=RL/Z0
7    x=XL/Z0
8    % Calculate B3,X3, and Element Values
9    B3=(x+sqrt(r*(r^2+x^2-r)))/(Z0*(r^2+x^2))
10   X3=Z0*sqrt((r^2+x^2-r)/r)
11   L2=X3/(2*pi*f)
12   L1=-1/(2*pi*f*B3)
```

Figure 5-23 Calculating the matching element values

Viewing Workspace Variables show that the shunt element is an inductor, $L1 = 5.008\ nH$ and the series element is another inductor, $L2 = 5.754\ nH$. To display the frequency response, setup a new Design with Schematic in Genesys and use the given element values. Simulate the schematic and display the input return loss and insertion loss in dB.

Power Transfer and Impedance Matching 241

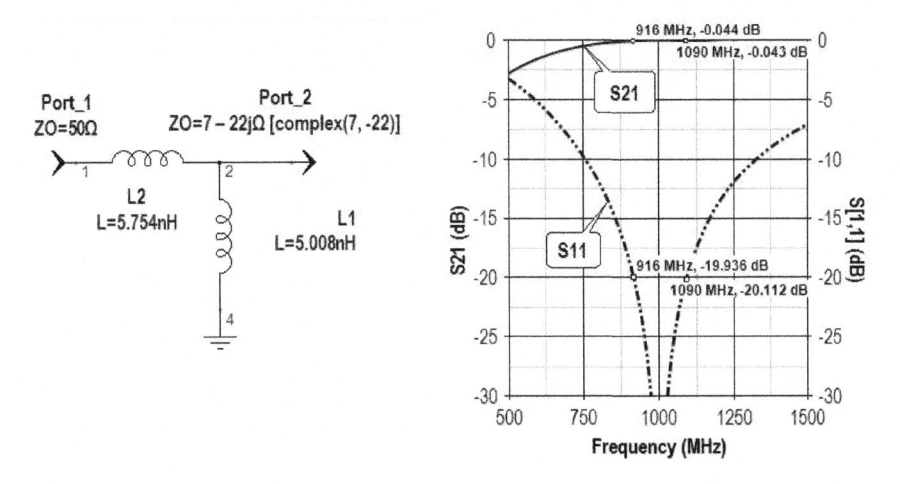

Figure 5-24 Schematic and response of the third solution

The simulated response in Figure 5-24 shows that the load is perfectly matched to the 50 Ω source resistor at 1 GHz and the measured fractional bandwidth at 20 dB return loss is about 17 %.

Fourth Solution: Use Equations (5-38) and (5-39) in an Equation Editor to calculate the matching element values.

```
1   %Enter Design Parameters
2   Z0=50
3   RL=7
4   XL=-22
5   f=1000e6
6   %Normalize Load Impedance
7   r=RL/Z0
8   x=XL/Z0
9   % Calculate B4,X4, and Element Values
10  B4=(x-sqrt(r*(r^2+x^2-r)))/(Z0*(r^2+x^2))
11  X4=-Z0*sqrt((r^2+x^2-r)/r)
12  C2=-1/(2*pi*f*X4)
13  L1=-1/(2*pi*f*B4)
```

Figure 5-25 Calculating the matching element values

Viewing the Workspace Variables show that the series element is an inductor, $L1 = 3.135\ nH$ and the shunt element is a capacitor, $C2 = 4.402\ pF$. To display the frequency response, setup a new Design with Schematic in Genesys and use the given element values. Simulate the schematic and display S11, and S21, in dB.

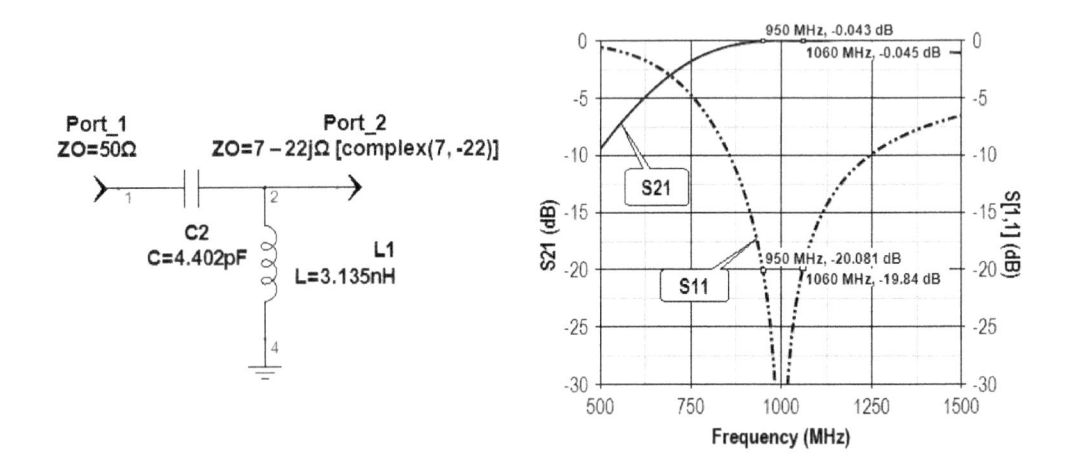

Figure 5-26 Schematic and response of the fourth solution

The simulated response in Figure 5-26 shows that the matching network matches the complex load impedance to a 50 Ω source resistor at 1 GHz and the measured fractional bandwidth at 20 dB return loss is about 11 %.

5.3.3 Matching a Real Load to a Real Source Impedance

When source and load impedances are both real the first matching configuration of Figure 5-7 is redrawn in Figure 5-27.

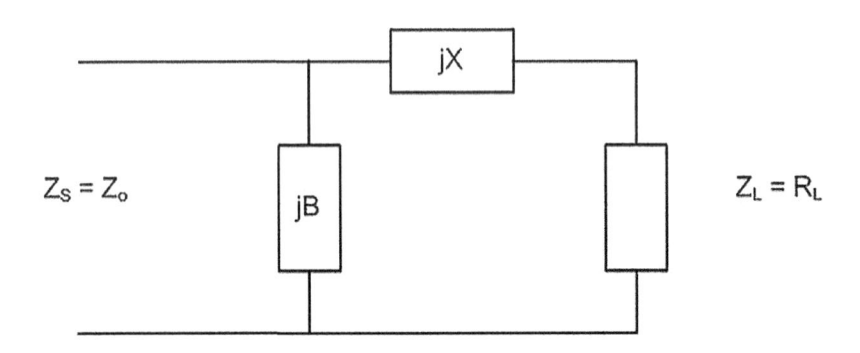

Figure 5-27 First matching configuration with $X_L = X_S = 0$

To design the matching network, apply the maximum power transfer condition and require that $Z_0^* = Z_{IN}$. Therefore,

Power Transfer and Impedance Matching 243

$$Z_0 = \cfrac{1}{jB + \left(\cfrac{1}{jX + R_L}\right)} \qquad (5\text{-}40)$$

Substituting $r = \dfrac{R_L}{Z_0}$ in Equation (5-40), we get:

$$Z_0 = \cfrac{1}{jB + \left(\cfrac{1}{jX + rZ_0}\right)} \qquad (5\text{-}41)$$

The solutions for B and X in Equation (5-41) can be obtained by substituting x = 0 in Equations (5-33) and (5-35).

$$B_1 = \frac{\sqrt{(1-r)/r}}{Z_0} \qquad (5\text{-}42)$$

$$X_1 = Z_0 \sqrt{r(1-r)} \qquad (5\text{-}43)$$

$$B_2 = -\frac{\sqrt{(1-r)/r}}{Z_0} \qquad (5\text{-}44)$$

$$X_2 = -Z_0 \sqrt{r(1-r)} \qquad (5\text{-}45)$$

Note that the two solutions given by Equations (5-42) through (5-45) are only valid if r is less than 1 or $R_L < Z_0$. If r > 1, the second matching network configuration in Figure 5-28 should be used.

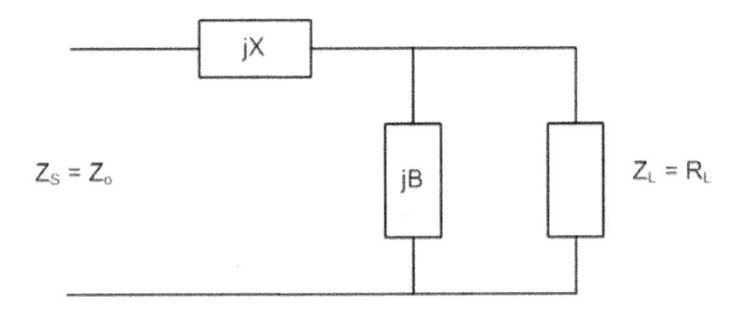

Figure 5-28 Second matching configuration with resistive load and source

To calculate the B and X values, we require that, $Z_0^* = Z_{IN}$, therefore,

$$Z_0^* = Z_0 = jX + \dfrac{1}{jB + \left(\dfrac{1}{Z_0 r}\right)} \quad (5\text{-}46)$$

The solutions for B and X in Equation (5-46) can be obtained by substituting x = 0 in Equations (5-36) through (5-39).

$$B_3 = \dfrac{\sqrt{r-1}}{Z_0 r} \quad (5\text{-}47)$$

$$X_3 = Z_0 \sqrt{r-1} \quad (5\text{-}48)$$

$$B_4 = \dfrac{-\sqrt{r-1}}{Z_0 r} \quad (5\text{-}49)$$

$$X_4 = -Z_0 \sqrt{r-1} \quad (5\text{-}50)$$

Note that the two solutions given by Equations (5-47) through (5-50) are only valid if r is greater than 1 or $R_L > Z_0$. The conditions for the validity of solutions are given below and summarized in Table 5-2.

Case 1 If r is less than 1 there are two L-networks that match the two impedances. The two solutions are given by Equations (5-42) through (5-45).

Case 2 If r is greater than 1 there are two L-networks that match the two impedances. The two solutions are given by Equations (5-47) through (5-50).

Case No	Condition	Solutions	Use Equations
1	$r < 1$	2	(5-42) to (5-45)
2	$r > 1$	2	(5-47) to (5-50)

Table 5-2 Impedance matching conditions and the number of solutions

Example 5.3-3: Design an L-network to match a 10 Ω load to a 50 Ω source resistor at 500 MHz.

First Solution: Since the load resistor is smaller than the source resistor, we get two solutions from Equations (5-42) through 5-45 in the Equation Editor.

```
1    %Enter Design Parameters
2    Z0=50
3    RL=10
4    f=500e6
5    %Normalize the Load Resistance
6    r=RL/Z0
7    %Calculate Matching Element Values
8    B1=sqrt((1-r)/r)/Z0
9    X1=Z0*sqrt(r*(1-r))
10   L1=X1/(2*pi*f)
11   C2=B1/(2*pi*f)
```

Figure 5-29 Calculation of the matching element values

To design the first matching L-network, create a new Design with Schematic in Genesys and select the element values by viewing the Workspace Variables. Simulate the schematic from 0 to 1000 MHz and display the response, as shown in Figure 5-30

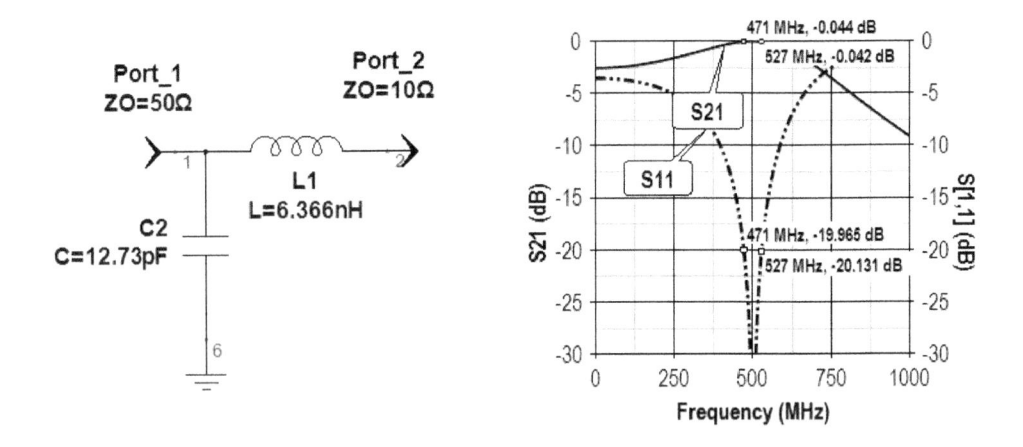

Figure 5-30 Schematic and response of the first matching L-network

The simulated response in Figure 5-30 shows that the matching L-network perfectly matches the 10 Ω load resistor to a 50 Ω source resistor at 500 MHz and the fractional bandwidth at 20 dB return loss is about 11 %.

Second Solution: To design the matching L-network, use Equations (5-44) and (5-45) in the Equation Editor to calculate the matching element values.

```
1    %Enter Design Parameters
2    Z0=50
3    RL=10
4    f=500e6
5    %Normalize the Load Resistance
6    r=RL/Z0
7    %Calculate Matching Element Values
8    B2=-sqrt((1-r)/r)/Z0
9    X2=-Z0*sqrt(r*(1-r))
10   C1=-1/(2*pi*f*X2)
11   L2=-1/(2*pi*f*B2)
```

Figure 5-31 Calculating the matching element values

To display the frequency response, setup a Design with Schematic in Genesys and select the element values by viewing the Workspace Variables. Simulate the schematic from 0 to 1000 MHz and display the response. The schematic and response of the second solution is shown in Figure 5-32.

Note that the load resistor is perfectly matched to the source at 500 MHz. The measured fractional bandwidth at 20 dB return loss is about 11 %.

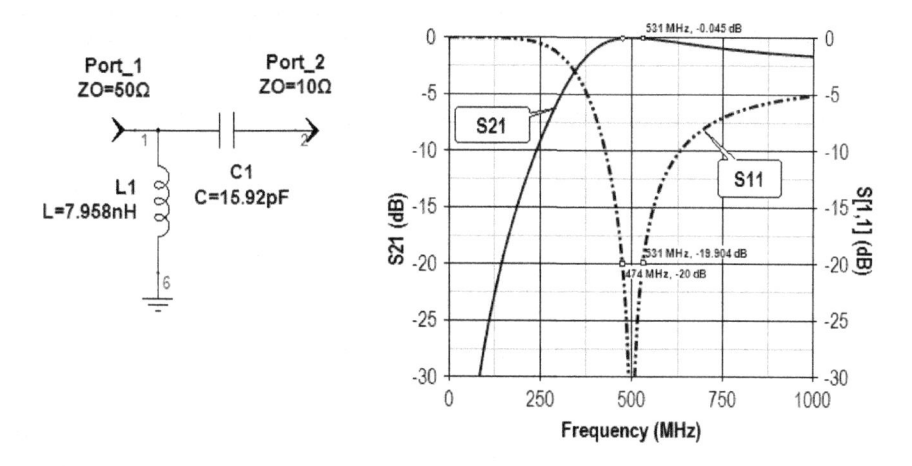

Figure 5-32 Schematic and response of the second matching L-network

5.4 Introduction to Broadband Matching Networks

In the previous sections, the L-section matching networks achieved an impedance match at a fixed frequency capable of producing a 20 dB return loss over a narrow fractional bandwidth of less than 20%. Broadband networks are generally considered to have greater than 20% fractional bandwidths. In this section it is demonstrated that the bandwidth of a matching network can be increased by cascading L-networks. It is demonstrated that by cascading L-networks of equal Q factor, the bandwidth of a network can be increased. The design of equal-Q matching networks is based on the selection of intermediate, or virtual, resistors not necessarily 50 Ω, and then matching the load and source impedance to the virtual resistors. Successively adding additional L-networks of equal Q will continue to extend the bandwidth of the overall circuit.

5.4.1 Analytical Design of Broadband Matching Networks

This section demonstrates the importance of the selection of the proper intermediate network resistance that will result in the best broadband return loss. In example 5.4-1 the complex source and load impedance are matched to one specific intermediate resistor thus creating two, equal-Q, L-networks. The purpose is to show that this method provides a broader matching bandwidth at 20 dB return loss compared to the case when we chose a different intermediate resistor.

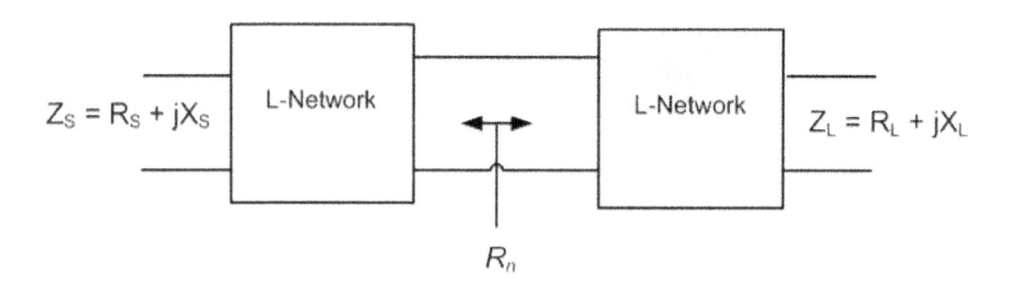

Figure 5-33 Cascaded L-networks with intermediate resistance, R_n

Example 5.4-1: A transmitter operates over a frequency range of 835 MHz to 1200 MHz. At its center frequency of 1 GHz the source impedance is $Z_S = 55 + j10\ \Omega$ while the antenna input impedance is $Z_L = 20 + j15\ \Omega$. Design a cascade of two L-networks that matches the transmitter output impedance to the antenna impedance across the full transmitter bandwidth.

Solution: The first step in the solution is to calculate the intermediate resistor, R_n. The optimum intermediate resistor value is equal to square root of the product of the real part of the source and load impedance.

$$R_n = \sqrt{(55)(20)} = 33.166\ \Omega$$

Use Equations (5-32) and (5-33) in an Equation Editor to calculate the element values of the load matching network. Viewing the Workspace Variables show that the series element is an inductor, $LS1 = 0.195$ nH and the shunt element is a capacitor, $CP2 = 3.893$ pF.

Power Transfer and Impedance Matching

```
1   %Enter Design Parameters
2   Z0=33.166
3   RL=20
4   XL=15
5   f=1000e6
6   %Normalize Load Impedance
7   r=RL/Z0
8   x=XL/Z0
9   % Calculate B1, B2, and Element Values
10  B1=sqrt((1-r)/r)/Z0
11  X1=Z0*sqrt(r*(1-r))-x*Z0
12  LS1=X1/(2*pi*f)
13  CP2=B1/(2*pi*f)
```

Figure 5-34 Calculating the matching element values

To display the frequency response of the L-network, set up a new Design with Schematic in Genesys and select the element values as given in Figure 5-35. Simulate the schematic from 0 to 2000 MHz and display the input return loss, S11, and the insertion loss, S21, in dB, as shown in Figure 5-35.

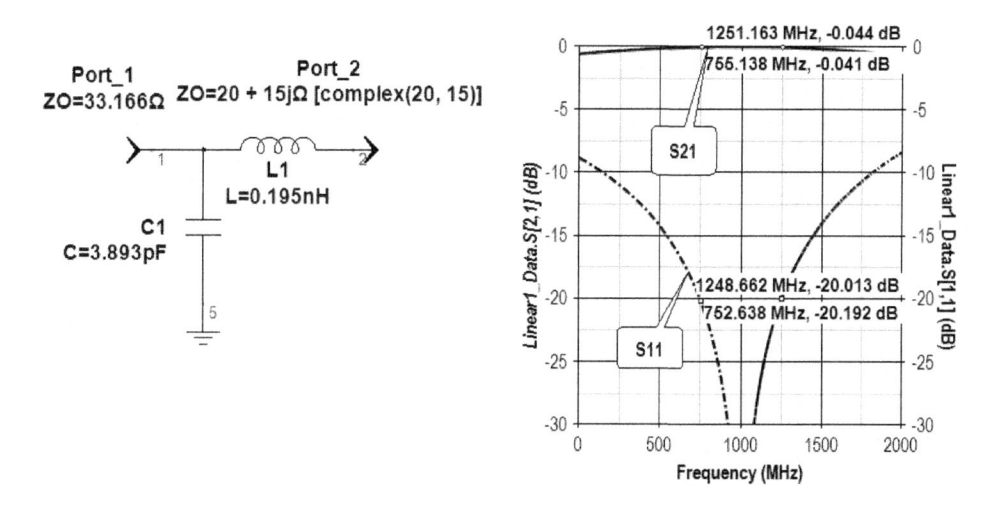

Figure 5-35 Schematic and response of the matching L-network

To match the source impedance to the intermediate resistor, we use Equations (5-36) and (5-37) in an Equation Editor and calculate the element values of the matching network. Viewing the Workspace Variables show that the series element is an inductor, $LS1$ = 4.458 nH and the shunt element is a capacitor, $CP2$ = 2.875 pF.

```
1    %Enter Design Parameters
2    Z0=33.166
3    RL=55
4    XL=10
5    f=1000e6
6    %Normalize Load Impedance
7    r=RL/Z0
8    x=XL/Z0
9    % Calculate B3,X3, and Element Values
10   B3=(x+sqrt(r*(r^2+x^2-r)))/(Z0*(r^2+x^2))
11   X3=Z0*sqrt((r^2+x^2-r)/r)
12   LS1=X3/(2*pi*f)
13   CP2=B3/(2*pi*f)
```

Figure 5-36 Calculating the matching element values

To display the frequency response, set up a new Design with Schematic in Genesys and simulate the schematic from 0 to 2000 MHz. The input return loss, S11, and the insertion loss, S21, in dB are shown in Figure 5-37.

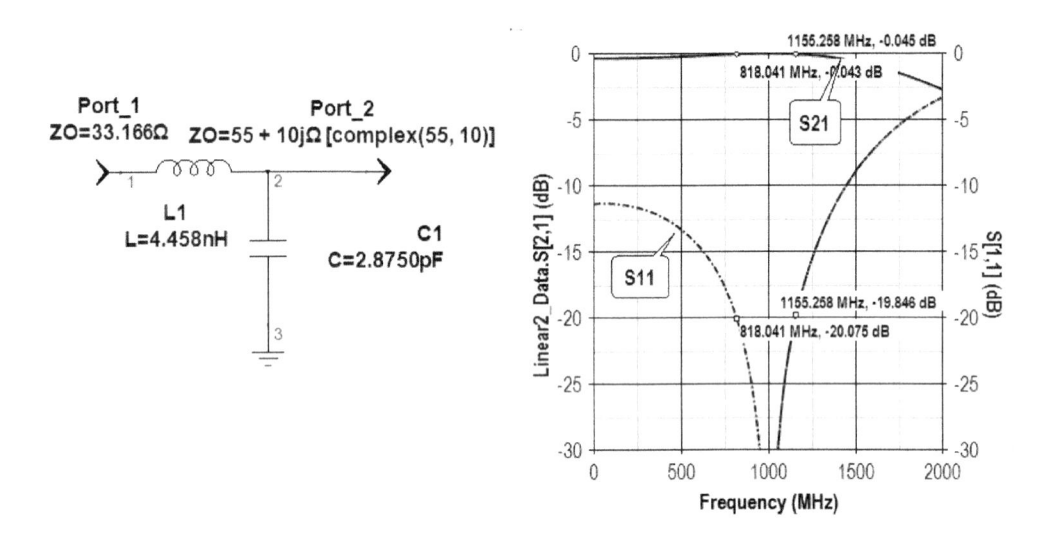

Figure 5-37 Schematic and response of the matching L-network

Finally, we cascade the two L-networks in Figure 5-38 and simulate the schematic to display the return loss and insertion loss in dB. To measure the matching bandwidth manually, place two markers at 20 dB return loss and record the marker readings. The schematic and simulated response of the cascaded network is shown in Figure 5-38.

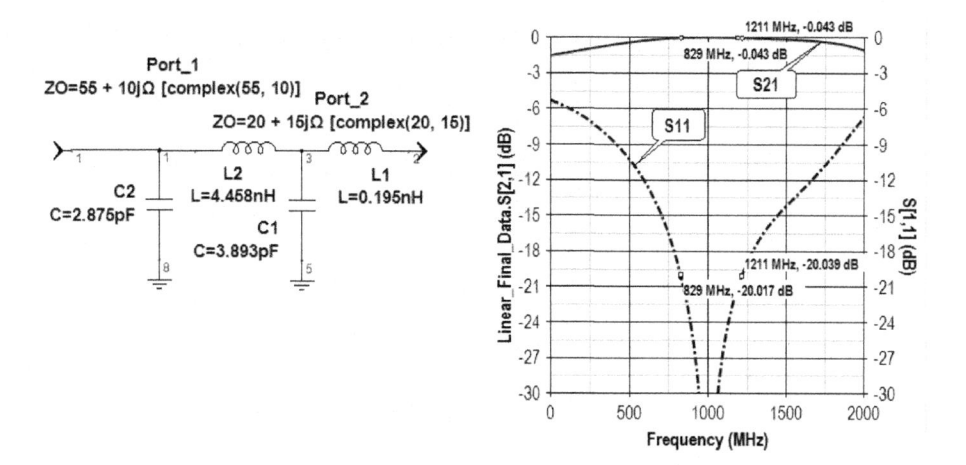

Figure 5-38 Cascaded L-networks schematic and response

Notice that the matching bandwidth at 20 dB return loss is:

$$BW_{20\,dB\ return\ loss} = 1211 - 829 = 382\ MHz$$

Therefore, the corresponding fractional bandwidths at 20dB return loss is:

$$FBW_{20\,dB\ return\,loss} = \frac{382}{\sqrt{(1211)(829)}} = 0.381 = 38.1\%$$

Notice that the higher fractional bandwidth is due to cascading two equal Q L-networks. The Q factor of a single L-network is calculated by the following equation [1]:

$$Q = \frac{1}{2}\sqrt{\frac{R_2}{R_1} - 1} \qquad R_2 > R_1$$

Where: R_2 and R_1 are the input and output resistors of the L-network.

For the Example 5.4-1 the Q factors of the two L-networks are equal:

$$Q_{source} = Q_{load} = \frac{1}{2}\sqrt{\frac{33.166}{20} - 1} = \frac{1}{2}\sqrt{\frac{55}{33.166} - 1} = 0.405$$

If the Q factors of the two cascading L-networks are not equal, the fractional bandwidth would be smaller. Example 5.4-2 proves this point.

Example 5.4-2: Redesign the matching network of Example 5.4-1 by matching the source and load impedances to an intermediate resistor of 50 Ω instead of 33.166 Ω. Compare the fractional bandwidth and Q factors of the two Examples.

Solution: To redesign the matching network, change the intermediate resistor in the Equation Editors to 50 Ω and recalculate the matching element values. The calculation results are shown in Figure 5-39.

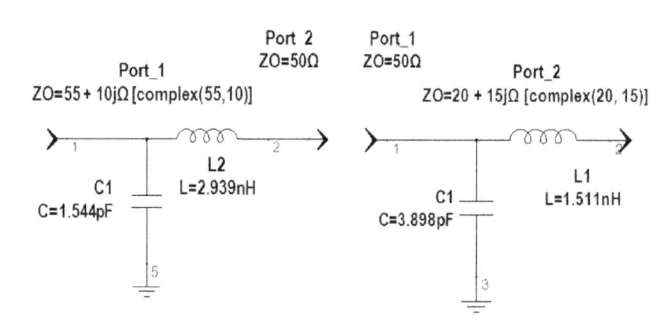

Figure 5-39 Load and source impedance matched to 50 Ω

Notice that the element values in both schematics have changed due to the change in the intermediate resistor. The cascaded schematic and simulated response are shown in Figure 5-40.

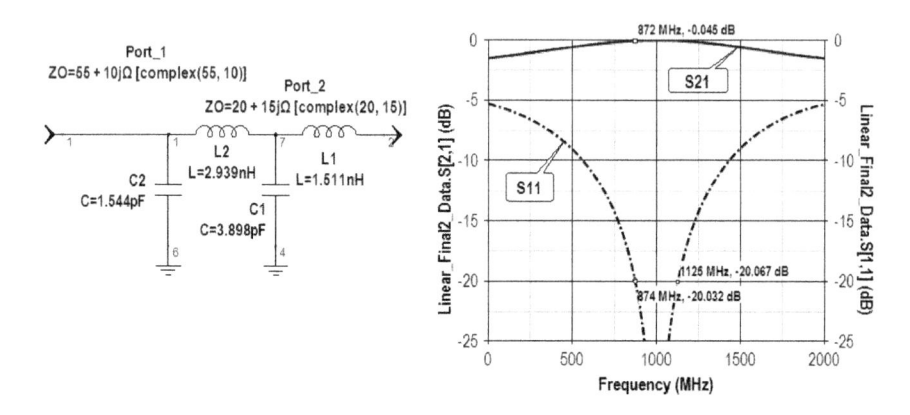

Figure 5-40 Complete matching network schematic and response

Power Transfer and Impedance Matching   253

The markers at 20 dB return loss show that the bandwidth is:

$$BW_{20dB\ return\ loss} = 1125 - 874 = 251\ MHz$$

Therefore, the corresponding fractional bandwidths at 20dB return loss is:

$$FBW_{20dB\ return\ loss} = \frac{251}{\sqrt{(1125)(874)}} = 0.253 = 25.3\ \%$$

A comparison of the fractional bandwidths for the two examples show that the matching network in Figure 5-38 provides over 50 % more matching bandwidth than the matching network in Figure 5-40.

Now we show that the less bandwidth is due to the fact that the source and load matching networks do not have the same Q factors.

$$Q_{source} = \frac{1}{2}\sqrt{\frac{55}{50}-1} = 0.158$$

$$Q_{load} = \frac{1}{2}\sqrt{\frac{50}{20}-1} = 0.612$$

The equal Q factors in Example 5.4-1 cause the cascaded network to transfer more power to the load than the unequal Q factors in example 5.4-2.

5.4.2 Broadband Impedance Matching Using N-Cascaded L-Networks

In Example 5.4-1 we showed that cascading two equal-Q matching L-networks increases the overall matching bandwidth. Therefore, we can further increase the matching bandwidth by increasing the number of equal-Q matching networks. In practice as the number of L-networks becomes greater than five, the element values become difficult to physically realize. In this section we show that by using a network of four equal-Q L-networks we can lower the individual Q factor and provide an even greater bandwidth than Example of 5.4-1.

Example 5.4-3: Redesign the matching network of Example 5.4-1 with four equal-Q L-networks. Compare the Q and bandwidth of Example 5.4-3 with Example 5.4-1.

Solution: This example uses the same steps developed in Example 5.4-1. The general equation for the calculation of any number of intermediate resistors is given by Equation (5-51).

$$R_n = R_S (r)^{n/N} \qquad n = 1, 2, 3, \text{ to } N-1 \qquad (5\text{-}51)$$

where:
  R_n is the intermediate resistor values
  R_S is the source resistor value
  r is the normalized load resistor, $r = R_L/R_S$
  N is the number of cascading networks

Equation Editor in Figure 5-41 is used to calculate 3 intermediate resistors.

```
1    %Enter Design Parameters
2    RS=55
3    RL=20
4    r=RL/RS
5    N=4
6    %Calculate Intermediate Resistors
7    R1=RS*r^(1/N)
8    R2=RS*r^(2/N)
9    R3=RS*r^(3/N)
```

Figure 5-41 Calculation of the intermediate resistors: R1, R2, and R3

Viewing the Workspace Variables show that $R1 = 42.71 \, \Omega$, $R2 = 33.166 \, \Omega$, and $R3 = 25.755 \, \Omega$. Now we design the four matching L-networks.

Power Transfer and Impedance Matching 255

Equatin Editor in Figure 5-42 is used to calculatet the matching L-network between R3 and the load impedance .

```
1    %Enter Design Parameters
2    Z0=25.755
3    RL=20
4    XL=15
5    f=1000e6
6    %Normalize Load Impedance
7    r=RL/Z0
8    x=XL/Z0
9    % Calculate Element Values
10   B1=sqrt((1-r)/r)/Z0
11   X1=Z0*sqrt(r*(1-r))-x*Z0
12   CS1=-1/(2*pi*f*X1)
13   CP2=B1/(2*pi*f)
```

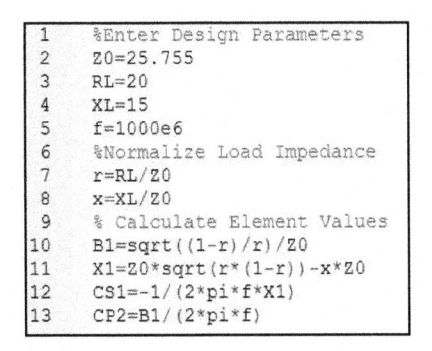

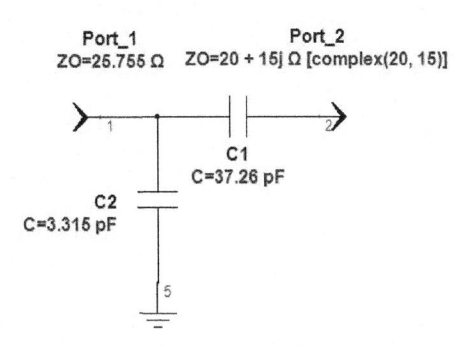

Port_1 ZO=25.755 Ω   Port_2 ZO=20 + 15j Ω [complex(20, 15)]

C1 C=37.26 pF
C2 C=3.315 pF

Figure 5-42 Calculation of the first matching L-network and schematic

Equatin Editor in Figure 5-43 is used to calculatet the matching L-network between R2 and R3.

```
1    %Enter Design Parameters
2    Z0=33.166
3    RL=25.755
4    f=1000e6
5    %Normalize Load Impedance
6    r=RL/Z0
7    % Calculate Element Values
8    B1=sqrt((1-r)/r)/Z0
9    X1=Z0*sqrt(r*(1-r))
10   LS1=X1/(2*pi*f)
11   CP2=B1/(2*pi*f)
```

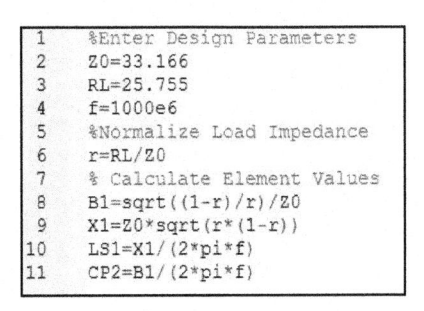

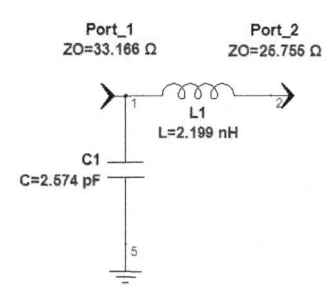

Port_1 ZO=33.166 Ω   Port_2 ZO=25.755 Ω

L1 L=2.199 nH
C1 C=2.574 pF

Figure 5-43 Calculation of the second matching L-network and schematic

Equatin Editor in Figure 5-44 is used to calculatet the matching L-network between R1 and R2.

```
1    %Enter Design Parameters
2    Z0=42.71
3    RL=33.166
4    f=1000e6
5    %Normalize Load Impedance
6    r=RL/Z0
7    % Calculate Element Values
8    B1=sqrt((1-r)/r)/Z0
9    X1=Z0*sqrt(r*(1-r))
10   LS1=X1/(2*pi*f)
11   CP2=B1/(2*pi*f)
```

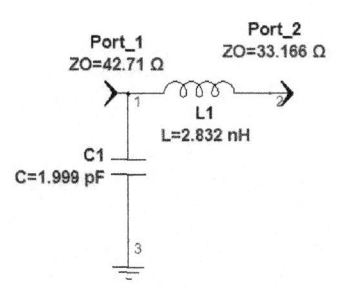

Port_1 ZO=42.71 Ω   Port_2 ZO=33.166 Ω

L1 L=2.832 nH
C1 C=1.999 pF

Figure 5-44 Calculation of the third matching L-network and schematic

Finally, Equatin Editor in Figure 5-45 is used to calculatet the matching L-network between R1 and the source impedance .

```
1    %Enter Design Parameters
2    Z0=42.71
3    RL=55
4    XL=10
5    f=1000e6
6    %Normalize Load Impedance
7    r=RL/Z0
8    x=XL/Z0
9    % Calculate B3,X3, and Element Values
10   B3=(x+sqrt(r*(r^2+x^2-r)))/(Z0*(r^2+x^2))
11   X3=Z0*sqrt((r^2+x^2-r)/r)
12   LS1=X3/(2*pi*f)
13   CP2=B3/(2*pi*f)
```

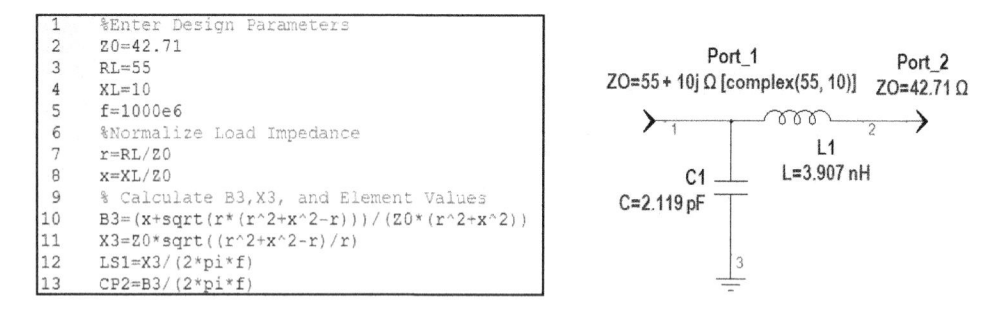

Figure 5-45 Calculation of the fourth matching L-network and schematic

Cascade the four L-networks and identify the nodes, from right to left, as shown in Figure 5-46.

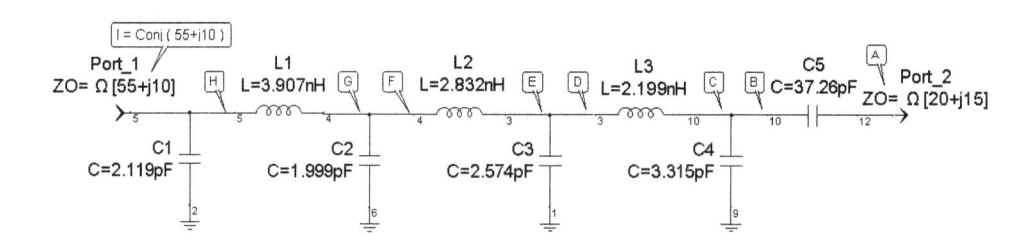

Figure 5-46 Schematic of the matching network using 4 L-networks

The simulated response shows that the matching bandwidth at 20 dB return loss is:

$$BW_{20dB\ return\ loss} = 1334 - 821 = 513\ MHz$$

Therefore, the fractional bandwidth at 20 dB return loss is:

$$FBW_{20dB\ return\ loss} = \frac{513}{\sqrt{(1334)(821)}} = 0.490 = 49.0\%$$

Power Transfer and Impedance Matching   257

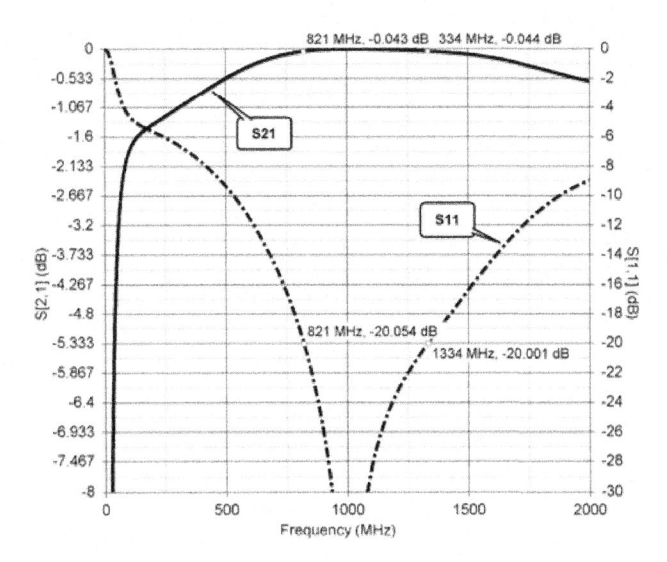

Figure 5-47 Simulated response of the matching network using 4 L-networks

The comparison between the fractional bandwidth of this example with the fractional bandwidth of example 5.4-1 shows that cascading four equal-Q matching networks increases the fractional bandwidth by more than 93 % over the two equal-Q matching networks. As we stated earlier, the reason for wider matching bandwidth is that all four individual matching networks in Figure 5-46 have the same Q factor as shown in the following calculations.

$$Q_{4L} = \frac{1}{2}\sqrt{\frac{55}{42.71}-1} = \frac{1}{2}\sqrt{\frac{42.71}{33.166}-1} = \frac{1}{2}\sqrt{\frac{33.166}{25.755}-1} = \frac{1}{2}\sqrt{\frac{25.755}{20}-1} = 0.268$$

5.4.3 Derivation of Equations for Q and the Number of L-Networks

The loaded Q of a single L-network matching two real resistors R_1 and R_2 was given in Section 5.4.1 by:

$$Q_1 = \frac{1}{2}\sqrt{\frac{R_2}{R_1}-1} \qquad\qquad R_2 > R_1$$

For a cascade of N equal-Q L-networks, the individual Q factor is given by:

$$Q_N = \frac{1}{2}\sqrt{\left(\frac{R_2}{R_1}\right)^{1/N} - 1} \quad N = 1,2,3 \qquad (5\text{-}52)$$

Notice that by increasing the number of L-networks the individual Q factor is decreased and, as a result, the matching bandwidth will be increased. Equation (5-52) can be used to calculate the number of L-networks needed for a given resistor ratio and Q factor. Consider the relationship between the resistance ratio and the Q factor.

$$1 + 4Q_N^2 = \left(\frac{R_2}{R_1}\right)^{1/N}$$

or

$$\left[1 + 4Q_N^2\right]^N = \left(\frac{R_2}{R_1}\right)$$

Taking the logarithm of both sides, the number of cascaded L-networks, N, is given by Equation (5-53).

$$N \geq \frac{\log\left(\frac{R_2}{R_1}\right)}{\log\left[1 + 4Q_N^2\right]} \qquad (5\text{-}53)$$

Notice for N to be a positive integer the resistor R_2 must be greater than R_1.

Example 5.4-4: For a load and source resistor ratio of 3, find the minimum number of cascaded L-networks to achieve a loaded Q of 0.3.

Solution: To calculate N, substitute the resistor ratio and Q into Equation (5-53).

$$N \geq \frac{\log(3)}{\log\left[1+4(0.3)^2\right]} = \frac{0.477}{0.133} = 3.58$$

Therefore, a minimum of four L-networks is needed to achieve a Q of 0.3 or less. For verification substitute the resistor ratio and N into Equation (5-52) to calculate the Q factor.

$$Q_4 = \frac{1}{2}\sqrt{\left(\frac{R_2}{R_1}\right)^{1/4} - 1} = 0.28$$

The same calculations are don in the Equation Editor of Figure 5-48.

```
1    %Enter Source and Loal Resistors
2    RS=60
3    RL=20
4    %Defin Resistor Ratio
5    r=RS/RL
6    %Enter Number of L Sections
7    N=4
8    %Calculate Q Given r and N
9    QN=0.5*sqrt(r^(1/N)-1)
10   %Calculate N Given r and Q
11   N1= log(r)/(log(1+4*QN^2))
```

Figure 5-48 Calculation of Q and N

Viewing the Workspace Variables show that QN=0.281 and N1=3.573

5.5 Designing with Q-Curves on the Smith Chart

We have seen that cascaded L-networks can be used to perform a conjugate match between any source and load impedances. Each successive section is matched to an intermediate resistance, R_n. The overall Q of the matching network is determined by the ratio of the source to load resistor and the number of cascading L-networks. In this process the Q factor is reduced as the number of equal-Q L-networks is increased. This cascade of successive L-networks of constant Q can also be accomplished on the Smith Chart.

Curves of constant Q can be drawn on the Smith Chart at every point at which $Q = X / R$. Such a plot of Q curves is shown in Figure 5-49 for Q values of 1, 2, and 3. A given Q curve is actually the external Q of a given L-network section. As such it does not account for loading on both ports of the network and is therefore two times the value of the Q as defined by Equation (5-52). The matching sequence progresses between the source impedance and the conjugate of the load impedance using the reactive element Smith Chart movements learned in Chapter 3. Each L-network should intersect a point on the real axis of the Smith Chart.

These points on the real axis represent the intermediate virtual resistance, R_n. The inductor and capacitor values must be chosen to stay within the boundary of the Q curve. Keeping each section equal distant to the Q curve results in a network in which each L-network has the same external Q factor. This is not to be confused with the overall loaded Q of the matching network. The Q Curve selection can be visualized as a compromise between the desired bandwidth and number of L-networks that may be required for a given impedance match. Using the smallest Q curve will result in the widest possible bandwidth. It can also be noted that for very high Q load impedances, it may not be possible to achieve a broadband match with reactive L-networks. For these types of loads it may be necessary to use transmission line transformer techniques [8]. Even in narrowband applications, engineers prefer to keep the matching network Q as low as possible in circuits that involve high power levels. High Q matching networks can create very high voltage or current peaks when excited with high RF power levels. The Q curve technique will be demonstrated by revisiting Example 5.4.1.

Power Transfer and Impedance Matching 261

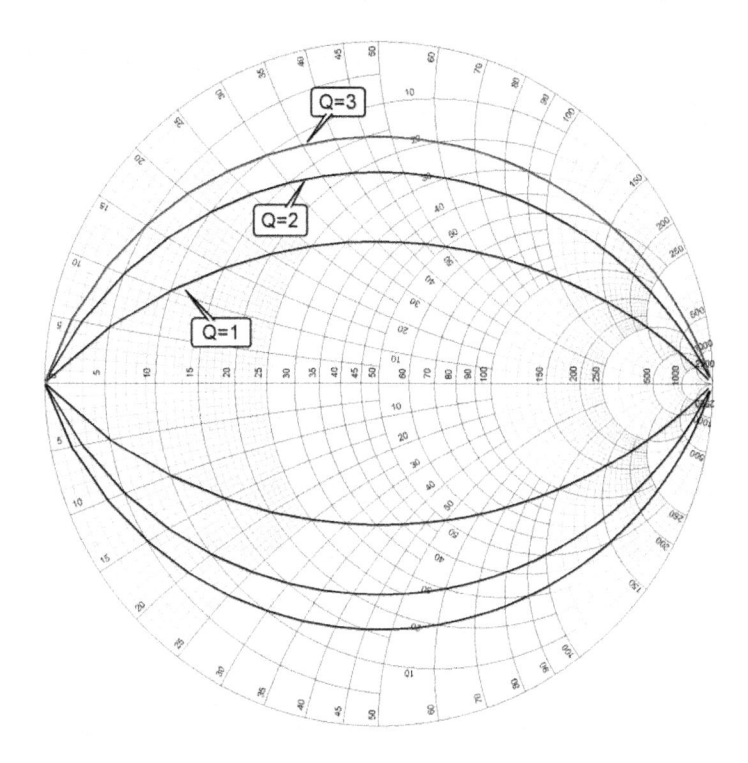

Figure 5-49 Smith Chart display of constant Q-curves

5.5.1 Q-Curve Matching Example

Example 5.5-1: A transmitter operates over a frequency range of 835 MHz to 1200 MHz. At its center frequency of 1 GHz the source impedance is $Z_S = 55 + j10\ \Omega$ while the antenna input impedance is $Z_L = 20 + j15\ \Omega$. Use the constant Q curve method in Genesys to implement the cascaded matching L-network.

Solution: A separate schematic and analysis is created in the Genesys Workspace to generate the Q curve. Create a schematic that consists of a one port impedance element. Make the resistance and reactance values tunable variables. The Linear analysis should be set as a fixed frequency in which any frequency can be used. Add a Parameter Sweep to the Workspace so that the resistance of the one port impedance can be swept over a range of 0.1 Ω to 4000 Ω in 5 Ω steps. Then create an Equation Editor as shown in Figure 5-50 to determine the impedance array that result in points of constant Q. These points can then be plotted on the Smith Chart to create the Q curve. The desired value of the Q Curve can be entered in the Equation Editor or in the Tune Window.

```
1   'Enter the desired Q Curve Value
2   Q=0.536
3   'Calculate R1 using X1 from the Parameter Sweep
4   X1=Sweep1_Data.IMP_IMPEDANCE_1_X_Swp_F
5   R1=X1./Q
6   'Define the complex impedance
7   Ztop=complex(R1,X1)
8   'Normalize the impedance to plot on Smith Chart
9   Ztopn=(Ztop-50)/(Ztop+50)
10  'Plot top circle
11  setindep("Ztopn","X1")
12  'Plot bottom half circle
13  'Define the complex impedance
14  Zbottom=complex(R1,-X1)
15  'Normalize the impedance to plot on Smith Chart
16  Zbottomn=(Zbottom-50)/(Zbottom+50)
17  'Plot bottom circle
18  setindep("Zbottomn","X1")
```

Figure 5-50 Equation Editor used to create the 0.536 Q curve

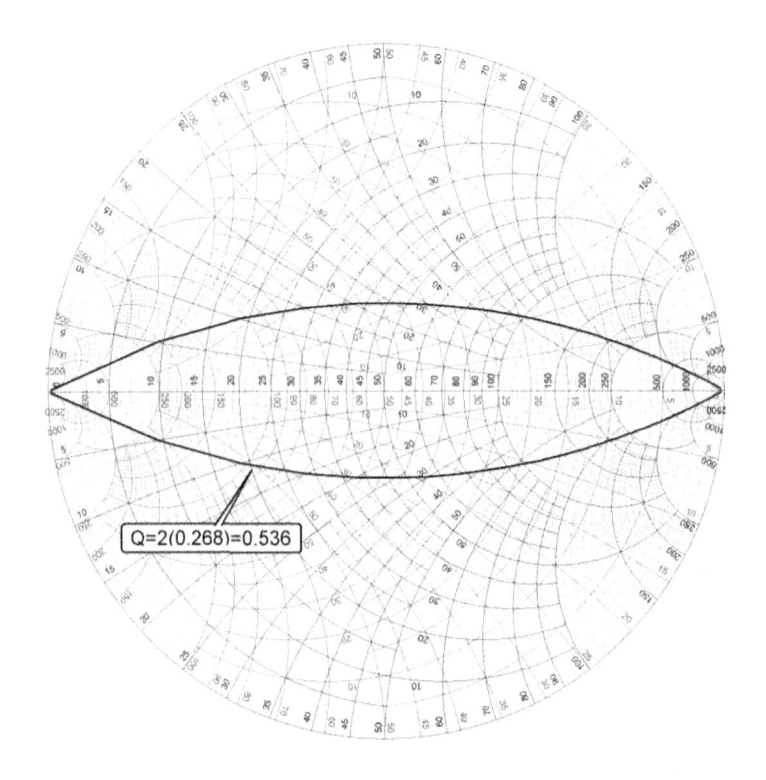

Figure 5-51 Q curve plot for Q=.536

Power Transfer and Impedance Matching 263

The Q curve for a Q value of 0.536 is shown in Figure 5-51. Because Q-curve plots do not consider loading on both ports of the network, we need to double the individual Q factor in order to obtain the external Q used in plotting the Q curve. For the network in Figure 5-46 the individual Q factor was 0.268, therefore, for plotting the Q curve on the Smith Chart we must use Q = 2 (0.268) = 0.536 on the Smith Chart and in the Equation Editor.

Impedance Match Progression: Using the lumped element movement techniques learned in Chapter 3 the matching network can be designed on the Smith Chart. On the Smith Chart we are matching from the load at Z_A to the conjugate of the source at Z_i. The Smith Chart of Figure 5-52 shows the lumped element movements on the constant conductance and resistance circles that maintain equal distant arcs within the specified Q curve. This ensures that each L-network maintains equal Q. The matching sequence between the source and load impedance progresses by using a series inductor from Z_A to intersect the Q curve at Z_B. Then add a shunt capacitor to move from the Q curve to the real axis at point Z_C. This process is continued until the last shunt capacitor intersects the impedance at Z_I.

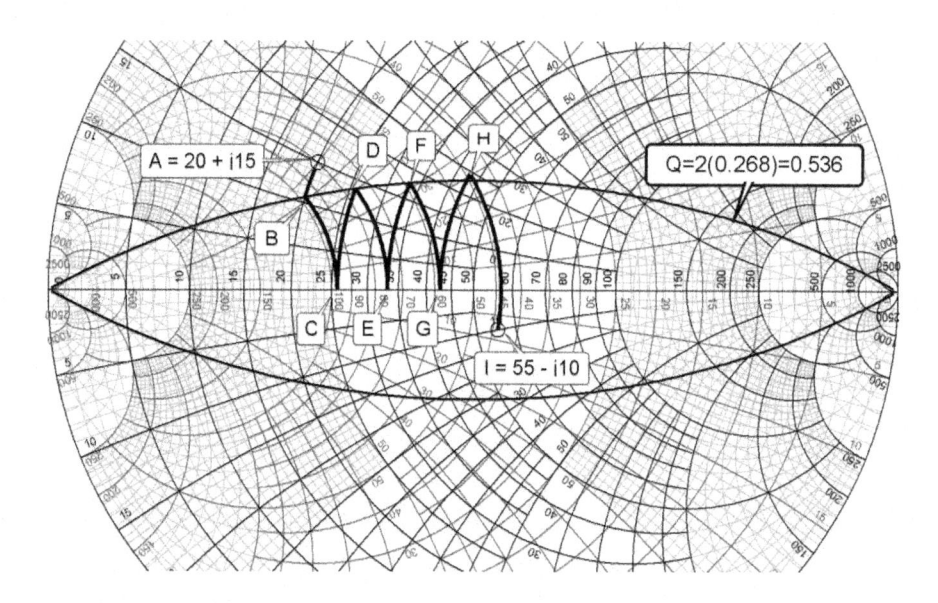

Figure 5-52 Q curve output for Example 5.4-3 (Q=0.536)

The Q curve plot of Figure 5-52 shows that all the node impedances fall on the same Q curve. To see the change in impedance as we move from the load impedance at 20 + j15 to the source impedance at 55 + j10 we have connected the points together indicating the addition of capacitors and inductors on the Smith Chart. The displayed node impedances on the Smith Chart show that three of the impedances fall on the real axis. These are the intermediate resistors. The intermediate virtual resistance values on the real axis are found to be:

$$Z_C = 25.777 \ \Omega \quad Z_E = 33.167 \ \Omega \quad Z_G = 42.715 \ \Omega$$

The impedances on the Q = 0.536 curve is measured as:

$$Z_B = 20 + j10.729 \ \Omega \qquad Z_D = 25.777 + j13.816 \ \Omega$$

$$Z_F = 33.167 + j17.797 \ \Omega \qquad Z_H = 42.715 + j24.55 \ \Omega$$

5.6 Limitations of Broadband Matching

From the previous example of section 5.5 we can deduce that the Q of the complex source and load impedance has an impact on the maximum bandwidth over which a good impedance match, low return loss, can be achieved. There is a finite limit on the achievable return loss for a given load impedance and circuit bandwidth known as Fano's limit [1]. Fano's Limit is the optimum reflection coefficient that can be achieved with a given load impedance. It is a theoretical limit that considers an infinite number of lossless matching elements. Fano's limit is defined by Equation (5-54).

$$|\Gamma| = e^{\frac{-\pi Q_L}{Q_{UL}}} \qquad (5\text{-}54)$$

where:
Q_{UL} = the unloaded Q or the ratio of the reactance to resistance of the load

Q_L = the loaded Q of the network or the ratio of the center frequency of the network divided by the 3 dB bandwidth as defined by Equation (4-11) in chapter 4 section 4.2.

Power Transfer and Impedance Matching   265

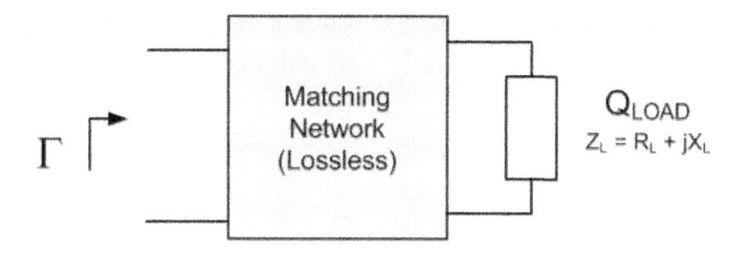

Figure 5-53 General configuration of impedance matching network

5.6.1 Example of Fano's Limit Calculation

In the following example we apply the Fano's limit to calculate the optimum reflection coefficient for a given network.

Example 5.6-1: Apply Fano's Limit to find the lowest reflection coefficient that can be achieved in the 3 dB range of the network shown in Figure 5-20.

Solution: From the network response in Figure 5-20 we notice that f_a=660 MHz and f_b=1233 MHz, therefore, the center frequency is $f_0 = \sqrt{(1233)(660)}$ = 902 MHz. The matching network loaded Q_L is defined as:

$$Q_L = \frac{f_0}{f_b - f_a} = \frac{902\,MHz}{1233\,MHz - 660\,MHz} = 1.574$$

For the load impedance, $Z_L = 7 - j22$, the unloaded Q_{UL} is defined as:

$$Q_L = \frac{X_L}{R_L} = \frac{22}{7} = 3.14$$

Fano's limit states that the best achievable Γ is given by:

$$|\Gamma| = e^{\frac{-\pi Q_L}{Q_{UL}}} = e^{\frac{-\pi(1.574)}{3.14}} = 0.20$$

From Table 2-1 we know that a reflection coefficient of 0.20 is equivalent to a return loss of 13.98dB or a 1.5 VSWR. We must remember that Fano's limit gives a strictly theoretical result for return loss assuming lossless components and an infinite number of elements. The reflection coefficient defined by Fano's limit is essentially a rectangular return loss characteristic which is not possible to achieve in practice. It can be used however to give a quick estimate of the difficulty of a particular matching network, particularly for high Q load impedances.

5.7 Matching Network Synthesis

5.7.1 Filter Characteristics of the L-networks

The impedance matching L-networks that we have designed take the form of either a low pass or high pass characteristic. Based on the location of the inductor and capacitor, the network, acts as a low pass or a high pass filter. The low pass network consists of a series inductor and a shunt capacitor while the high pass network consists of a series capacitor and shunt inductor. Figure 5-54 shows two low pass configurations that are used to match a source and load impedance based on the ratio of the real component of the source and load impedance. With the low pass configuration at very low frequencies, the inductive reactance is very small (short circuit) and the output is close to input. At very high frequencies the inductor has a high reactance and the output is near zero.

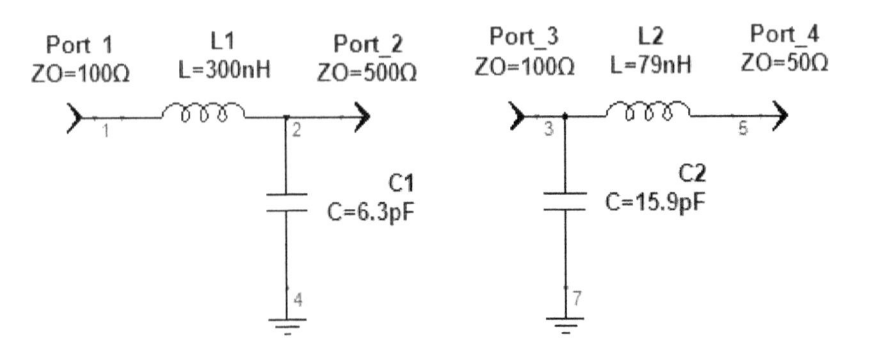

Figure 5-54 Low pass matching L-networks

Power Transfer and Impedance Matching 267

The high pass network consists of a shunt inductor and a series capacitor. Figure 5-55 shows two high pass configurations that are used to match a source and load impedance based on the ratio of the real component the source and load impedance. At very low frequencies the inductor has a very small reactance, (short circuit), and the output is close to zero. At high frequencies the inductor has a high reactance, (like an open), and the output is nearly equal to input. At very low frequencies the capacitor has a very high reactance, (open circuit), and the output is close to zero while at high frequencies the capacitor has a small reactance, (short circuit), and the output is nearly equal to the input.

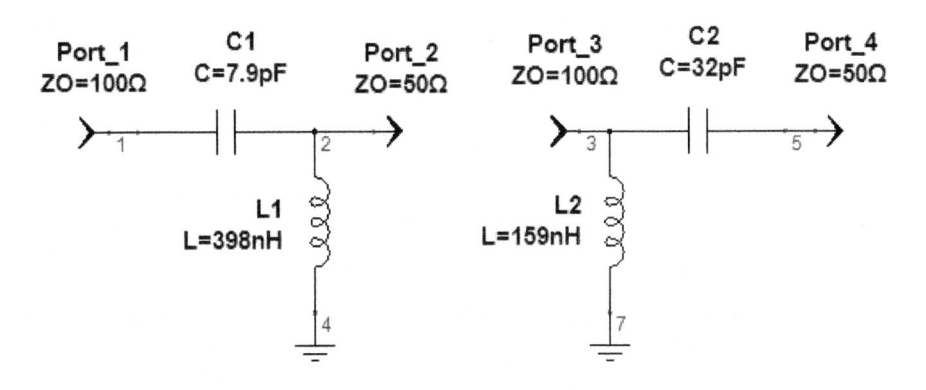

Figure 5-55 High pass matching L-networks

5.7.2 L-Network Impedance Matching Utility

The analytical impedance matching techniques introduced in this chapter form the mathematical basis for network synthesis. We can use the VBScript capability of Genesys to create a simple L-network impedance matching utility to solve the equations for arbitrary source and load impedance values. Equations (5-7) through (5-10) and (5-12) through (5-15) are programmatically solved in the hosted VBScript language. Each matching solution can be quickly calculated and the schematic response displayed. The VBScript capability within Genesys is a very powerful tool that the reader is encouraged to explore. The schematic for the L-Network impedance matching utility is shown in Figure 5-56.

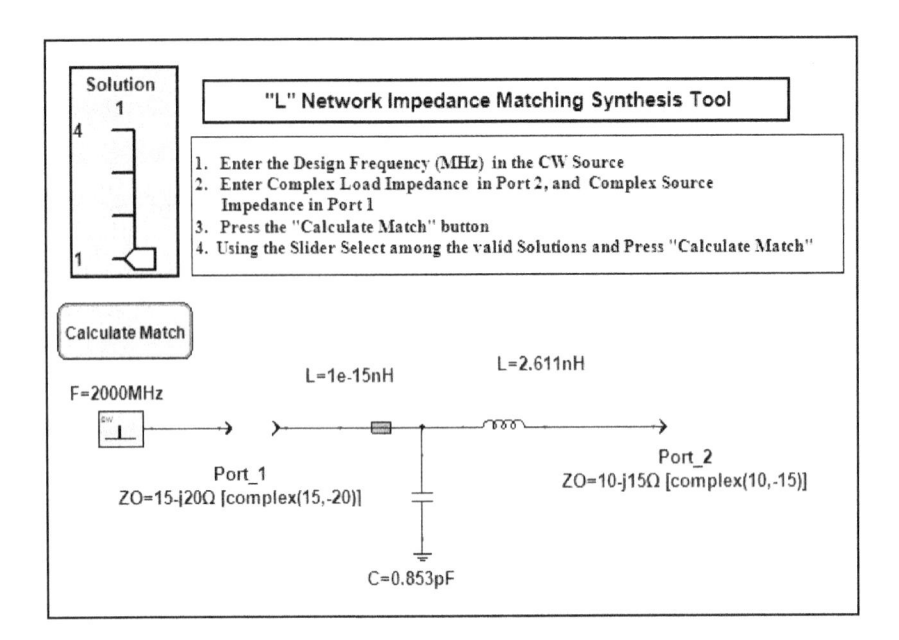

Figure 5-56 L-network impedance matching utility

VBScript provides the designer with the capability of changing parts and their properties on the schematic. A three element Tee network is created with three schematic elements. Depending on the network topology the appropriate two-element L-Network that matches the source and load impedance will be displayed. The small bead symbol is used to represent a short circuit of a given series element position. This reduces the three element schematic to a two element L-network. Note the extremely small inductance value that is assigned to the element to signify that it is a short circuit. The user inputs are the source and load impedance and the frequency for which the network is to be calculated. The port impedance can be edited by entering the desired impedance in the Properties window of the input and output port. The impedance should always be entered in a complex format. If it is desired to have real impedance for the source or load simply enter an extremely small value such as 0.001 for the reactive component. A value of zero will result in a division by zero error. The frequency is entered by editing the Properties window of the CW power source. The VBScript program retrieves the values from the input and output ports as well as the CW power source element. A command button and slider control is also placed on the schematic.

Power Transfer and Impedance Matching 269

```
55     '              Define Network Configurations
56     Sub ShuntC
57     My_L_Network=WsDoc.Designs.Sch1.PartList.X1.ChangeSymbol("CAPACITOR")
58     My_L_Network=WsDoc.Designs.Sch1.PartList.X1.ChangeModel("CAP")
59     My_L_Network=WsDoc.Designs.Sch1.PartList.X1.ParamSet.C.SetValue(B)
60     end sub
61
62     Sub ShuntL
63     My_L_Network=WsDoc.Designs.Sch1.PartList.X1.ChangeSymbol("INDUCTOR")
64     My_L_Network=WsDoc.Designs.Sch1.PartList.X1.ChangeModel("IND")
65     My_L_Network=WsDoc.Designs.Sch1.PartList.X1.ParamSet.L.SetValue(B)
66     end sub
```

Figure 5-57 VBScript code for changing the schematic element

The VBScript code is actually loaded in the command button as shown in Figure 5-58. Pressing the command button will initiate execution of the VBScript code. This enables the execution of the script when the user presses the command button. The slider control is programmatically updated to display the number of valid L-network solutions that exist for the source to load impedance match. The slider control is linked to the variable, Solutions, which is the calculated number of valid solutions for the impedance matching L-Section. The user can scroll through the valid solutions by selecting a solution on the slider control and then pressing the command button.

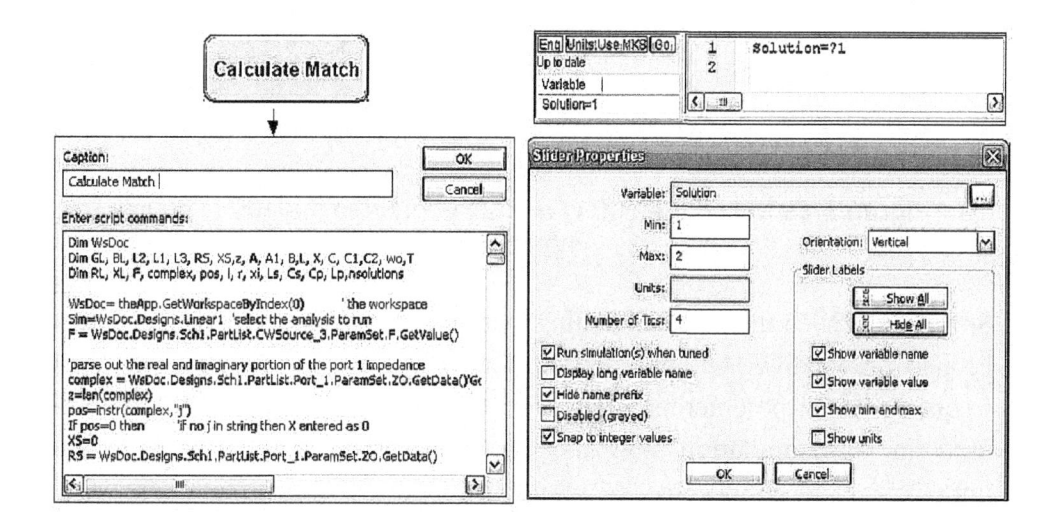

Figure 5-58 VBScript code loaded into the command button and the slider properties linked to the variable 'solution' in the Equation Editor

The VBScript will update the schematic with the calculated component values and display the insertion loss, S21, and return loss, S11. The complete VBScript program listing is given in Appendix B.

5.7.3 Network Matching Synthesis Utility in Genesys

Chapter 4 section 4.8 introduced the filter synthesis utility for the design of filter networks. Genesys also has a powerful impedance matching utility integrated into the software. In a Genesys workspace right click on the Designs folder and select Synthesis followed by Add Impedance Match. Name the circuit or accept the default name and the synthesis window will open. The following example shows the typical use of the Genesys Impedance Matching Utility.

Example 5.7-1 Use the Genesys Impedance Matching Utility to match a source impedance of 100 Ω to a load that is comprised of a 500 Ω resistor in parallel with a 2 pF capacitance as shown in Figure 5-59. It is required that a 20 dB return loss and greater than -1 dB insertion loss be achieved over a frequency range of 60 MHz to 164 MHz.

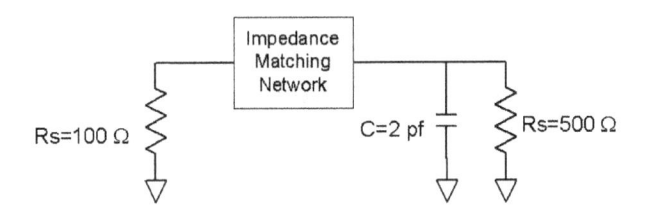

Figure 5-59 Match the 500 Ω and 2 pF loads to the 100 Ω source impedance

Solution: The impedance matching parameters are defined in the Match Properties window. On the Setting Tab, the lower and upper passband frequencies can be entered. Also enter the desired number of points to be used in the simulation. On the Sections Tab, the source and load impedances are defined by the selection of the network type as shown in Figure 5-60. Also on the Sections Tab, the type of network type and order can be selected. Figure 5-61 shows the variety of network types that can be selected. Basic LC tee networks can be used to model the L-networks as

Power Transfer and Impedance Matching 271

designed in section 5.3. Because the goal of this design is to have a bandpass characteristic the LC Bandpass network type is selected. To keep the network realized of all LC elements, select 'no transformers' to be used in the matching network. Finally select the order of the network and press Calculate Match button. A schematic of the synthesized network along with its response will be automatically generated. It is very easy to perform 'what-if' analysis using the Matching Utility. Simply enter a higher network order and select 'Calculate' to quickly compare the return loss.

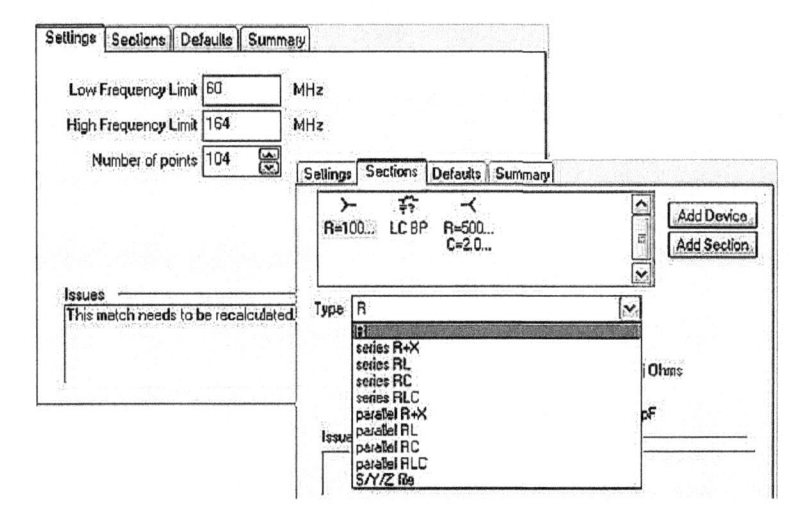

Figure 5-60 Specify the frequency band and input and output impedance

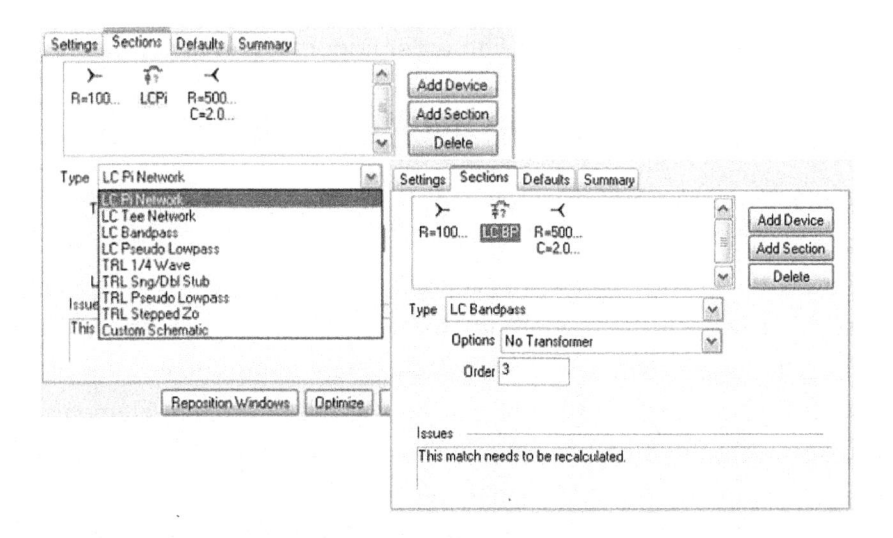

Figure 5-61 Specify the network type and order

Figure 5-62 shows the return loss comparison between a third order and fifth order matching network. To meet the 20 dB return loss specification the fifth order network should be chosen.

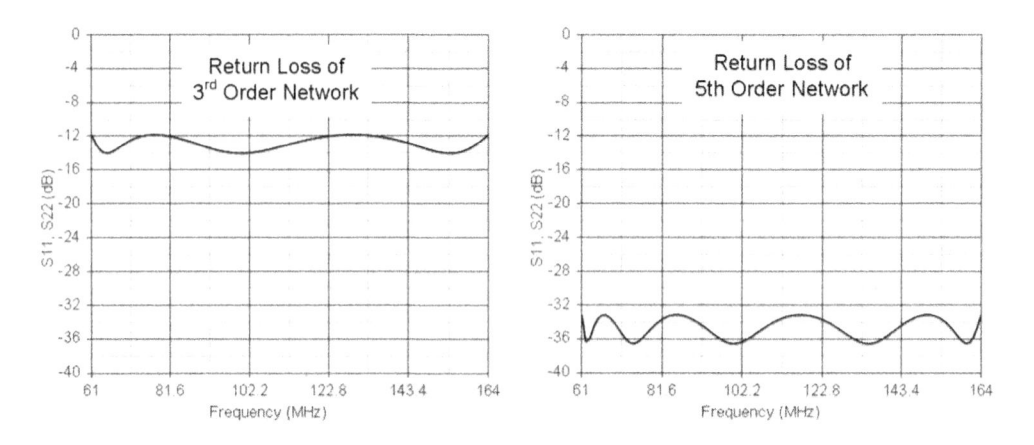

Figure 5-62 Return loss of third and fifth order matching networks

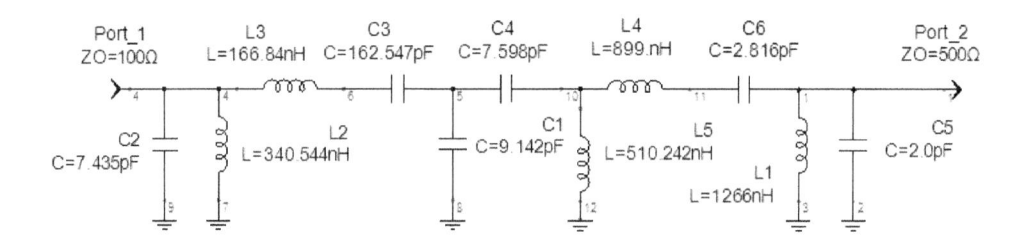

Figure 5-63 schematic of the fifth order impedance matching network

5.7.4 Effect of Finite Q on the Matching Networks

To complete the matching circuit design, the components need to be converted to their physical equivalents by accounting for the finite Q factor of the capacitors and inductors. This can be accomplished by using either S parameter or Modelithic models for the elements. A quick approach is to model each element with a Q factor that closely approximates the physical components. Note that a Q factor can be specified for each element in the Matching synthesis utility. In this design we'll demonstrate a technique to assign a common variable name representing the Q factor of each component. This allows the evaluation of the necessary Q factor of the components required to meet the specifications of the matching network. The variable name, Qcap, is assigned to all capacitors in the matching network. Similarly the variable name, Qind, is assigned to all of the inductors. The frequency at which the Q factor is defined is also defined as a variable, Qf. This allows the Q factor of all inductors and all capacitors to be tuned simultaneously. Figure 5-64 shows the revised schematic with the Q factor defined for all components. The variables are then initialized in the Equation Editor as shown in Figure 5-65. Note the use of the (=?) which enables the variable to be tuned.

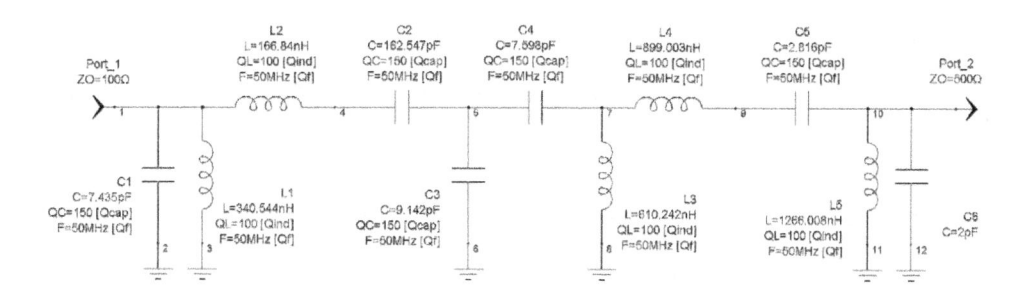

Figure 5-64 Fifth order matching schematic with Q factor variables

```
1    Qcap=?150
2    Qind=?100
3    Qf=?5e+007
```

Figure 5-65 Q factor variables defined in the Equation Editor

The network response with the initial Qcap and Qind values is shown in Figure 5-66. We can see that the insertion loss (>-1dB) and return loss (<20 dB) specifications have been met. Using the Tune control in Genesys, the inductor and capacitor Q can be decreased to determine the minimum component Q factor that is required to meet the design specification. By using the common variable name, all capacitor Q's are changed simultaneously. It is not necessary to tune each individual capacitor Q in the matching network. As Figure 5-67 shows, a capacitor Q of 50 and an inductor Q of 40 results in an insertion loss that falls short of meeting the insertion loss specification. The response with Qcap=50 and Qind=40 is shown in Figure 5-67.

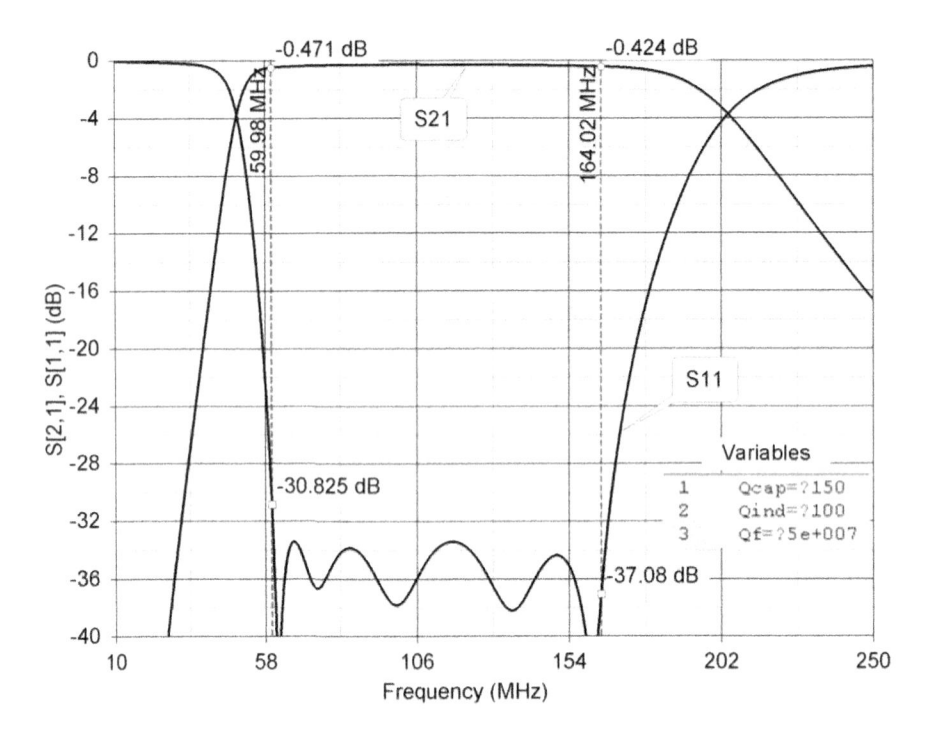

Figure 5-66 Response with Qcap=150 and Qind=100

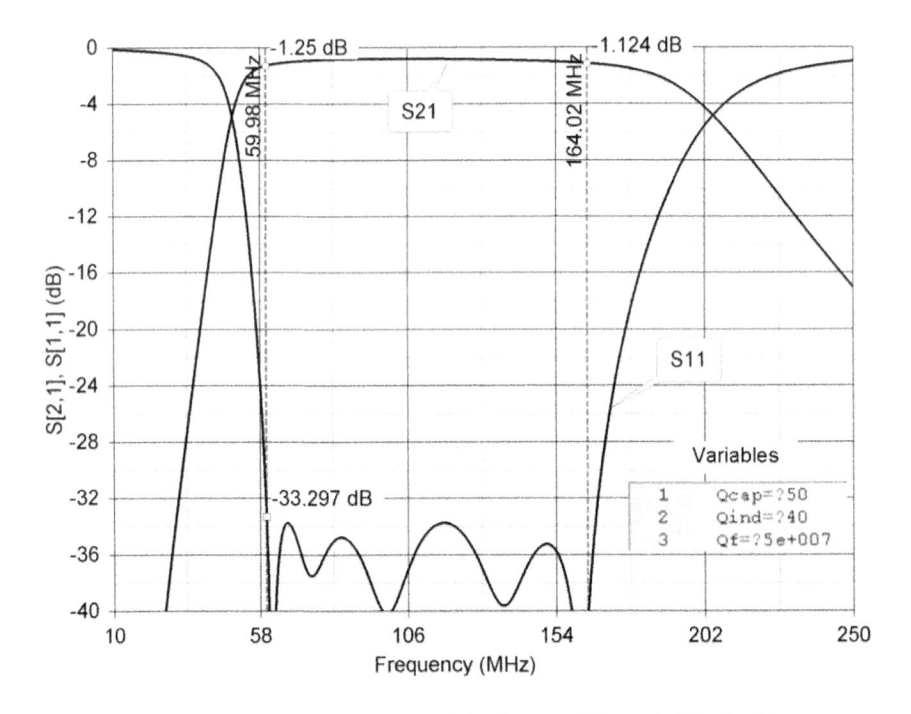

Figure 5-67 Response with Qcap=50 and Qind=40

References and Further Reading

[1] Ali A. Behagi and Stephen D. Turner, *Microwave and RF Engineering*, BT Microwave LLC, State College, PA, 2011

[2] Keysight Technologies, *Genesys 2014.03, Users Guide*, www.keysight.com

[3] Guillermo Gonzales, *Microwave Transistor Amplifiers – Analysis and Design*, Second Edition, Prentice Hall Inc., Upper Saddle River, NJ.

[4] Randy Rhea, *The Yin-Yang of Matching: Part 2 – Practical Matching Techniques*, High Frequency Electronics, March 2006

[5] Steve C. Cripps, *RF Power Amplifiers for Wireless Communications*, Artech House Publishers, Norwood, MA. 1999.

[6] David M. Pozar, *Microwave Engineering*, Fourth Edition, John Wiley & Sons, New York, 2012

[7] R. Ludwig, P. Bretchko, *RF Circuit Design*, Theory and Applications, Prentice Hall, Upper Saddle River, NJ, 2000

[8] Jerry Sevick, *Transmission Line Transformers*, The American Radio Relay League, Newington, CT., 1990.

Problems

5-1. For a load of 75 Ω with a 10 pF series capacitance we want to have maximum power transfer at 1 GHz. Find a source inductance that cancels the reactance of the load at this frequency.

5-2. Analytically design all the L-networks that will match a complex load impedance, $Z_L = 15 - j10$ Ω, to a complex source impedance,

$Z_S = 25 + j20$ Ω, at a frequency of 1 GHz. Verify all the solutions by plotting the response of the matching networks.

5-3. Analytically design all the L-networks that will match a source impedance, $Z_S = 30$ Ω, to a complex load impedance, $Z_L = 25 + j20$ Ω, at a frequency of 1 GHz. Verify all the solutions by plotting the response of the matching networks.

5-4. Analytically design all the L-networks that will match a load impedance, $Z_L = 15$ Ω, to a source impedance, $Z_S = 75$ Ω, at a frequency of 1.5 GHz. Verify all the solutions by plotting the response of the matching networks.

5-5. The output impedance of a transmitter operating at a frequency of 2 GHz is: $Z_S = 30 + j10$ Ω. Analytically design all the matching L-networks such that the maximum power is delivered to the antenna whose input impedance is $Z_L = 10 + j15$ Ω.

5-6. Redesign the network of Problem 5-5 by matching the source and load impedances to a 50 Ω line. Compare fractional bandwidth and Q factors for examples 5.5 and 5.6.

5-7. Redesign the matching network in Example 5.5 with four equal-Q L-networks. Compare the Q and bandwidth of the two networks.

5-8 For a load and source resistor ratio of 5, find the minimum number of cascaded L sections needed to achieve a loaded Q of 0.5.

5-9. Use the Genesys Impedance Matching Utility to solve a complex load matching example with $R_L = 1000$ Ω in parallel with a 3 pF capacitance. For the source select $R_S = 50$ Ω, and for the load select the parallel RC with R = 1000 Ω and C = 3 pF. Compare the Pi and Tee Matching Network response.

Chapter 6

Analysis and Design of Distributed Matching Networks

6.1 Introduction

Distributed networks are comprised of transmission line elements rather than discrete resistors, inductors, and capacitors. These transmission lines can take the form of the various transmission lines covered in chapter 2. At RF and microwave frequencies, where the wavelengths of the signals become comparable to the physical dimensions of the components, even lumped elements behave like distributed components. At microwave frequencies the distributed network is a more realizable form of a matching network than the lumped element versions in the previous chapter. As a general rule, any electrical part larger than one tenth of the signal wavelength should be analyzed as distributed element. In an impedance matching network, operating at RF and microwave frequencies, reflections from short lengths of wire can create effects that are not predictable by the lumped element analysis. In this chapter both narrowband and broadband distributed matching networks are analyzed. Several examples are given to show how the matching networks are analytically and graphically designed. Distributed matching networks, discussed in this chapter, include quarter-wave and single-stub matching networks. For the quarter-wave matching networks the matching bandwidth, power loss and Q factor are analytically calculated. For the single-stub matching networks a matching utility program in Genesys is developed that automatically calculates the line and stub lengths.

6.2 Quarter-Wave Matching Networks

The quarter-wave transformer is a useful network for impedance matching between two resistors. At RF and microwave frequencies impedance matching between a resistive source and load impedance can easily be achieved by a 90 degree transmission line known as a Quarter-Wave Transformer or Quarter-Wave Network. In this section the quarter-wave matching network is defined and its properties are investigated.

6.2.1 Analysis of Quarter-Wave Matching Networks

In chapter 2, Equation (2-41), we showed that the input impedance of a quarter-wave network with characteristic impedance Z_0 terminated in a resistive load R_L is given by:

$$Z_{IN} = \frac{Z_0^2}{R_L}$$

This equation can be written as:

$$Z_O = \sqrt{R_L Z_{IN}} \qquad (6\text{-}1)$$

In chapter 5, section 5.2.1, we also showed that maximum power transfer from a source with resistance R_S to a network terminated in a resistive load R_L is achieved only if the input impedance of the network is equal to the source resistance:

$$Z_{IN} = R_S$$

Therefore, the characteristic impedance of quarter-wave matching network, as shown in Figure 6-1, must satisfy the following equation.

$$Z_O = \sqrt{R_S R_L} \qquad (6\text{-}2)$$

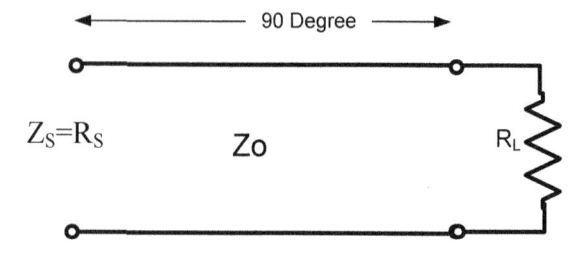

Figure 6-1 Quarter-wave network terminated in resistor R_L

Equation (6-2) states that the characteristic impedance of the quarter-wave network, matching R_L to R_S, must be equal to the square root of the product of source and load resistors.

If we normalize the load resistor R_L with respect to R_S,

$$r = \frac{R_L}{R_S} \tag{6-3}$$

The characteristic impedance of the quarter-wave network becomes a function of R_S and r.

$$Z_O = R_S \sqrt{r} \tag{6-4}$$

Notice that for $R_L > R_S$, the characteristic impedance Z_0 is greater than R_S while for $R_L < R_S$, the characteristic impedance Z_0 is less than R_S. The fractional bandwidth of a network, FBW, is defined in Equation (6-5) where f_H and f_L are the upper and lower frequencies of the bandwidth and the center frequency f_0 is equal to $\sqrt{f_H f_L}$, respectively.

$$FBW = \frac{f_H - f_L}{\sqrt{f_H f_L}} \tag{6-5}$$

The fractional bandwidth of a quarter-wave matching network is given in Equation (6-6) [5]:

$$FBW = 2 - \frac{4}{\pi} \cdot \cos^{-1}\left(\frac{2\Gamma_m \sqrt{r}}{\sqrt{1-\Gamma_m^2}\,|1-r|}\right) \tag{6-6}$$

where, Γ_m is the magnitude of the reflection coefficient.

Equation (6-6) shows that the fractional bandwidth of a quarter-wave matching network depends upon the magnitude of the input reflection coefficient, Γ_m, and the mismatch ratio, r. The solutions to Equation (6-6) are only valid if,

$$\frac{2\Gamma_m \sqrt{r}}{\sqrt{1-\Gamma_m^2}\,|1-r|} \leq 1$$

At 3 dB return loss the reflection coefficient is $\Gamma_m = 0.707$, therefore, Equation (6-6) reduces to:

$$FBW_{3dB} = 2 - \frac{4}{\pi} \cdot \cos^{-1}\left(\frac{2\sqrt{r}}{|1-r|}\right) \qquad (6\text{-}7)$$

At 3 dB return loss Equation (6-7) has valid solutions only if,

$$\frac{2\sqrt{r}}{|1-r|} \leq 1 \quad or \quad r^2 - 6r + 1 \geq 0 \qquad (6\text{-}8)$$

Similarly, if we define the bandwidth at $\Gamma_m = 0.1$, corresponding to 20 dB return loss, as a good matching bandwidth it is insightful to evaluate the fractional bandwidth associated with this 20 dB return loss. From Equation (6-6) the fractional bandwidth at $\Gamma_{in} = 0.1$, is:

$$FBW_{20dB} = 2 - \frac{4}{\pi} \cdot \cos^{-1}\left(\frac{0.2\sqrt{r}}{\sqrt{0.99}\,|1-r|}\right) \qquad (6\text{-}9)$$

At 20 dB return loss Equation (6-9) has valid solutions only if,

$$\frac{0.2\sqrt{r}}{\sqrt{0.99}\,|1-r|} \leq 1 \quad or \quad 99r^2 - 202r + 99 \geq 0 \qquad (6\text{-}10)$$

The loaded quality factor, Q_L, of the quarter-wave matching network is defined as the inverse of the fractional bandwidth at 3 dB return loss; therefore, the loaded Q factor can be calculated from Equation (6-11).

Analysis and Design of Distributed Matching Networks

$$Q_L = \frac{1}{FBW_{3dB}} = \frac{1}{2 - \frac{4}{\pi} \cdot \cos^{-1}\left(\frac{2\sqrt{r}}{|1-r|}\right)} \qquad (6\text{-}11)$$

Equation (6-11) shows that the validity condition in Equation (6-8) for the 3 dB fractional bandwidth is the same for the Q factor except that whenever the 3 dB fractional bandwidth tends towards infinity the Q factor tends towards zero. The Q factor given in Equation (6-11) is not to be confused with the unloaded transmission line Q factor as covered in chapter 4. This is actually an external Q factor as it relates to the loaded Q of the overall network. If f is the center (design) frequency, the bandwidth of the circuit is then calculated from Equation (6-12).

$$BW = (f) \cdot (FBW) \qquad (6\text{-}12)$$

6.2.2 Analytical Design of Quarter-Wave Matching Networks

In this subsection, based on the equations developed in section 6.2.1, two quarter-wave matching networks for $R_L = 2\ \Omega$ and $R_L = 150\ \Omega$ are designed. For each R_L the loaded Q factor and bandwidth is calculated at both 20 dB and 3 dB return loss.

Example 6.2-1 Design a quarter-wave network intended to match a 50 Ω source to a 2 Ω load at 100 MHz, as shown in Figure 6-2. Calculate the Q factor and the fractional bandwidths at 3 dB and 20 dB return loss. Compare the calculations with simulation in Genesys.

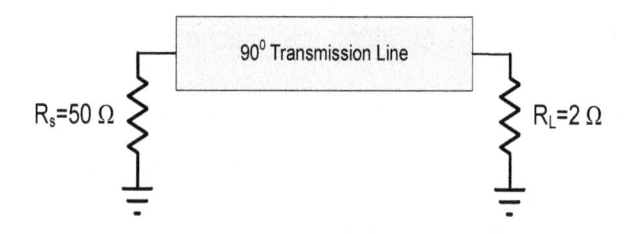

Figure 6-2 Matching quarter-wave transformer ($R_L < R_s$)

Solution: Using Equation (6-2) the characteristic impedance of the quarter wave matching network is:

$$Zo = \sqrt{(2)(50)} = 10 \ \Omega$$

The schematic of quarter-wave matching network is shown in Figure 6-3.

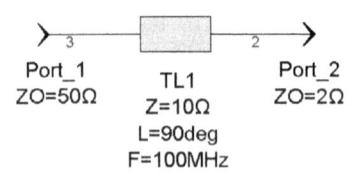

```
>─3───┬──────┬──2──>
   Port_1   TL1    Port_2
   ZO=50Ω   Z=10Ω  ZO=2Ω
            L=90deg
            F=100MHz
```

Figure 6-3 Schematic of the quarter-wave matching network

Using the equations developed in section 6.2.1, the characteristic impedance, loaded Q factor, and the bandwidths of the quarter wave matching network at 3 dB and 20 dB return loss are calculated in an Equation Editor as shown in Figure 6-4.

```
1   %Enter Design Parameters RS, RL, and Frequency
2   RS=50
3   RL=2
4   f=100e6
5   %Normalize the Load Impedance
6   r=RL/RS
7   %Calculate Electrical Length of the Line
8   Z1=RS*sqrt(r)
9   %Calculate Fractional Bandwidth at 20dB Retun Loss
10  FBW20dB=2-(4/PI)*acos((0.2*sqrt(r))/(sqrt(0.99)*(abs(1-r))))
11  BW20dB=(f*FBW20dB)
12  %Calculate Fractional Bandwidth at 3dB Return Loss
13  FBW3dB=2-(4/PI)*acos((2*sqrt(r))/(abs(1-r)))
14  BW3dB=(f*FBW3dB)
15  %Calculate Qe
16  Qe=1/FBW3dB
```

Figure 6-4 Calculating Z1, Q factor, and fractional bandwidths ($R_L < R_S$)

Viewing Workspace Variables show that the loaded Q factor is 1.827 and the bandwidths at 3dB and 20dB return loss are 54.72 MHz and 5.333 MHz, respectively. The fractional bandwidth at 20 dB return loss is only 5.333 % indicating a narrowband matching network. This is characteristic of a narrowband matching network with a load resistor that is much smaller than the source resistor. Next we analyze the matching network in Genesys to measure the same parameters. Set up a Linear Analysis to simulate the schematic of Figure 6-3. Display the input reflection coefficient, S11, and forward transmission, S21, from 50 MHz to 150 MHz. Use markers to measure the simulated bandwidth at 3 dB and 20 dB return loss as shown in Figure 6-5.

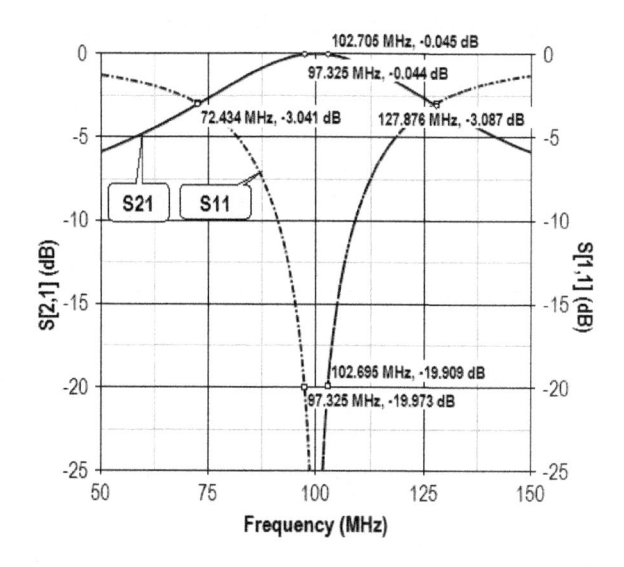

Figure 6-5 Response of the quarter-wave matching network ($R_L < R_S$)

Reading marker frequencies at 3 and 20 dB return loss, the corresponding bandwidths are measured.

$$BW_{3dB} = 127.4 - 72.6 = 54.8 \text{ MHz}$$

$$BW_{20dB} = 102.7 - 97.3 = 5.4 \text{ MHz}$$

The loaded Q of the matching network is measured by using Equation (6-9),

$$Q_L = \frac{1}{FBW_{3dB}} = \frac{\sqrt{(127.4)(72.6)}}{127.4 - 72.6} = 1.75$$

Note that the measured Q factor and bandwidths are in close agreement with the calculated values in Figure 6-4.

Example 6.2-2 Design a quarter-wave network intended to match a 50 Ω source to a 150 Ω load. The design frequency is 100 MHz. Compare the calculated Q factor and the fractional bandwidths, at 3 and 20 dB return loss, with the measurements.

Solution: The schematic of the quarter-wave matching network is shown in Figure 6-6.

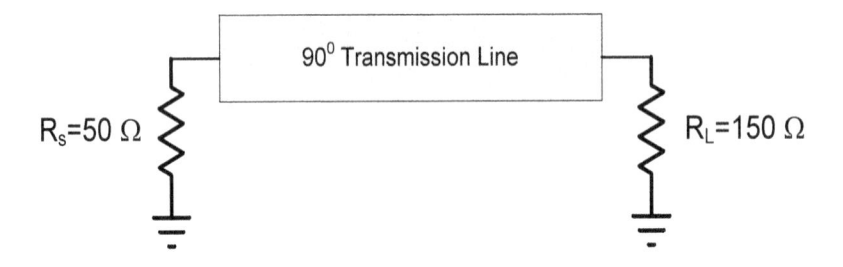

Figure 6-6 Quarter-wave matching network ($R_L > R_S$)

Using Equation (6-2) the characteristic impedance of the quarter-wave matching network is:

$$Z_1 = \sqrt{(50)(150)} = 86.6 \, \Omega$$

The Genesys schematic of the matching network is shown in Figure 6-7.

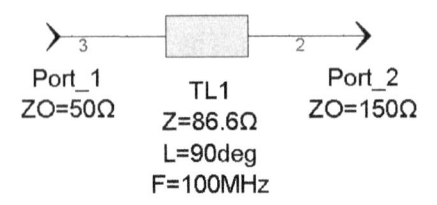

Figure 6-7 Schematic of the matching quarter-wave network

Using Equations developed in section 6.2.1, the characteristic impedance, the loaded Q factor, and the bandwidths of the matching network at 3 and 20 dB return loss are calculated in Figure 6-8.

Analysis and Design of Distributed Matching Networks 285

```
1    'Enter Design Parameters RS, RL, and Frequency
2    RS=50
3    RL=150
4    f=100e6
5    'Normalize the Load Impedance
6    r=RL/RS
7    'Calculate Electrical Length of the Line
8    Z1=RS*sqrt(r)
9    'Calculate Fractional Bandwidth at 20dB Retun Loss
10   FBW20dB=2-(4/PI)*acos((0.2*sqrt(r))/(sqrt(0.99)*(abs(1-r))))
11   BW20dB=(f*FBW20dB)
12   'Calculate Fractional Bandwidth at 3dB Return Loss
13   FBW3dB=2-(4/PI)*acos((2*sqrt(r))/(abs(1-r)))
14   BW3dB=(f*FBW3dB)
15   'Calculate Qe
16   Qe=1/FBW3dB
```

Figure 6-8 Calculating Z1, bandwidths, and Q factor ($R_L > R_S$)

Viewing Workspace Variables show that the bandwidth at 20 dB return loss is 22.28 MHz. This is over four times the bandwidth that was achieved with the 2 Ω load impedance. Notice that the overall loaded Q of the network has significantly increased due to higher load impedance. Next analyze the matching network in Genesys by simulating the quarter wave transformer. Create a Linear Analysis and simulate the schematic of Figure 6-7. Display the input reflection coefficient, S11, and forward transmission, S21, from 50 to 150 MHz. Place markers at the 20 dB return loss as shown.

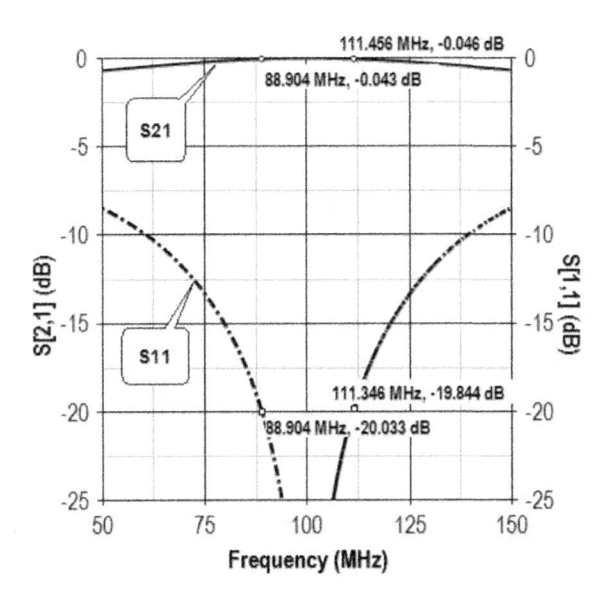

Figure 6-9 Response of the quarter-wave matching network

As Figure 6-9 shows the measured bandwidth at 20 dB return loss is:

$$111.3 - 88.9 = 22.4 \text{ MHz}$$

This agrees well with the result of calculation in the Equation Editor.

6.3 Quarter-Wave Matching Network Bandwidth

Equation (6-6) shows that the achievable bandwidth in a quarter-wave matching network is related to the ratio of the load to the source impedance (mismatch ratio) as well as the value of the input reflection coefficient. It is insightful to examine the relationship between these quantities when one of the parameters is swept in value.

6.3.1 Effect of Load Impedance on Matching Bandwidth

To investigate the effect of load variation on the matching bandwidth, create a new schematic of the quarter-wave matching network by making the output impedance of Port 2 a tunable value. Similarly make the impedance and frequency of the quarter wave variable by assigning the names Z1 for the impedance and F for the frequency. The new schematic is shown in Figure 6-10. These variables will be defined in a new Equation Editor as shown in Figure 6-11. Due to variable load impedance, the quarter-wave characteristic impedance must be recalculated for each load resistance value. The fractional bandwidth corresponding to a 20 dB return loss is also calculated in the Equation Editor. A new Linear Analysis should be setup for a fixed frequency analysis at 100 MHz.

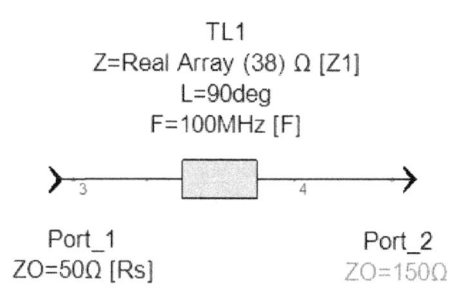

Figure 6-10 Set up for the variable load impedance

Analysis and Design of Distributed Matching Networks

```
1   'Enter Source Resistance and Frequency
2   Rs=50
3   F=100e6
4   'Take the Load Resistance from the Parameter Sweep Dataset
5   RL=Sweep1_Data.Port_2_Z0_Swp_F
6   r=RL/Rs
7   Z1=Rs*sqrt(r)
8   FBW20dB=2-(4/PI)*acos((0.2*sqrt(r))/(sqrt90.99)*(abs(1-r))))
9   FBW20dB=FBW20dB*(PI/180)
10  BW20dB=F*FBW20dB
11  setindep("Z1","RL")
12
```

Figure 6-11 Bandwidth calculation with variable load resistance

A Parameter Sweep is added to the Workspace to vary the load impedance. The Parameter Sweep Properties window is shown in Figure 6-12.

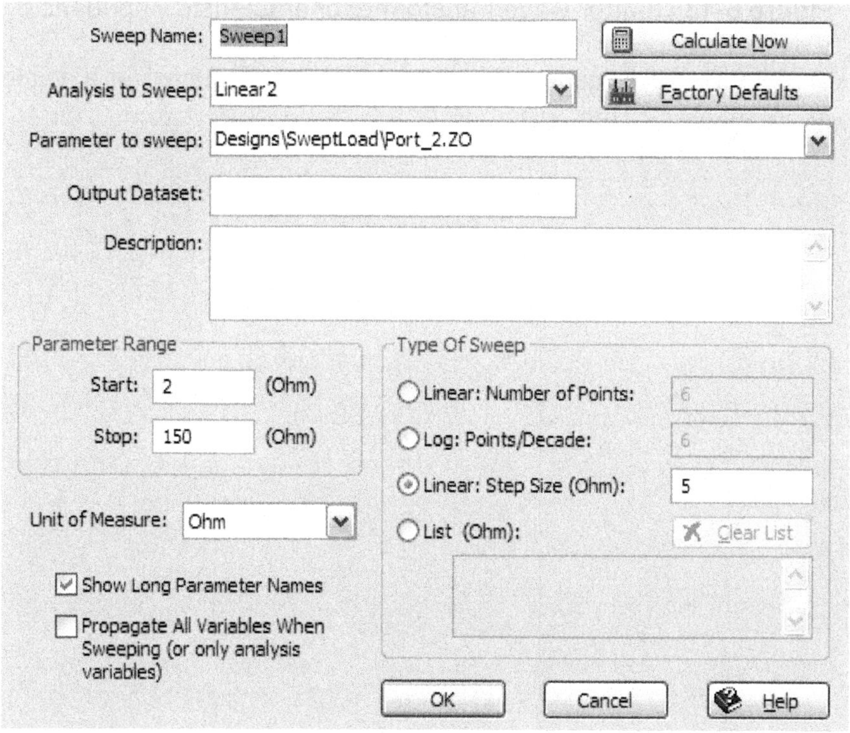

Figure 6-12 Parameter sweep for varying the load impedance

Note the setindep ("Z1", RL") line in the Figure 6-11. This line allows the transmission line impedance Z1 to be plotted as an independent variable versus the dependent variable RL, as shown in Figure 6-13.

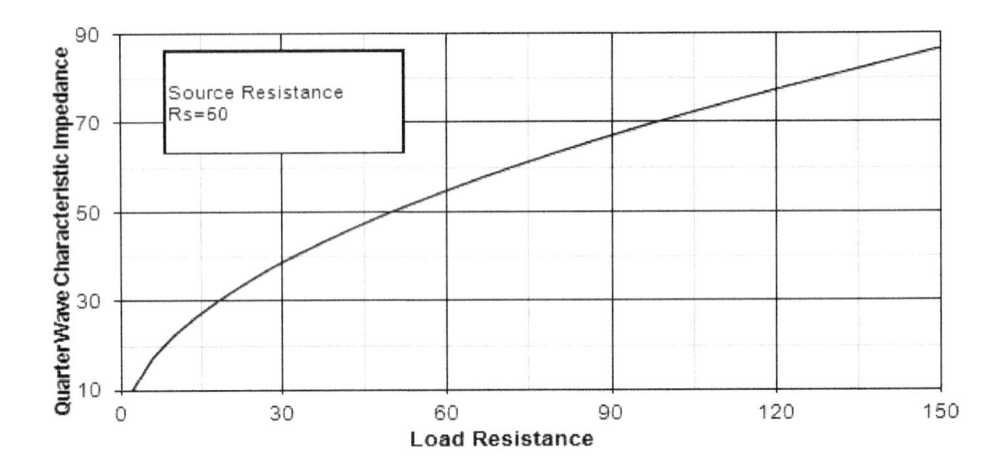

Figure 6-13 Quarter-wave transformer characteristic impedance

The variables in the Equation Editor can also be displayed in a Table for analysis, as shown in Table 6-1.

RL (Ω)	Equation_SweptLoad.FBW20dB...	Equation_SweptLoad.BW20dB...
2	0.053	5.333e+6
7	0.111	11.15e+6
12	0.165	16.54e+6
17	0.227	22.73e+6
22	0.306	30.61e+6
27	0.416	41.62e+6
32	0.59	58.96e+6
37	0.926	92.63e+6
42	1.#QO	1.#QO
47	1.#QO	1.#QO
52	1.#QO	1.#QO
57	1.#QO	1.#QO
62	1.53	153e+6
67	0.96	95.96e+6
72	0.739	73.87e+6
77	0.611	61.14e+6
82	0.527	52.7e+6
87	0.467	46.66e+6
92	0.421	42.09e+6
97	0.385	38.51e+6
102	0.356	35.61e+6
107	0.332	33.22e+6
112	0.312	31.2e+6
117	0.295	29.48e+6
122	0.28	27.99e+6
127	0.267	26.68e+6
132	0.255	25.53e+6
137	0.245	24.5e+6
142	0.236	23.57e+6
147	0.227	22.74e+6
150	0.223	22.28e+6

Table 6-1 Load resistance versus fractional bandwidth

Table 6-1 shows the variation of the fractional bandwidth, for 20 dB return loss, and the corresponding bandwidth as the load resistor varies from R_L=2 Ω to R_L=150 Ω. Note that the bandwidth with R_L=2 Ω and R_L=150 Ω is the same as that which was singly calculated in Section 6.2. The symbol 1.#QO in Table 6-1 indicates that for the load resistances equal to 42, 47, 52, and 57 Ω, Equation (6-10) is not satisfied, therefore, the bandwidth value is not a number.

6.3.2 Quarter-Wave Network Matching Bandwidth and Power Loss in Genesys

The fractional bandwidth and power loss of a quarter-wave matching network can be calculated in an Equation Editor as a function of the input reflection coefficient, Γ_{IN}, and the normalized load resistor, r. The procedure is listed here.

1. Enter equations for input reflection coefficient, power loss, and the conversion of reflection coefficient to return loss in dB.
2. Enter the desired values for the source and load resistors
3. Normalize the load resistor with respect to source resistor
4. Use Equation (6-6) to calculate the fractional bandwidth.

Example 6.3-1: For a 50 Ω source and 2 Ω load resistors, calculate the fractional bandwidth and power loss from $\Gamma = 0.1$ to $\Gamma = 0.707$.

Solution: The fractional bandwidth and power loss is calculated in the following Equation Editor.

```
1    'Enter Reflection Coefficient, Powerloss, and Return Loss Equations
2    vswr=Sweep1_Data.VSWR1
3    ReflCoef=(vswr-1)/(vswr+1)
4    Powerloss=(1-(1-(abs(ReflCoef)^2)))*100
5    RLdB=-20*log(abs(ReflCoef))
6    'Enter the Source and Load Resistor Values
7    RS=50
8    RL=2
9    'Normalize the Load Resistor
10   r=RL/RS
11   'Calculate the Fractional Bandwidth
12   FBW=(2-(4/PI)*acos((2*(ReflCoef)*sqrt(r))/(sqrt((1-(ReflCoef)^2))*abs(1-r))))
```

Figure 6-14 Fractional bandwidth and power loss calculations for $R_L=2\ \Omega$,

Sweep the parameters to display the fractional bandwidth and power loss for reflection coefficients from 0.1 to 0.707, as shown in Table 6-2.

Analysis and Design of Distributed Matching Networks

	ReflCoef	FBW	RLdB	Powerloss
1	0.1	0.053	20	1
2	0.11	0.059	19.172	1.21
3	0.12	0.064	18.417	1.44
4	0.13	0.07	17.721	1.69
5	0.14	0.075	17.077	1.96
6	0.15	0.081	16.478	2.25
7	0.16	0.086	15.917	2.56
8	0.17	0.092	15.391	2.89
9	0.18	0.097	14.895	3.24
10	0.19	0.103	14.425	3.61
11	0.2	0.108	13.979	4
12	0.3	0.167	10.458	9
13	0.4	0.233	7.959	16
14	0.5	0.309	6.021	25
15	0.6	0.405	4.437	36
16	0.707	0.547	3.012	49.985

Table 6-2 Fractional bandwidth and power loss for $R_L = 2\ \Omega$

Table 6-2 shows that the fractional bandwidths at 0.1 reflection coefficient, corresponding to 20 dB return loss, is 5.3 % with 1% power loss while at 0.707 reflection coefficient, corresponding to 3 dB return loss, the fractional bandwidth is 54.7 % with 49.985 % power loss. Notice that the fractional bandwidths, at 3 and 20 dB return loss, are the same as calculated in Figure 6-4.

Example 6.3-2: For a 50 Ω source and 150 Ω load resistors, calculate the fractional bandwidth and power loss from $\Gamma = 0.1$ to $\Gamma = 0.707$.

Solution: Following the procedure, the calculations are shown in Figure 6-15. Sweep the parameters and display the result, as shown in Table 6-3.

```
1    'Enter Reflection Coefficient, Powerloss, and Return Loss Equations
2    vswr=Sweep1_Data.VSWR1
3    ReflCoef=(vswr-1)/(vswr+1)
4    Powerloss=(1-(1-(abs(ReflCoef)^2)))*100
5    RLdB=-20*log(abs(ReflCoef))
6    'Enter the Source and Load Resistor Values
7    RS=50
8    RL=150
9    'Normalize the Load Resistor
10   r=RL/RS
11   'Calculate the Fractional Bandwidth
12   FBW=(2-(4/PI)*acos((2*(ReflCoef)*sqrt(r))/(sqrt((1-(ReflCoef)^2))*abs(1-r))))
```

Figure 6-15 Fractional bandwidth and power loss calculation, $R_L = 150\ \Omega$

	ReflCoef	FBW (rad)	RLdB	Powerloss
1	0.1	0.223	20	1
2	0.11	0.246	19.172	1.21
3	0.12	0.269	18.417	1.44
4	0.13	0.292	17.721	1.69
5	0.14	0.315	17.077	1.96
6	0.15	0.339	16.478	2.25
7	0.16	0.362	15.917	2.56
8	0.17	0.386	15.391	2.89
9	0.18	0.411	14.895	3.24
10	0.19	0.435	14.425	3.61
11	0.2	0.46	13.979	4
12	0.3	0.733	10.458	9
13	0.4	1.091	7.959	16
14	0.5	2	6.021	25
15	0.6	1.#QO	4.437	36
16	0.707	1.#QO	3.012	49.985

Table 6-3 Fractional bandwidth and power loss measurements for $R_s=50\ \Omega$ and $R_L=150\ \Omega$

Table 6-3 shows that the fractional bandwidth at $\Gamma = 0.1$, corresponding to 20 dB return loss, is 22.3 % with a 1% power loss. The fractional bandwidth is the same as calculated in Figure 6-8. The higher fractional bandwidth in this example, compared to its value in Example 6.3-1, is due to the lower ratio of the load to source resistor.

6.4 Single-Stub Matching Networks

The single-stub, also known as line and stub, matching network is a popular narrowband transmission line technique used to match real or complex load impedance to real source impedance. This technique is frequently used in distributed matching circuit designs using microstrip or stripline. The single-stub matching network, shown in Figure 6-16, consists of a series transmission line connected directly to the load impedance and a shunt stub (short-circuited or open-circuited transmission line) attached to the source impedance. The characteristic impedance of both matching elements has the same value as the source impedance, R_s. This section demonstrates both analytical and graphical techniques to design fixed frequency single-stub matching networks.

Analysis and Design of Distributed Matching Networks 293

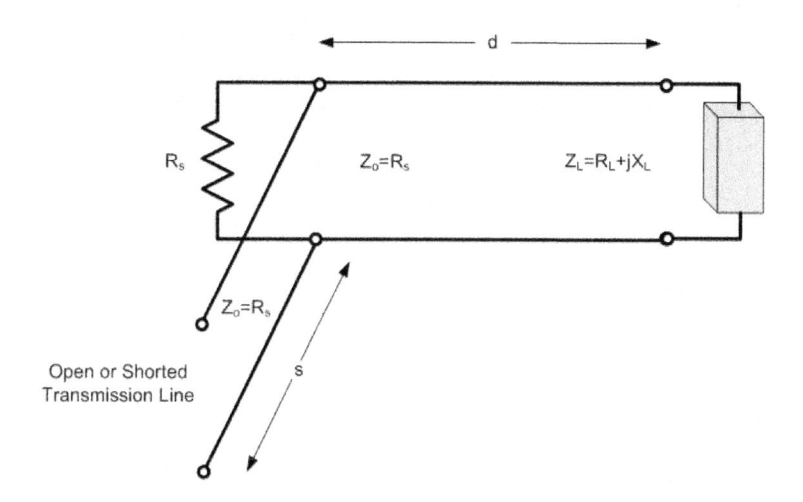

Figure 6-16 Single-stub matching network

To start the design of the single-stub matching network, attach a transmission line section, with characteristic impedance equal to R$_S$, to the load and determine its electrical length in such a way that the normalized admittance at a distance d from the load falls on the unit conductance circle of the Smith Chart. Calculation of the electrical lengths for the series line and shunt stub are given separately.

6.4.1 Analytical Design of Series Transmission Line

The input impedance of a lossless transmission line of length d and characteristic impedance Z_0 terminated in an arbitrary load Z_L, was given in chapter 2 in Equation (2-46) and repeated here for convenience.

$$Z_{IN} = Z_o \frac{Z_L + jZ_o \tan \beta d}{Z_o + jZ_L \tan \beta d}$$

Setting $Z_0 = R_S$ and $\tan \beta d = t$, the input admittance of the network can be written as:

$$Y_{IN} = \frac{1}{Z_{IN}} = \frac{R_S + jZ_L t}{R_S(Z_L + jR_S t)} = G_{IN} + jB_{IN} \qquad (6\text{-}13)$$

Substituting the normalized load impedance, $Z_L/R_s = r + jx$ into Equation (6-13) and separating its real and imaginary parts we get:

$$G_{IN} = \frac{r(1+t^2)}{R_S(r^2 + x^2 + t^2 + 2xt)} \quad (6\text{-}14)$$

and,

$$B_{IN} = \frac{xt^2 + (r^2 + x^2 - 1)t + x}{R_S(r^2 + x^2 + t^2 + 2xt)} \quad (6\text{-}15)$$

The value of d, which implies t, can be obtained by setting the input conductance, G_{IN}, equal to source conductance:

$$\frac{r(1+t^2)}{R_S(r^2 + x^2 + t^2 + 2xt)} = \frac{1}{R_S} \quad (6\text{-}16)$$

Equation (6-16) can be rearranged as:

$$(r-1) \cdot t^2 - 2xt - (r^2 + x^2 - r) = 0 \quad (6\text{-}17)$$

Notice that the quadratic Equation (6-17) has two solutions for t. For a wider bandwidth and lower loss usually the smaller value of t is selected. The two solutions for t are:

$$t_1 = \frac{x + \sqrt{r(r^2 + x^2 - 2r + 1)}}{r - 1} \quad (6\text{-}18)$$

and,

$$t_2 = \frac{x - \sqrt{r(r^2 + x^2 - 2r + 1)}}{r - 1} \quad (6\text{-}19)$$

With $\tan \beta d = t$, and $\beta \lambda = 2\pi$, we have $d = \frac{\lambda}{2\pi} \tan^{-1} t$ and the two solutions for d are:

$$d_1 = \frac{\lambda}{2\pi} \tan^{-1} t_1 \qquad t_1 \geq 0 \qquad (6\text{-}20)$$

and

$$d_2 = \frac{\lambda}{2\pi} \tan^{-1} t_2 \qquad t_2 \geq 0 \qquad (6\text{-}21)$$

To specify the lengths of d_1 and d_2 in electrical degrees, we get:

$$d_1 = \frac{360}{2\pi} \tan^{-1} t_1 \qquad t_1 \geq 0 \qquad (6\text{-}22)$$

and

$$d_2 = \frac{360}{2\pi} \tan^{-1} t_2 \qquad t_2 \geq 0 \qquad (6\text{-}23)$$

Because at every half wavelength the input impedance of a transmission line repeats, there are an infinite number of transmission line lengths that matches the load to source impedance. Usually the shorter length is selected to improve the matching bandwidth. If t_1 or t_2 is negative, we add half a wavelength to each line to get positive d_1 and d_2.

$$d_1 = \frac{360(\pi + \tan^{-1} t_1)}{2\pi} \qquad t_1 < 0 \qquad (6\text{-}24)$$

and

$$d_2 = \frac{360(\pi + \tan^{-1} t_2)}{2\pi} \qquad t_2 < 0 \qquad (6\text{-}25)$$

6.4.2 Analytical Design of Shunt Transmission Line

To calculate the electrical length of the shunt stub, first substitute t_1 and t_2 in Equation (6-15) to determine B_1 and B_2.

$$B_1 = \frac{xt_1^2 + (r^2 + x^2 - 1)t_1 + x}{R_S(r^2 + x^2 + t_1^2 + 2xt_1)} \qquad (6\text{-}26)$$

$$B_2 = \frac{xt_2^2 + (r^2 + x^2 - 1)t_2 + x}{R_S(r^2 + x^2 + t_2^2 + 2xt_2)} \qquad (6\text{-}27)$$

Then, the electrical lengths of the open circuited stubs are found by setting the susceptance of the stubs equal to the negative of the input susceptance.

$$so_1 = \frac{-\lambda\left(tan^{-1}(R_S B_1)\right)}{2\pi} \qquad (6\text{-}28)$$

$$so_2 = \frac{-\lambda\left(tan^{-1}(R_S B_2)\right)}{2\pi} \qquad (6\text{-}29)$$

If either stub length in Equation (6-28) or (6-29) is negative, add one half wavelength to obtain a positive stub length. For short-circuited stubs, the two solutions are:

$$ss_1 = \frac{\lambda\left(tan^{-1}\left(\frac{1}{R_S B_1}\right)\right)}{2\pi} \qquad (6\text{-}30)$$

$$ss_2 = \frac{\lambda\left(tan^{-1}\left(\frac{1}{R_S B_2}\right)\right)}{2\pi} \qquad (6\text{-}31)$$

6.4.3 Single-Stub Matching Design Example

Example 6.4-1: Design a single-stub network to match a load resistance Z_L = 2 - j5 Ω to a resistive source R_S = 50 Ω at 100 MHz. Calculate the fractional bandwidths of the matching network at 3 and 20 dB return loss. Display the simulated response and compare the calculations with measurements.

Analysis and Design of Distributed Matching Networks

Solution: The design example is shown in Figure 6-17.

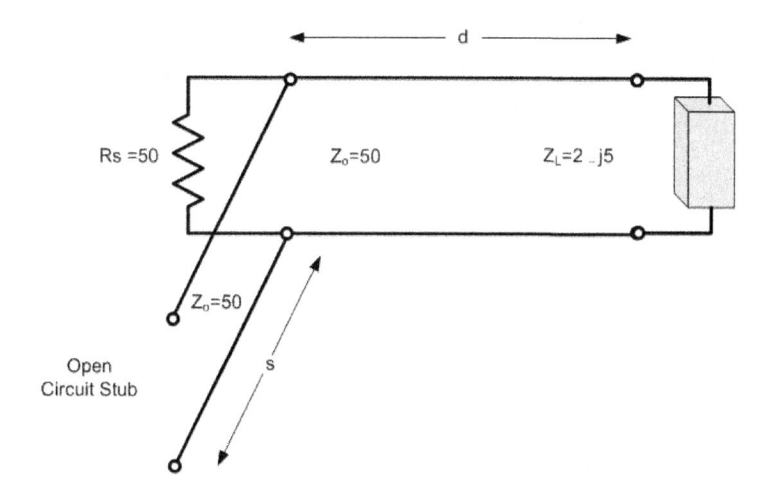

Figure 6-17 Complex load to resistive source matching network

Equation Editor of Figure 6-18 calculates the electrical length of the line and stub matching network.

```
1    %Enter frequency, souce and load impedances
2    RS=50
3    RL=2
4    XL=-5
5    f=100e6
6    %Normalize the load impedance
7    r=RL/RS
8    x=XL/RS
9    %Calculte t1 and t2
10   t1=(x+sqrt(r*(r^2+x^2-2*r+1)))/(r-1)
11   t2=(x-sqrt(r*(r^2+x^2-2*r+1)))/(r-1)
12   %Calculate d1 and d2
13   d1=360*(atan(t1))/(2*pi)
14   d2=360*(atan(t2))/(2*pi)
15   %Calculate B1 and B2
16   B1=(x*t1^2+(r^2+x^2-1)*t1-x)/(RS*(r^2+x^2+t1^2+2*x*t1))
17   B2=(x*t2^2+(r^2+x^2-1)*t2-x)/(RS*(r^2+x^2+t2^2+2*x*t2))
18   %Calculate so1 and so2
19   so1=360*(pi-atan(RS*B1))/(2*pi)
20   so2=-360*atan(RS*B2)/(2*pi)
```

Figure 6-18 Calculation of line and stub lengths

The calculations in Figure 6-18 show that the problem has two solutions. Either solution can be used in the design of the single-stub matching network. For both line and stub usually the shorter line lengths are chosen.

For this example we select the shorter electrical lengths in the second solution; d2 for the line and so2 for the open-circuited stub. The Genesys schematic of the single-stub matching network is shown in Figure 6-19. The simulated response of Figure 6-21 shows that the 3-dB bandwidth is about 11 MHz while the 20-dB bandwidth is about 1 MHz.

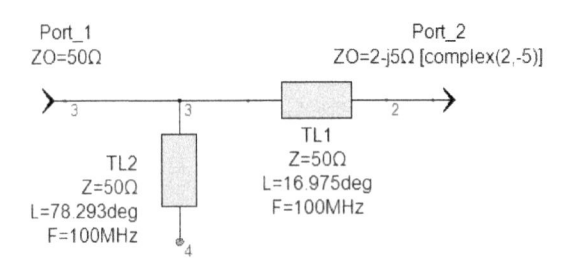

Figure 6-19 Schematic of the single-stub matching network

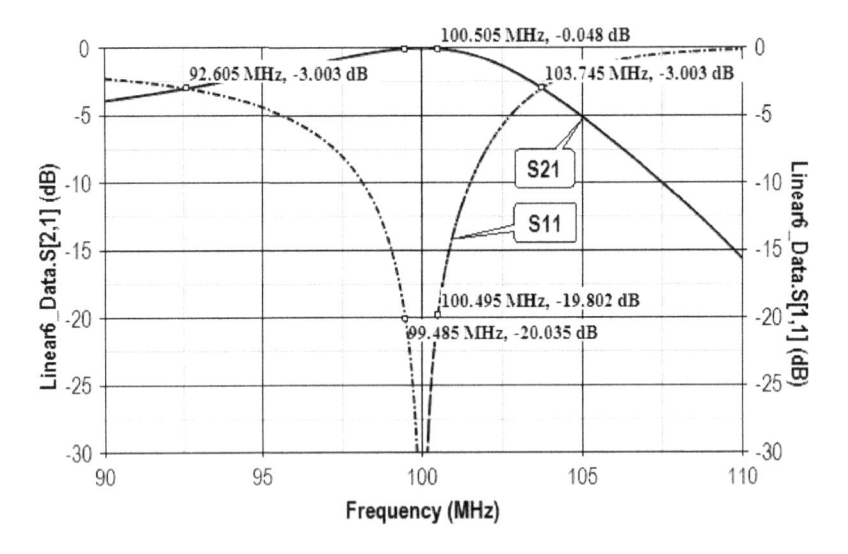

Figure 6-20 Simulated response of the single-stub matching network

6.4.4 Automated Calculation of Line and Stub Lengths

The conditional IF-THEN statements can be used in the Equation Editor to choose the proper solution. For the series transmission line, the shorter line length is chosen. For the shunt transmission line, the non-negative solution

Analysis and Design of Distributed Matching Networks 299

is chosen. The IF-THEN statement is used to redefine the line and stub calculations as shown in the Equation Editor of Figure 6-21.

```
1    'Enter the source and load impedances
2    RS=50
3    RL=2
4    XL=-5
5    r=RL/RS
6    x=XL/RS
7    'Calculate t1 and t2 and electrical lengths
8    t1=(x+sqrt(r*(r^2+x^2-(2*r)+1)))/(r-1)
9    t2=(x-sqrt(r*(r^2+x^2-(2*r)+1)))/(r-1)
10   If t1>0 then
11   d1=(360/(2*PI))*atan(t1)
12   else
13   d1=360*(PI+atan(t1))/(2*PI)
14   endif
15   If t2>0 then
16   d2=(360/(2*PI))*atan(t2)
17   else
18   d2=360*(PI+atan(t2))/(2*PI)
19   endif
20   If d1<d2 then
21   SeriesLineLength=d1
22   else
23   SeriesLineLength=d2
24   endif
25   Calculate Shunt Line Parameters B1 and B2
26   B1=(x*t1^2+(r^2+x^2-1)*t1-x)/(RS*(r^2+x^2+t1^2+2*x*t1))
27   B2=(x*t2^2+(r^2+x^2-1)*t2-x)/(RS*(r^2+x^2+t2^2+2*x*t2))
28   so1=-360*(atan(RS*B1))
29   so2=-360*atan(RS*B2)/(2*PI)
30   If so1<0 then
31   OpenShuntStub=so2
32   else
33   OpenShuntStub=so1
34   endif
35   'short circuit stubs
36   ss1=(360*atan(1/(RS*B1)))/(2*PI)
37   ss2=(360*atan(1/(RS*B2)))/(2*PI)
38   If ss1<0 then
39   ShortedShuntStub=ss2
40   else
41   ShortedShuntStub=ss1
42   endif
```

Figure 6-21 Equation Editor calculating line and stub line lengths

6.4.5 Development of Single-Stub Matching Utility

Because the single-stub matching network design is frequently encountered in microwave circuit design it is helpful to develop a utility program in VBScript to handle such repetitive calculations. An example of such a "Line & Stub Impedance Matching Utility" in Genesys is shown in Figures 6-22 and 6-23. In this program a slider control is used to select between an open circuit stub solution and a short circuit stub solution to the impedance matching problem. A Genesys output port is used to terminate the shunt transmission line. The symbol for the port is changed in VBScript to take on the symbol of a Ground or a TRL_END depending on whether the short circuit or open circuit solution is selected. For a short circuit termination the impedance of the port is set to a very small, non-zero, value such as 0.0001 Ω. For an open circuit termination the port impedance is made an extremely high value such as 100 MΩ. The VBScript code can be copied into the 'Calculate Match' button so that the code is executed when the button is pressed. Therefore the operation of the utility begins with setting the design frequency into the CW source. Then choose between solution 1 or solution 2 and press the 'Calculate Match' button. Make sure that the Linear Analysis frequency range covers the design frequency and the network will automatically be swept and the response plotted on the graph. This provides a quick and convenient means of comparing the frequency response of the two solutions. The transmission lines can then be easily copied into another schematic where the Advanced TLine program can be used to convert the ideal transmission lines into physical transmission lines. The complete program listing for the Line & Stub matching tool is listed in Appendix D.

Example 6.4-2: Design a single-stub network to match a complex impedance of: $Z_L = 150 - j40$ Ω to a resistive source $R_S = 50$ Ω at 100 MHz.

Solution: The single-stub matching utility is used here to calculate and select the line and short-circuited stub matching components, as shown.

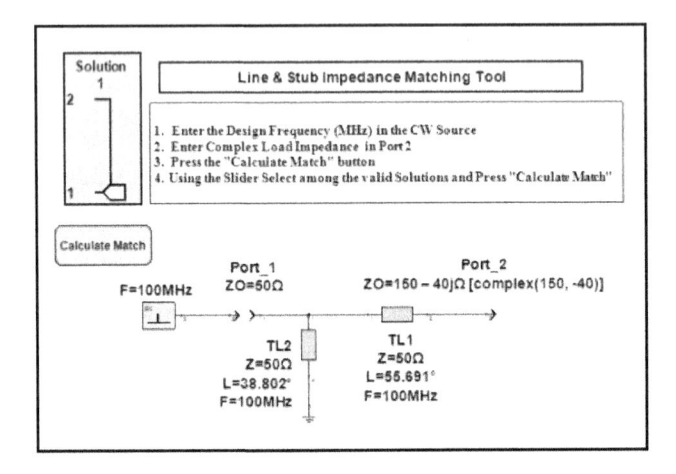

Figure 6-22 Short-circuited line and stub matching utility

In Figure 6-23, the single-stub matching utility is used to calculate the line and open-circuited stub matching components.

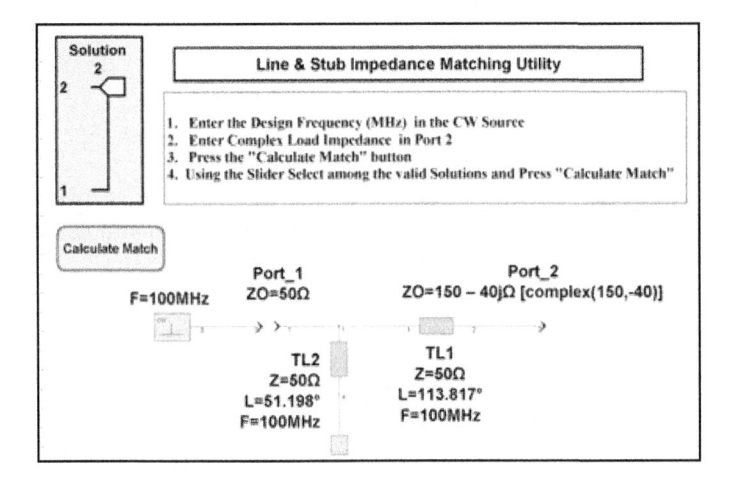

Figure 6-23 Open-circuited line and stub matching utility

6.5 Graphical Design of Single-Stub Matching Networks

Building on the techniques covered in chapter 3.7 this section demonstrates the ease in which the single-stub matching network can be graphically designed using the Smith Chart. The Smith Chart combined with tunable elements in a Genesys schematic is a very powerful tool in which matching networks can be approximately designed without the need to solve the exact Equations of section 6.4.1.

6.5.1 Smith Chart Design Using an Open Circuit Stub

In this section a 50 Ω source will be matched to a 5 - 25j Ω load at a frequency of 1000 MHz. Both open-circuited and short-circuited shunt stubs will be synthesized. Consider the 5 - j25 Ω impedance shown at position A in Figure 6-24. Add a 50 Ω series transmission line to move this impedance to intersect the unit conductance circle on the top half of the Smith Chart to point B. Because we intend to place a shunt element after the series transmission line, enable the admittance circles on the Smith Chart.

As the schematic inset shows, an electrical length of 42.5° places the impedance on the unit conductance circle at B.

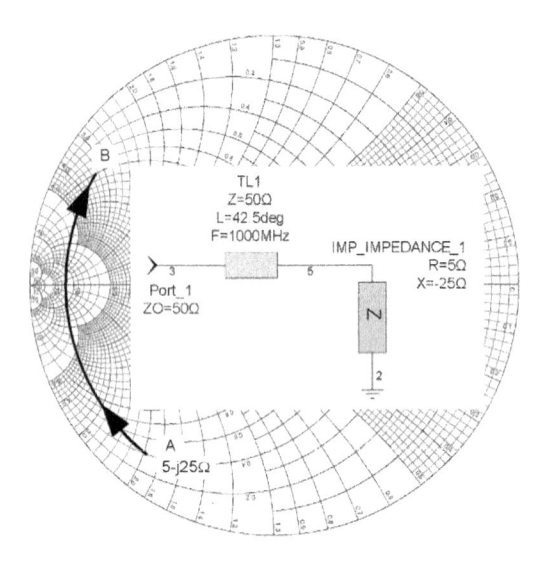

Figure 6-24 Adding series transmission line (electrical length=42.5°)

Then add an open-circuited shunt transmission line and tune the length until the impedance moves to the center of the chart (50Ω). As the schematic of Figure 6-25 shows, a 73° length of transmission line would be required.

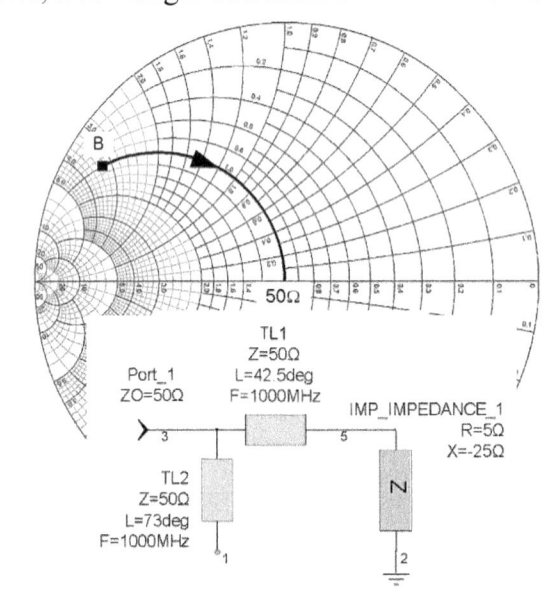

Figure 6-25 Adding open-circuited shunt transmission line (73°) to network

6.5.2 Smith Chart Design Using a Short Circuit Stub

A short-circuited shunt transmission line can be used to perform the function of the shunt stub. Using a short-circuited shunt transmission line, the series transmission line should intersect the unit conductance circle on the bottom half of the Smith Chart. Add a series transmission line of 11° electrical length to move the impedance at point A to intersect the unit conductance circle at point B as shown in Figure 6-26. Then add the short-circuited shunt transmission line as shown in Figure 6-27. Tune the length to 17° to move the impedance to the center of the Smith Chart.

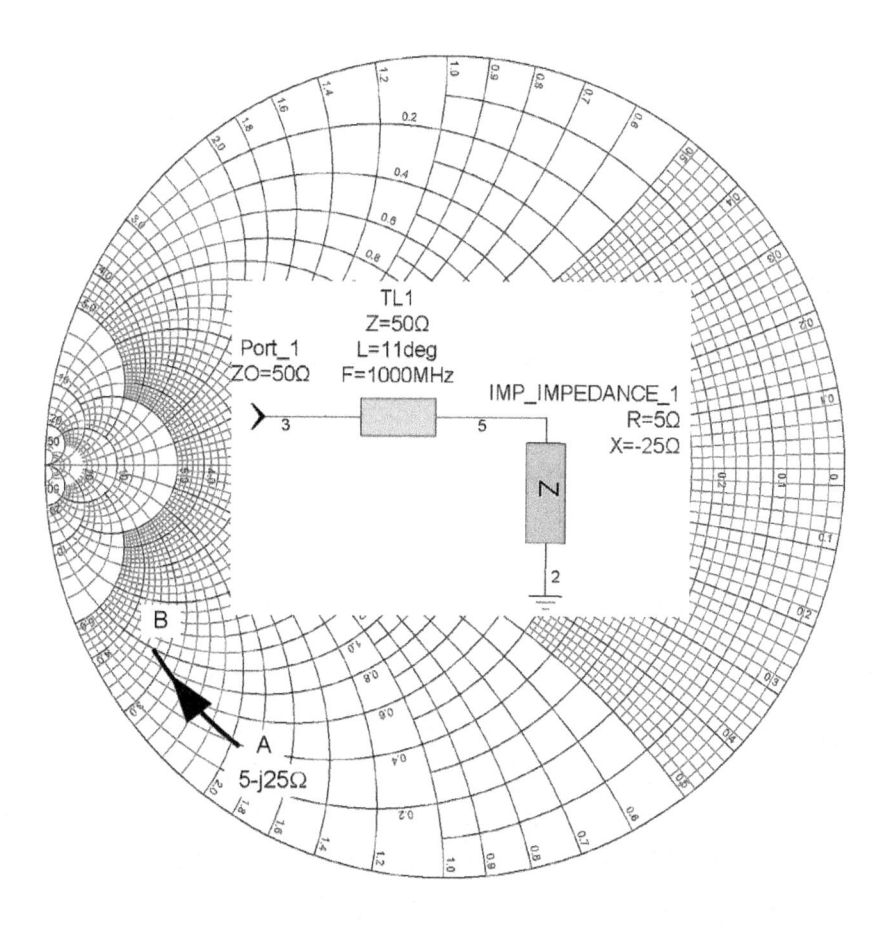

Figure 6-26 Adding series transmission line (11°) to 5-j25 Ω

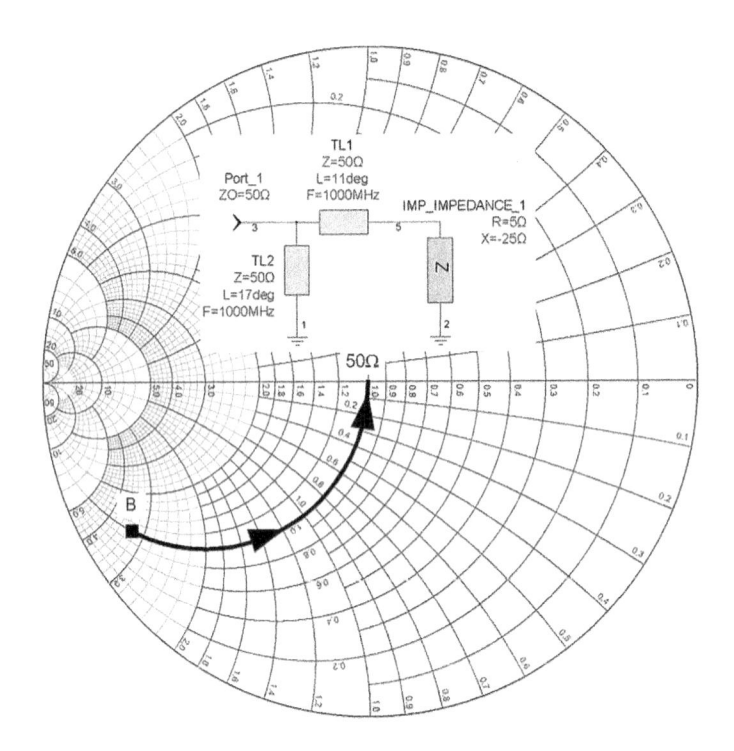

Figure 6-27 Adding short-circuited shunt transmission line (17°)

6.6 Design of Cascaded Single-Stub Matching Networks

When the impedance of the load and the source are both complex, we can define a virtual resistor and design single-stub networks to match the complex impedances to the virtual resistor. The final matching network is obtained by cascading the two single-stub matching networks. The procedure is demonstrated in the following example.

Example 6.6-1: Design single-stub networks to match a complex load $Z_L = 10 - j5$ Ω to a complex source $Z_S = 50 - j15$ Ω at 100 MHz. Calculate the electrical lengths of the lines and the fractional bandwidths of the matching network at 3 dB and 20 dB return loss. Display the simulated response and verify the calculations with simulation in Genesys.

Solution: The following procedure is used to match the two complex impedances

1. Find the intermediate resistor, $R_1 = \sqrt{R_L R_S}$
2. Design a single-stub matching network between $R1$ and the source impedance
3. Design a second matching network between $R1$ and the load impedance
4. Cascade the two single-stub networks

First Solution: For this example we match the source to R_1, where: $R_1 = \sqrt{(10)(50)} = 22.36\ \Omega$. There are two matching networks in each case.

```
1   %Enter Design Parameters
2   RS=22.36
3   RL=50
4   XL=-15
5   f=100e6
6   %Normalize the load impedance
7   r=RL/RS
8   x=XL/RS
9   %Calculte t1 and t2
10  t1=(x+sqrt(r*(r^2+x^2-2*r+1)))/(r-1)
11  t2=(x-sqrt(r*(r^2+x^2-2*r+1)))/(r-1)
12  %Calculate d1 and d2
13  d1=360*(atan(t1))/(2*pi)
14  d2=360*(pi+atan(t2))/(2*pi)
15  %Calculate B1 and B2
16  B1=(x*t1^2+(r^2+x^2-1)*t1-x)/(RS*(r^2+x^2+t1^2+2*x*t1))
17  B2=(x*t2^2+(r^2+x^2-1)*t2-x)/(RS*(r^2+x^2+t2^2+2*x*t2))
18  %Calculate the electrical lenghts of the open-ended stubs
19  so1=360*(pi-atan(RS*B1))/(2*pi)
20  so2=-360*atan(RS*B2)/(2*pi)
```

Figure 6-28 Calculation of the first line and stub matching network

First Solution: For the first solution we select, from the Workspace Variables, d2=114.019 and so2=43.245 as shown in Figure 6-29.

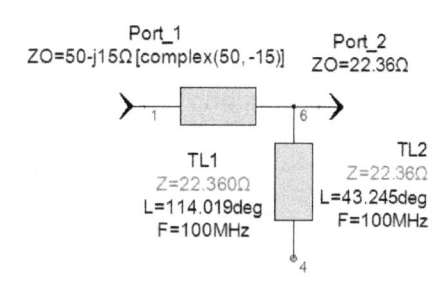

Port_1
ZO=50-j15Ω [complex(50,-15)]

Port_2
ZO=22.36Ω

TL1
Z=22.360Ω
L=114.019deg
F=100MHz

TL2
Z=22.36Ω
L=43.245deg
F=100MHz

Figure 6-29 Schematic of the first single-stub matching network

Second Solution: For the second solution we select, from the Workspace Variables, d2=48.39 and so2=41.722 degrees as shown in Figure 6-31.

```
1    %Enter Design Parameters
2    RS=22.36
3    RL=10
4    XL=-5
5    f=100e6
6    %Normalize the load impedance
7    r=RL/RS
8    x=XL/RS
9    %Calculte t1 and t2
10   t1=(x+sqrt(r*(r^2+x^2-2*r+1)))/(r-1)
11   t2=(x-sqrt(r*(r^2+x^2-2*r+1)))/(r-1)
12   %Calculate d1 and d2
13   d1=360*(pi+atan(t1))/(2*pi)
14   d2=360*(atan(t2))/(2*pi)
15   %Calculate B1 and B2
16   B1=(x*t1^2+(r^2+x^2-1)*t1-x)/(RS*(r^2+x^2+t1^2+2*x*t1))
17   B2=(x*t2^2+(r^2+x^2-1)*t2-x)/(RS*(r^2+x^2+t2^2+2*x*t2))
18   %Calculate the electrical lenghts of the open-ended stubs
19   so1=360*(pi-atan(RS*B1))/(2*pi)
20   so2=-360*atan(RS*B2)/(2*pi)
```

Figure 6-30 Calculation of the second single-stub matching network

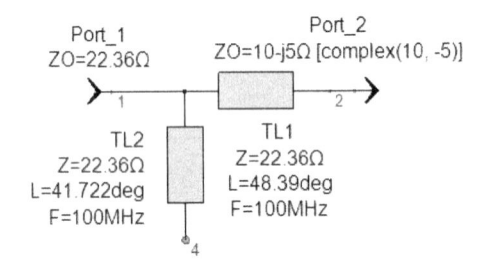

Figure 6-31 Schematic of the second single-stub matching network

Now connect both line and stub matching networks in cascade to obtain the complete matching network, as shown in Figure 6-32.

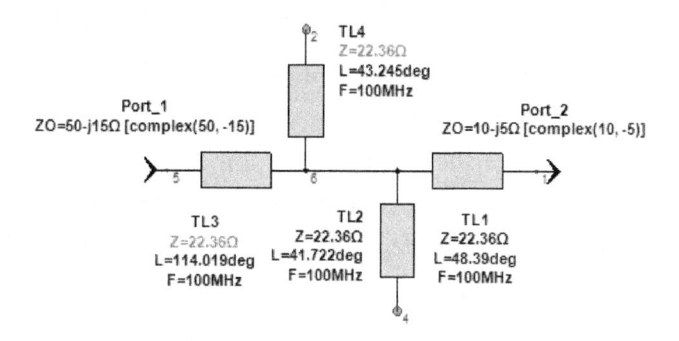

Analysis and Design of Distributed Matching Networks 307

Figure 6-32 Cascading single-stub matching networks

The simulated response of the complete network is shown in Figure 6-33. Notice that the cascaded single-stub matching network has about 100 % fractional bandwidth at the 3 dB return loss and 47% fractional bandwidth at the 12 dB return loss.

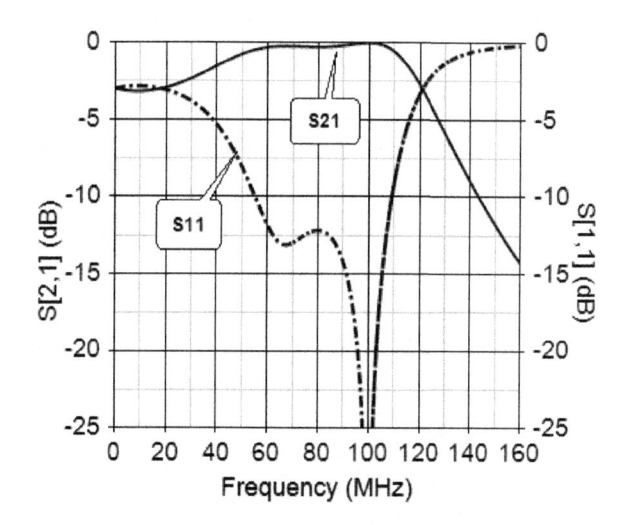

Figure 6-33 Response of the cascaded matching network

6.7 Broadband Quarter-Wave Matching Network Design

In Examples 6.2-1 and 6.2-2, the bandwidth of a single quarter-wave transformer matching network with a large reflection coefficient is less than 10 % which is considered to be a narrowband matching network. We can increase the bandwidth by cascading two or more quarter-wave transformers to achieve a broadband matching network. To analytically design a broadband matching network with N quarter-wave transformers first use Equation (6-32) in an Equation Editor to calculate the characteristic impedance of each quarter-wave transformer then cascade all the sections into one matching network. Let Rs and R_L be the source and load impedances to be matched by the N quarter-wave transformation network. The characteristic impedance of each section can be calculated from Equation (6-32).

$$Z_n = R_S(r)^{(2n-1)/2N} \qquad n = 1, 2, \ldots, N \qquad (6\text{-}32)$$

Where $r = R_L/R_S$ is the normalized load resistance and N is the number of quarter-wave transformers.

Example, 6.7-1 Design a three-section quarter-wave transformer network to match a load resistance $R_L = 2\ \Omega$ to a resistive source $R_S = 50\ \Omega$ at 100 MHz.

(a) Calculate the characteristic impedance of the quarter-wave transformers, the Q factor and the fractional bandwidths of the matching network at 3 and 20 dB return loss.

(b) Display the simulated response and compare the measurements with the measurements of single quarter-wave transformer matching network.

Solution: (a) Using Equation (6-32) the characteristic impedances of the 3-section quarter-wave transformers are:

$$Z1 = R_S(r)^{1/2N} = 50(0.04)^{1/6} = 29.24$$

$$Z2 = R_S(r)^{3/2N} = 50(0.04)^{3/6} = 10.00$$

$$Z3 = R_S(r)^{5/2N} = 50(0.04)^{5/6} = 3.42$$

The intermediate resistors, R_n, can be obtained from Equation (6-33).

Analysis and Design of Distributed Matching Networks 309

$$R_n = R_S (r)^{\frac{n}{N}} \quad n = 1, 2, 3, \ldots, N-1 \qquad (6\text{-}33)$$

Therefore, the two intermediate resistors are:

$$R_1 = 50 (0.04)^{\frac{1}{3}} = 17.1 \ \Omega$$

$$R_2 = 50 (0.04)^{\frac{2}{3}} = 5.848 \ \Omega$$

These calculations are solved in the Equation Editor of Figure 6-34. The Genesys schematic of the three-section quarter-wave matching network is shown in Figure 6-35.

```
1   %Enter Given Design Parameters
2   RS=50
3   RL=2
4   f=100e6
5   %Normalize Load Impedance
6   r=RL/RS
7   %Select Number of Quarter-Wave Lines
8   N=3
9   %Calculate Characteristic Impedances
10  Z1=RS*r^(1/(2*N))
11  Z2=RS*r^(3/(2*N))
12  Z3=RS*r^(5/(2*N))
13  %Calculate Intermediate Impedances
14  R1=RS*r^(1/N)
15  R2=RS*r^(2/N)
```

Figure 6-34 Calculation of characteristic impedances and intermediate resistors for Example 6.7-1

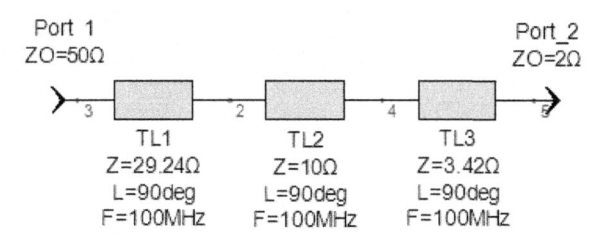

Port 1
ZO=50Ω

Port_2
ZO=2Ω

TL1
Z=29.24Ω
L=90deg
F=100MHz

TL2
Z=10Ω
L=90deg
F=100MHz

TL3
Z=3.42Ω
L=90deg
F=100MHz

Figure 6-35 Three-section quarter-wave matching network

The simulated response of the matching network is shown in Figure 6-36.

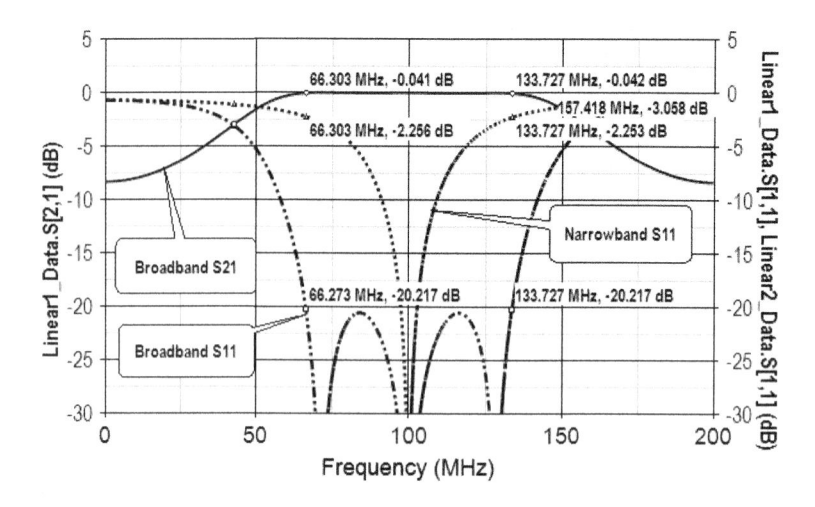

Figure 6-36 Simulated response of the cascaded matching network

Figure 6-36 shows that the bandwidth simulated at the 3 dB return loss is:

$$BW_{3dB} = 157.6 - 42.4 = 115.2 \text{ MHz}$$

$$FBW_{3dB} = \frac{115.2 \times 100}{\sqrt{(157.6).(42.4)}} = 140.8\,\%$$

Similarly the simulated bandwidth at the 20 dB return loss is:

$$BW_{20dB} = 133.8 - 66.2 = 67.6 \quad \text{MHz}$$

$$FBW_{20dB} = \frac{67.6 \times 100}{\sqrt{(133.8).(66.2)}} = 71.8\,\%$$

The measured Q factor for the matching network is,

$$Q = \frac{1}{1.408} = 0.71$$

The wide bandwidth and lower Q factor is an indication that by adding two more quarter-wave sections the Q factor has reduced to less than one half and the 3-dB bandwidth has more than doubled compared to a single quarter-wave matching network.

Example 6.7-2: Design a broadband quarter-wave network to match a complex load, $Z_L = 10 - j5\ \Omega$ to $50\ \Omega$ source impedance at 100 MHz. Calculate the bandwidth at 20 dB return loss and compare with the bandwidth of a single-stub matching network.

Solution: To design a broadband matching network between a resistive source and complex load impedance; first design a 5 section matching network between the real source and the real part of the load impedance. Then replace the quarter-wave transformer adjacent to the load with a single-stub that matches the complex load to the real resistor. Calculation of the broadband matching network between R_L and R_S is shown in the Equation Editor of Figure 6-37. The Genesys schematic of the five-section matching network is shown in Figure 6-38.

```
1    %Enter Design Parameters
2    RS=50
3    RL=10
4    f=100e6
5    r=RL/RS
6    N=5
7    %Characteristic Impededance
8    Z1=RS*r^(1/(2*N))
9    Z2=RS*r^(3/(2*N))
10   Z3=RS*r^(5/(2*N))
11   Z4=RS*r^(7/(2*N))
12   Z5=RS*r^(9/(2*N))
13   %Intermediate Resistors
14   R1=RS*r^(1/N)
15   R2=RS*r^(2/N)
16   R3=RS*r^(3/N)
17   R4=RS*r^(4/N)
```

Figure 6-37 Calculating 4 resistors and 5 characteristic impedances

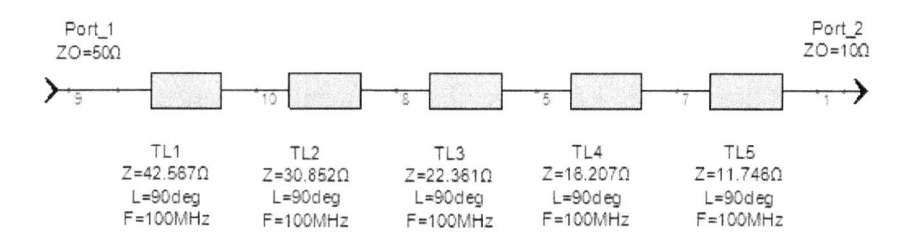

```
Port_1                                                                Port_2
ZO=50Ω                                                                ZO=10Ω
   ─┤ ├──□──┤ ├──□──┤ ├──□──┤ ├──□──┤ ├──□──┤ ├─
     9      10      8       5       7       1
        TL1       TL2      TL3      TL4      TL5
     Z=42.567Ω  Z=30.852Ω Z=22.361Ω Z=16.207Ω Z=11.746Ω
     L=90deg   L=90deg   L=90deg   L=90deg   L=90deg
     F=100MHz  F=100MHz  F=100MHz  F=100MHz  F=100MHz
```

Figure 6-38 Five-section quarter-wave transformer matching network

Next replace the quarter-wave section adjacent to the load with a single-stub matching network. The design of the single-stub matching network is shown in the following Equation Editor. Note that the source resistor, R4 = 13.797 Ω, was calculated in the Equation Editor of Figure 6-39. The single-stub matching network is shown in Figure 6-40.

```
1   %Enter Design Parameters
2   RS=13.797
3   RL=10
4   XL=-5
5   f=100e6
6   %Normalize the load impedance
7   r=RL/RS
8   x=XL/RS
9   %Calculte t1 and t2
10  t1=(x+sqrt(r*(r^2+x^2-2*r+1)))/(r-1)
11  t2=(x-sqrt(r*(r^2+x^2-2*r+1)))/(r-1)
12  %Calculate Electrical Lengths of the Lines, d1 and d2
13  d1=360*(pi+atan(t1))/(2*pi)
14  d2=360*(atan(t2))/(2*pi)
15  %Calculate B1 and B2
16  B1=(x*t1^2+(r^2+x^2-1)*t1-x)/(RS*(r^2+x^2+t1^2+2*x*t1))
17  B2=(x*t2^2+(r^2+x^2-1)*t2-x)/(RS*(r^2+x^2+t2^2+2*x*t2))
18  %Calculate Electrical Lengths of Open Stubs, so1 and so2
19  so1=360*(pi-atan(RS*B1))/(2*pi)
20  so2=-360*atan(RS*B2)/(2*pi)
```

Figure 6-39 Second line and stub calculations

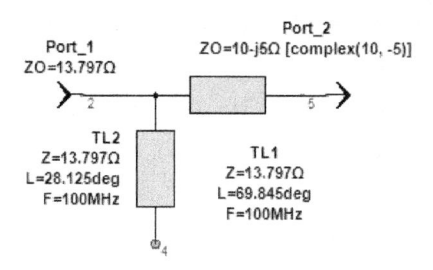

Analysis and Design of Distributed Matching Networks 313

Figure 6-40 Schematic of the single-stub matching network

Now cascade the line and stub with four quarter-wave networks to form the final design of the matching network in Figure 6-41. The simulated response of the matching network is shown in Figure 6-42.

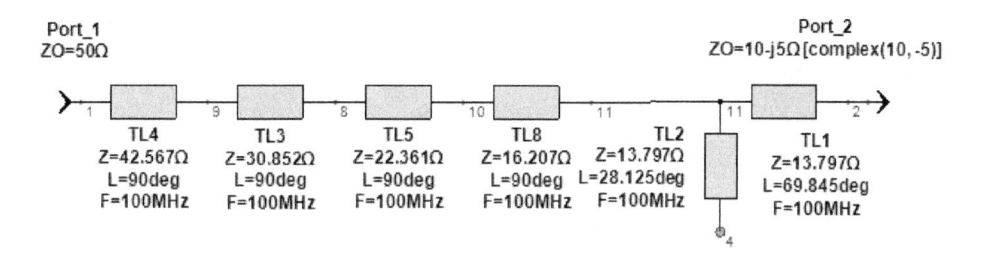

Figure 6-41 Schematic of the broadband matching network

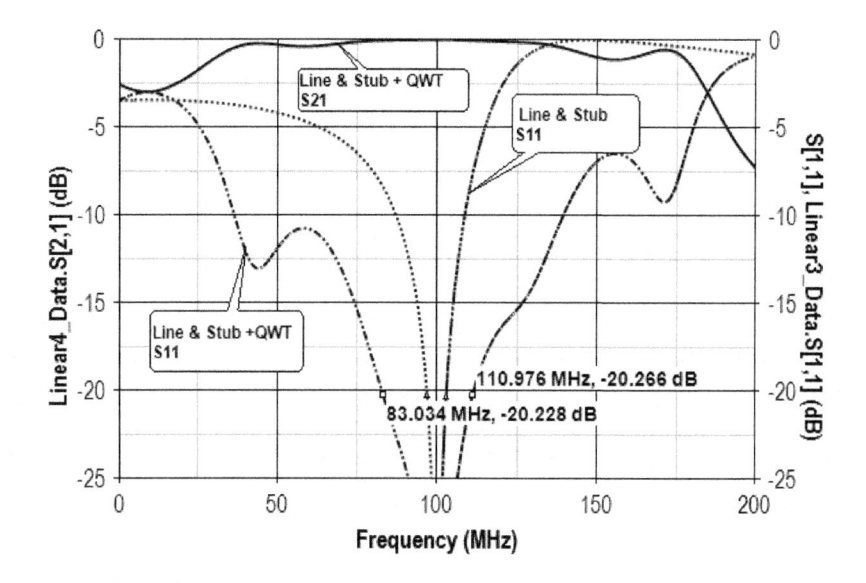

Figure 6-42 Simulated response of the broadband matching network

Figure 6-48 shows that the bandwidth at 20 dB return loss is:

$$BW_{20dB} = 110 - 83 = 27 \text{ MHz}$$

And the fractional bandwidth is:

$$FBW_{20dB} = \frac{27}{\sqrt{(110).(83)}} = 28.2\%$$

The same numbers for the single-stub matching network are:

$$BW_{20dB} = 102.8 - 96.8 = 6 \quad \text{MHz}$$

$$FBW_{20dB} = \frac{6 \times 100}{\sqrt{(102.8).(96.8)}} = 6\%$$

Notice that fractional bandwidth for the broadband at 20 dB return loss is 29.7 % as opposed to only 6 % for the narrowband single-stub matching network. Therefore, we have increased the matching bandwidth at 20 dB return loss by nearly five times over the single-stub matching network.

Example 6.7-3: Design a broadband network to match a complex load, $Z_L = 150 - j30$ to 50 Ω source impedance at 100 MHz. Display the simulated response and measure the bandwidth at 20 dB return loss. Compare the results with the singe-stub matching network.

Solution: Use the same method as in Example 6.6.1 to design the broadband matching network. The design of the broadband quarter-wave transformer matching network with five quarter-wave sections is shown in Figure 6-43.

Analysis and Design of Distributed Matching Networks 315

```
1    %Enter Design Parameters
2    RS=50
3    RL=150
4    f=100e6
5    r=RL/RS
6    N=5
7    %Calculate Characteristic Impededance
8    Z1=RS*r^(1/(2*N))
9    Z2=RS*r^(3/(2*N))
10   Z3=RS*r^(5/(2*N))
11   Z4=RS*r^(7/(2*N))
12   Z5=RS*r^(9/(2*N))
13   %Calculate Intermediate Resistors
14   R1=RS*r^(1/N)
15   R2=RS*r^(2/N)
16   R3=RS*r^(3/N)
17   R4=RS*r^(4/N)
```

Figure 6-43 Calculating 5 characteristic impedances and 4 intermediate resistors

The Genesys schematic of the five-section quarter-wave matching network is shown in Figure 6-44. The simulated response of the quarter-wave matching network is shown in Figure 6-45.

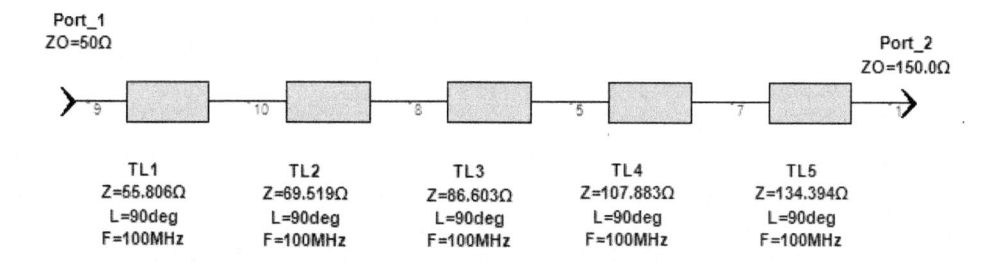

Port_1
ZO=50Ω

Port_2
ZO=150.0Ω

TL1	TL2	TL3	TL4	TL5
Z=55.806Ω	Z=69.519Ω	Z=86.603Ω	Z=107.883Ω	Z=134.394Ω
L=90deg	L=90deg	L=90deg	L=90deg	L=90deg
F=100MHz	F=100MHz	F=100MHz	F=100MHz	F=100MHz

Figure 6-44 Five-section quarter-wave matching network

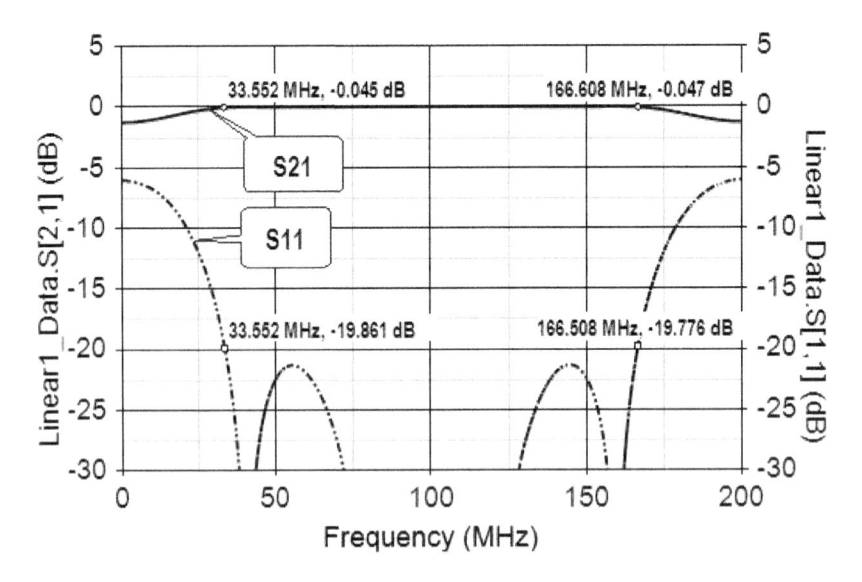

Figure 6-45 Simulated response of the matching network

Next replace the quarter-wave transformer adjacent to the load with a single-stub matching network. The design of the single-stub matching network is shown in the following Equation Editor. Note that the source resistor, R4 = 13.797 Ω, was calculated in the Equation Editor of Figure 6-46.

```
1    %Enter frequency, souce and load impedance
2    RS=120.411
3    RL=150
4    XL=-30
5    f=100e6
6    %Normalize the load impedance
7    r=RL/RS
8    x=XL/RS
9    %Calculate t1 and t2
10   t1=(x+sqrt(r*(r^2+x^2-2*r+1)))/(r-1)
11   t2=(x-sqrt(r*(r^2+x^2-2*r+1)))/(r-1)
12   %Calculate Line Lengths
13   d1=360*(atan(t1))/(2*pi)
14   d2=360*(pi+atan(t2))/(2*pi)
15   %Calculate B1 and B2
16   B1=(x*t1^2+(r^2+x^2-1)*t1-x)/(RS*(r^2+x^2+t1^2+2*x*t1))
17   B2=(x*t2^2+(r^2+x^2-1)*t2-x)/(RS*(r^2+x^2+t2^2+2*x*t2))
18   %Calculate Open-Stub Lengths
19   so1d=360*(pi-atan(RS*B1))/(2*pi)
20   so2d=-360*atan(RS*B2)/(2*pi)
```

Figure 6-46 Equation Editor showing the second line and stub calculations

The single-stub matching network is shown in Figure 6-47.

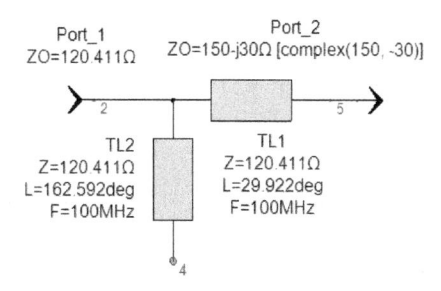

Figure 6-47 Schematic of the single-stub matching network

Now cascade the two matching networks to form the final design of broadband matching network, as shown in Figure 6-48. The simulated response of the quarter-wave matching network is shown in Figure 6-49.

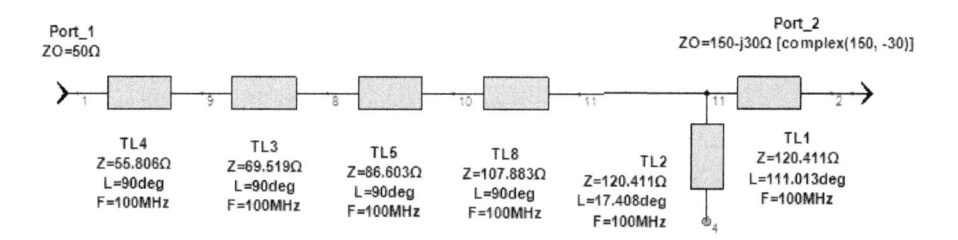

Figure 6-48 Schematic of the broadband matching network

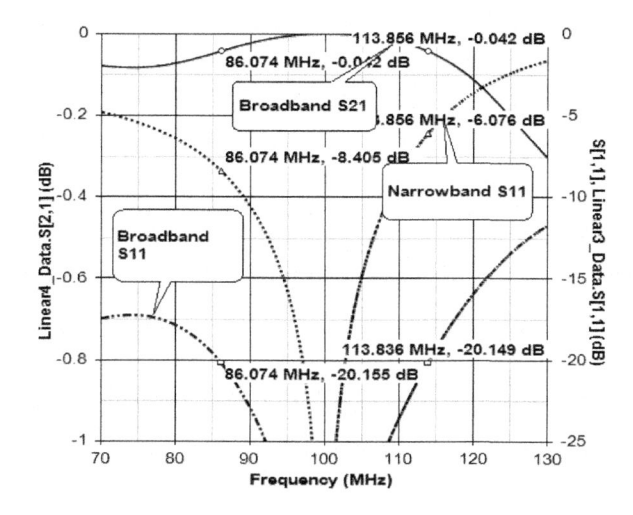

Figure 6-49 Simulated response of the broadband matching network

Figure 6-49 shows that the bandwidths of the broadband matching network at 20 dB return loss are:

$$BW_{20dB} = 113.8 - 86.0 = 27.8 \text{ MHz}$$

$$FBW_{20dB} = \frac{27.8 \times 100}{\sqrt{(113.8).(86.0)}} = 28.1\%$$

The same numbers for the narrowband single-stub matching network are:

$$BW_{20dB} = 102.8 - 97.1 = 5.7 \text{ MHz}$$

$$FBW_{20dB} = \frac{5.7 \times 100}{\sqrt{(102.8).(97.1)}} = 5.7\%$$

Notice that the fractional bandwidth at 20 dB return loss is about 28.1 % as opposed to only 5.7 % for the narrowband matching network. This is an indication that, by adding four quarter-wave sections to the single-stub matching network, we have increased the matching bandwidth at 20 dB return loss nearly five times over a single-stub matching network.

References and Further Reading

[1] Keysight Technologies, *Genesys 2014.03, Users Guide*, www.keysight.com

[2] Guillermo Gonzales, *Microwave Transistor Amplifiers – Analysis and Design*, Second Edition, Prentice Hall Inc., Upper Saddle River, NJ.

[3] Randy Rhea, *The Yin-Yang of Matching: Part 1 – Basic Matching Concepts*, High Frequency Electronics, March 2006

[4] Steve C. Cripps, *RF Power Amplifiers for Wireless Communications*, Artech House Publishers, Norwood, MA. 1999

[5] David M. Pozar, *Microwave Engineering*, Fourth Edition, John Wiley & Sons, New York, 2012

[6] Ali A. Behagi and Stephen D. Turner, *Microwave and RF Engineering*, BT Microwave LLC, State College, PA, 2011

Problems

6-1. Design a quarter-wave transmission line to match a load resistance $R_L = 5\ \Omega$ to a resistive source $R_S = 75\ \Omega$ at 500 MHz. Calculate the characteristic impedance of the quarter-wave line, the Q factor and the fractional bandwidths of the matching network at 3 and 20 dB return loss. Display the simulated response and compare the calculations with measurements. For the quarter-wave matching network, calculate the fractional bandwidth and power loss from $\Gamma = 0.1$ to $\Gamma = 0.707$.

6-2. Design a quarter-wave transmission line to match a load resistance $R_L = 100\ \Omega$ to a resistive source $R_S = 25\ \Omega$ at 600 MHz. Calculate the characteristic impedance of the quarter-wave line, the Q factor and the fractional bandwidths of the matching network at 3 dB and 20 dB return loss. Display the simulated response and compare the calculations with measurements. For the quarter-wave matching network, calculate the fractional bandwidth and power loss from $\Gamma = 0.1$ to $\Gamma = 0.707$.

6-3. Design a single-stub network to match a load resistance $Z_L = 5 - j5\ \Omega$ to a resistive source $R_S = 75\ \Omega$ at 700 MHz. Calculate the electrical lengths of the matching line and stub and the fractional bandwidths of the matching network at 3 and 20 dB return loss. Display the simulated response and compare the calculations with measurements.

6-4. Design a single-stub network to match a load resistance $Z_L = 100 + j20\ \Omega$ to a resistive source $R_S = 40\ \Omega$ at 800 MHz. Calculate the electrical lengths of the matching line and stub and the fractional bandwidths of the matching network at 3 dB and 20 dB return loss. Display the simulated response and compare the calculations with measurements.

6-5. Design a single-stub network to match a complex load $Z_L = 20 + j5\ \Omega$ to a complex source $Z_S = 50 + j20\ \Omega$ at 1000 MHz. Calculate the electrical lengths of the matching line and stub and the fractional bandwidths of the matching network at 3 dB and 20 dB return loss. Display the simulated response and verify the calculations with measurements.

6-6. Design a three-section quarter-wave network to match a load resistance $R_L = 5\ \Omega$ to a resistive source $R_S = 50\ \Omega$ at 100 MHz. Calculate the characteristic impedance of the quarter-wave line, the Q factor and the fractional bandwidths of the matching network at 3 and 20 dB return loss. Display the simulated response and compare the measurements with the singe quarter-wave matching network.

6-7. Design a three-section quarter-wave network to match a load resistance $R_L = 100\ \Omega$ to a resistive source $R_S = 25\ \Omega$ at 900 MHz. Calculate the fractional bandwidths of the matching network at 3 and 20 dB return loss. Display the simulated response and compare the measurements with the singe quarter-wave matching network.

6-8. Design a broadband network to match a complex load, $Z_L = 25 + j10\ \Omega$ to 75 Ω source impedance at 300 MHz. Measure the bandwidth at 20 dB return loss and compare it with the results of singe-stub matching network.

Chapter 7

Single Stage Amplifier Design

7.1 Introduction

Modern amplifier design involves the use of high frequency transistors to use DC power to control and amplify RF energy. These transistors are fundamentally of the bipolar and field effect transistor (FET) variety. There are many subtypes of bipolar and FET devices that the reader is encouraged to explore [6]. For the many types of transistors the impedance matching techniques covered in chapters 5 and 6 are very useful for the design and simulation of linear transistor amplifiers. There are many different ways in which transistor amplifiers are impedance matched depending on the function or purpose of the amplifier circuit. In the design of cascaded amplifier networks the output of each stage must be matched to the input of the next stage. In this chapter we design four single stage amplifiers that require four different impedance matching techniques.

1. Maximum Gain Amplifier Design
2. Specific Gain Amplifier Design
3. Low Noise Figure Amplifier Design
4. Power Amplifier Design

Most amplifier designs actually involve selective mismatching of the transistor to its source and load impedance to accomplish its intended purpose. Only the maximum gain amplifier requires a conjugate matching network design. The specific matching techniques for each category of amplifiers are listed as:

1. **Maximum Gain Amplifier:**
   In a maximum gain amplifier the input and the output are conjugate matched to achieve maximum gain.

2. **Specific Gain Amplifier:**
   A specific gain amplifier is mismatched at either the input or output to achieve a specified value of gain less than its maximum.

3. **Low Noise Figure Amplifier:**

In a low noise amplifier the transistor's input is mismatched from 50 Ω to achieve a specific Noise Figure.

4. **Power Amplifier:**

In a power amplifier the transistor's output is mismatched from 50 Ω for a specific amount of output power.

7.2 Maximum Gain Amplifier Design

This section covers the design of the maximum gain amplifier at 2.35 GHz using the SHF-0189 transistor. The SHF-0189 is a high performance Hetrostructure FET (HFET) housed in a surface-mount plastic package (SOT-89) as shown in Figure 7-1. The maximum gain amplifier is conjugate matched at the input and output resulting in very good return loss over a narrow bandwidth. A device can only be conjugate matched if it is unconditionally stable [1]. If the device is potentially unstable in the band of interest, a conjugate impedance match cannot be realized unless we add additional circuitry to stabilize the device.

Figure 7-1 GaAs HFET specifications (courtesy of RF Micro Devices)

7.2.1 Transistor Stability Considerations

Most Microwave transistors are potentially unstable at some frequency. This does not necessarily make them undesirable for use as an amplifier. A potentially unstable transistor does not mean that it will definitely oscillate in a circuit. Referring to the device in Figure 7-2 there may exist some combination of input or output reflection coefficient, Γ_S or Γ_L, that, when presented to the transistor, may indeed make the device oscillate. An oscillation condition is indicated by $|\Gamma_{IN}| > 1$ or $|\Gamma_{OUT}| > 1$. This can also be viewed as a positive value for the Return Loss, S11 or S22, in dB. This is referred to as negative resistance and is actually a design goal in the design of microwave oscillators. The conditions for unconditional stability for an RF transistor are given by equations (7-1) through (7-4) [2].

$$|\Gamma_S| < 1 \tag{7-1}$$

$$|\Gamma_L| < 1 \tag{7-2}$$

$$|\Gamma_{IN}| = \left| S_{11} + \frac{S_{12}S_{21}\Gamma_L}{1 - S_{22}\Gamma_L} \right| < 1 \tag{7-3}$$

$$|\Gamma_{OUT}| = \left| S_{22} + \frac{S_{12}S_{21}\Gamma_S}{1 - S_{11}\Gamma_S} \right| < 1 \tag{7-4}$$

Typically there exists a select set of reflection coefficient values for Γ_S and Γ_L that will give the device this negative resistance characteristic. One method of dealing with this problem is to select values for Γ_S and Γ_L that avoid these unstable regions of impedance. This technique is demonstrated in Section 7.3. Another technique is to introduce a network that forces the device to be unconditionally stable for any values of Γ_S and Γ_L. This technique is often preferable as it will be shown that a transistor can be made unconditionally stable over a very wide frequency range. This is important because even though a device may be stable in its desired frequency band, it may become unstable at some frequency outside of the band of operation. A simultaneous numerical solution of Equations (7-1) through (7-4) for all values of Γ_S and Γ_L can prove that there are two conditions for unconditional stability. These entities are the stability factor, K, and stability measure, B1, and are given by the following Equations [3].

$$K = \frac{1 - |S_{11}|^2 - |S_{22}|^2 + |\Delta|^2}{2|S_{12} \cdot S_{21}|} > 1 \quad (7\text{-}5)$$

$$B_1 = 1 + |S_{11}|^2 - |S_{22}|^2 - |\Delta|^2 > 0 \quad (7\text{-}6)$$

where,

$$|\Delta| = |S_{11} \cdot S_{22} - S_{12} \cdot S_{21}| \quad (7\text{-}7)$$

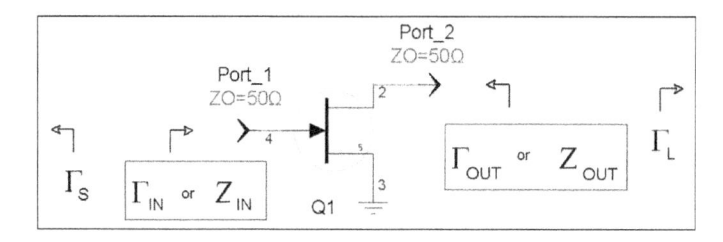

Figure 7-2 Transistor input and output reflection coefficients

When the transistor is unconditionally stable a simultaneous conjugate match can be defined. The simultaneous conjugate match exists when $\Gamma_{IN} = \Gamma_S^*$ and $\Gamma_{OUT} = \Gamma_L^*$. The simultaneous conjugate match reflection coefficients are often referred to as: Γ_{MS} and Γ_{ML}. The maximum gain that can be obtained with a simultaneous conjugate matched amplifier is determined from the transducer gain equation, G_{Tmax}.

$$G_{Tmax} = \frac{\left(1 - |\Gamma_{MS}|^2\right)|S_{21}|^2 \left(1 - |\Gamma_{ML}|^2\right)}{\left|(1 - S_{11}\Gamma_{MS})(1 - S_{22}\Gamma_{ML}) - S_{12}S_{21}\Gamma_{ML}\Gamma_{MS}\right|^2} = \frac{|S_{21}|}{|S_{12}|}\left(K - \sqrt{K^2 - 1}\right) \quad (7\text{-}8)$$

When K is exactly equal to one, G_{Tmax} becomes the Maximum Stable Gain, G_{MSG}. G_{MSG} is the maximum value that G_{Tmax} can achieve.

$$G_{MSG} = \frac{|S_{21}|}{|S_{12}|} \quad (7\text{-}9)$$

7.2.2 Stabilizing the Device in Genesys

Example 7.2-1 Design the stability network for the SHF-0189 device from 100 to 8000 MHz.

Solution: As equations (7-5) through (7-7) show, the stability factors are a function of the transistor's S parameters. An amplifier design begins with the placement of the device S parameters in a schematic design workspace as shown in Figure 7-2. The Genesys software has built in functions to determine the stability factor and stability measure as well as Γ_{MS} and Γ_{ML}, the conjugate match reflection coefficients. Sweep the frequency range of the device over the entire range of frequencies contained in the S parameter file. Plot the parameters K, B1, and GMAX for the SHF-0189 device.

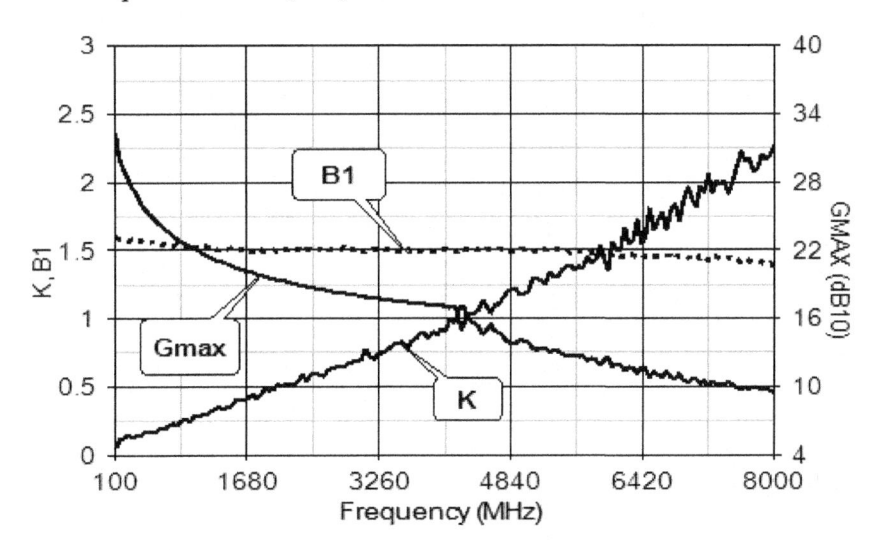

Figure 7-3 Plot of stability parameters and GMAX for the SHF-0189 device

Note that stability factor k is less than 1 from 100 MHz up to about 4500 MHz indicating that the transistor is potentially unstable over most of its frequency range including the design frequency of 2350 MHz. An effective stabilization network for medium and high power transistors employs a parallel RC circuit on the input of the device as shown in Figure 7-4. Make both the R and C values tunable with a starting value of 10 Ω and 10 pf respectively. Tune the resistance and capacitor values until K > 1 for frequencies above 500 MHz. The values shown in Figure 7-4 provide sufficient stability above 500 MHz.

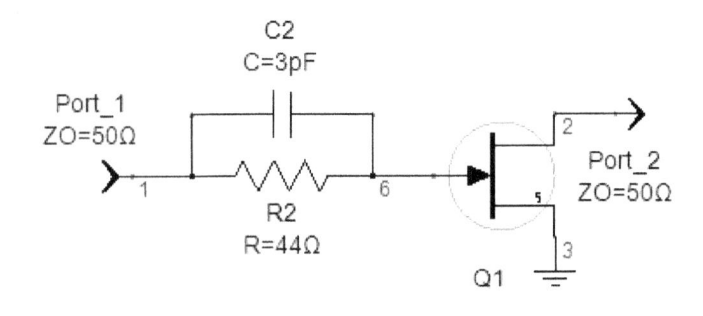

Figure 7-4 Adding parallel RC stability network

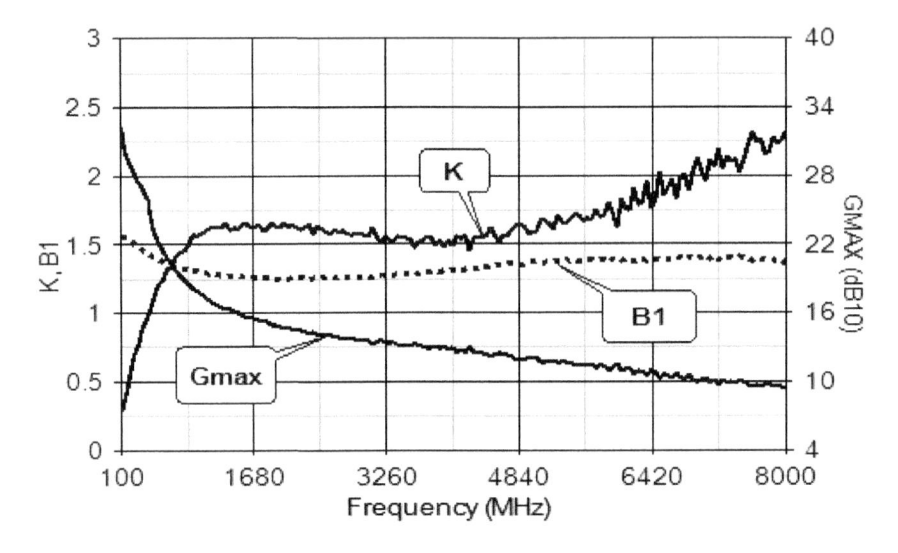

Figure 7-5 Stability parameters and GMAX with parallel RC network

As seen in Figure 7-5, the stability measure, B1, is greater than zero throughout the entire frequency range. However the stability factor, K, goes below one at frequencies from less than 427 MHz. We want to make sure that the device is stable at all frequencies so that there is no possibility of any value of Γ_{ML} or Γ_{MS} that could make the device unstable at these out-of-band frequencies. Therefore add an R-L network in shunt with the input of the transistor for additional low frequency stability. This R-L network can be absorbed into the bias feed for the gate voltage of the transistor. Choose a fixed value of 50 Ω for the resistor. Select a starting value of 100 nH for the inductor and make this value tunable. This R-L network will load the previous R-C network so its values will need to be changed. The inductor's

Single Stage Amplifier Design

value can be reduced which in turn reduces the low frequency gain while not significantly reducing the in band gain at 2.35 GHz. Figure 7-6 shows the final stabilization network for the device.

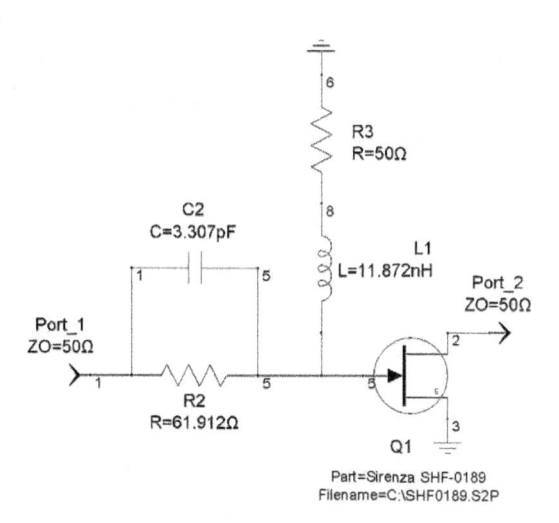

Figure 7-6 Stabilization network for the SHF-0189 transistor

As shown in Figure 7-7, GMAX = 15.975 dB and K = 1.271 at the design frequency of 2.35 GHz. The device is now unconditionally stable for all frequencies from 50 MHz to 8 GHz. Now that the device is unconditionally stable, we can determine the simultaneous conjugate match source and load impedance, Γ_{MS} and Γ_{ML}.

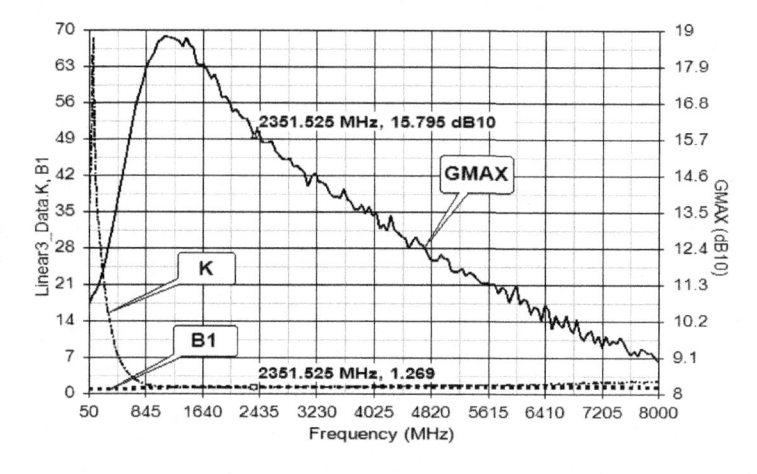

Figure 7-7 Stability parameters and G_{MAX} of stabilized device

7.2.3 Finding Simultaneous Match Reflection Coefficients and Impedances

Genesys has built in functions for the calculation of the Γ_{MS} and Γ_{ML} reflection coefficients as well as the corresponding impedance Z_{MS} and Z_{ML}. Using the schematic of Figure 7-6 create a tabular output of the simultaneous conjugate source and load parameters as shown in Figure 7-8.

Measurement	Label (Optional)	Units	Complex Format	Hide?
GMAX		dB10	(default)	
GM1		dB	Mag abs+Angle	
GM2		dB	Mag abs+Angle	
ZM1			Real+Imaginary	
ZM2			Real+Imaginary	
			(default)	

F (MHz)	GMAX (dB10)	mag (GM1) (dB)	ang (GM1)	mag (GM2) (dB)	ang (GM2)	re(ZM1) ()	im (ZM1)	re(ZM2) ()	im (ZM2)
2350	15.791	0.793	131.771	0.605	141.09	6.918	22.023	13.72	16.471

Figure 7-8 Tabular output functions showing G_{MAX}, Γ_{MS}, Γ_{ML}, Z_{MS}, and Z_{ML}

Note the equivalence between the Genesys function notation and the respective reflection coefficient and impedance.

$$GM1 = \Gamma_{MS} = 0.793 \angle 131.7 \qquad ZM1 = Z_{MS} = 6.92 + 22.02 \; \Omega$$

$$GM2 = \Gamma_{ML} = 0.605 \angle 141.1 \qquad ZM2 = Z_{ML} = 13.72 + 16.47 \; \Omega$$

7.3 Analytical and Graphical Impedance Matching Techniques

There are different methods to complete the task of matching the 50 Ω source or load impedance to the Γ_{ML} or Γ_{MS}. These techniques are summarized as graphical (using the Smith Chart), or analytical (by numerical computation). Because we are given a reflection coefficient that represents the conjugate of the impedance that the device represents, it is recommended to progress through the input and output match by moving from the 50 Ω point on the Smith Chart to the Γ_{ML} or Γ_{MS} point on the same Smith Chart. Theoretically this impedance match can be realized with any number of reactive elements. For single frequency applications or narrow bandwidths, < 10%, the impedance match is often realized with only two elements. Sometimes the engineer may prefer to use three or more elements for single frequency matching because it will make the design less sensitive to variation in the individual inductance or capacitance value. This becomes an important concern in manufacturing as practical lumped element components typically have a tolerance placed on their nominal value of 5 % to 20 %. In addition to the nominal tolerance, all components have a temperature coefficient of their nominal value. That is to say that the component's inductance or capacitance value may change with its temperature.

7.3.1 Analytical Design of the Input Matching Network

Example 7.3-1: Design the SHF-0189 input matching network.

Solution: We will use the analytical impedance matching techniques developed in chapter 5 and the L-Network Synthesis Utility to generate the input and output matching networks. When we are designing the input and output matching networks, we are matching the 50 Ω source and load impedance to the impedance looking into the transistor input or output. Therefore the complex impedances that we are matching are given as the conjugate of Z_{MS} and Z_{ML}.

$$Z1 = 6.92 - j22.02 \text{ } \Omega$$
$$Z2 = 13.72 - j16.47 \text{ } \Omega$$

To match the complex load impedance to a resistive source first obtain r and x by normalizing the load impedance with respect to the source resistor, and then determine the matching networks on the basis of the following conditions.

- If $r < 1$ and $r^2 + x^2 - r < 0$ there exists only two matching networks obtained from Equations (5-21) through (5-24).

- If $r < 1$ and $r^2 + x^2 - r > 0$ there exists four matching networks obtainable from Equations (5-21) through (5-24) and Equations (5-26) through (5-29).

- If $r > 1$ there exists only two matching networks obtained from Equations (5-26) through (5-29).

For the input matching network $r = 0.1384$ and $x = -0.44$, therefore, $r < 1$ and $r^2 + x^2 - r = 0.074 > 0$. According to the above conditions, there exists four L-networks that can be used for the input matching network. To calculate the element values of the first matching network we can utilize the Equation Editor in Genesys and follow the procedure:

1. Enter the source and load impedance and design frequency
2. Normalize the load impedance
3. Write the appropriate equations to calculate B and X
4. Based on the positive or negative values of B and X, calculate the element values of the matching network as in Figure 7-9.

```
1    %Enter Design Parameters
2    Z0=50
3    RL=6.918
4    XL=-22.03
5    f=2.35e9
6    %Normalize Load Impedance
7    r=RL/Z0
8    x=XL/Z0
9    %Write Equations
10   B2=-sqrt((1-r)/r)/Z0
11   X2=-Z0*(sqrt(r*(1-r))+x)
12   %Calculate Matching Elements
13   L1=X2/(2*pi*f)
14   L2=-1/(2*pi*f*B2)
```

Figure 7-9 Equation Editor calculating matching element values

To display the response of the input matching network, add a Schematic in Genesys and place the matching elements with the values in the Figure 7-10. Add a Linear Analysis from 1350 MHz to 3350 MHz to analyze the response of the matching network, S11, and S21 both in dB. The swept insertion loss and return loss are also shown in Figure 7-10.

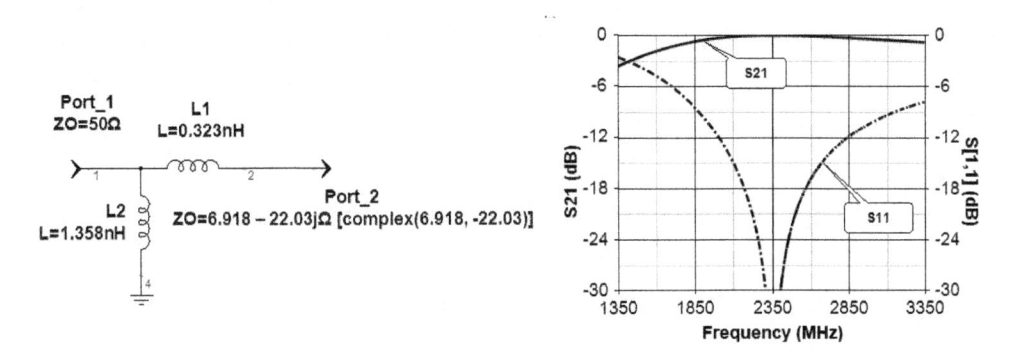

Figure 7-10 Impedance matching network for the input match

7.3.2 Synthesis Based Input Matching Networks

As an alternative to writing the equations in an equation editor, a circuit synthesis technique such as the L Network matching utility as created in Chapter 5 can be used to quickly evaluate all potential solutions. This allows the designer to quickly evaluate all potential solutions and choose the one that best fits the desired response and topology. Figure 7-11 shows the L Network Synthesis tool created with VBScript to design a two element matching network between arbitrary source and load impedances. Using the slider bar we can quickly view both the component values and the response for each of the four potential solutions for the impedance matching L network. Invalid networks are identified as those solutions that return negative or complex component values.

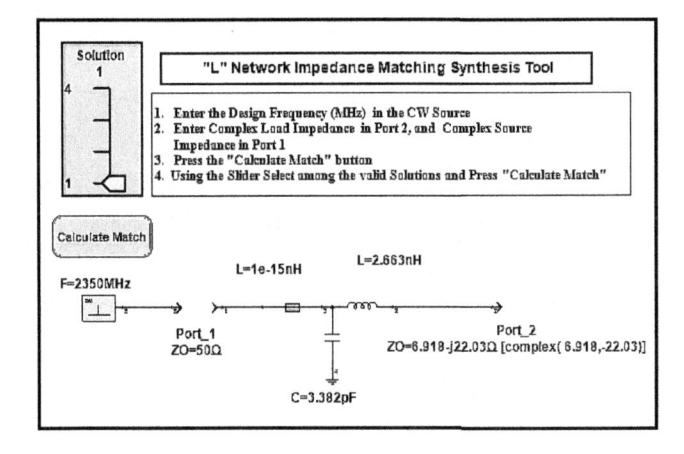

Figure 7-11 L-network impedance matching synthesis utility

Enter Z1 for the complex impedance at Port 2 in the L-Network Impedance Matching Synthesis Tool. Find all four of the impedance matching circuits by selecting Solution #1 through #4 with the slider control. After selecting a topology with the slider, press the "Calculate Match" button to display the L-network. Repeating for all four possible L Networks we can discard the solutions that return capacitor or inductor values with negative values. For this impedance match we find that there are four valid topologies that can match 50 Ω to Z1. These circuits are shown in Figure 7-12. Any of the four networks shown in Figure 7-12 could be chosen for the matching circuit. There is a variety of reasons why the circuit designer may chose one network over another. One reason could be the transmission bandwidth characteristic. Figure 7-12c appears to have the broadest bandwidth. The solutions of Figure 7-12a and 7-12d have a tendency toward a high pass frequency response. Figure 7-12b has a tendency toward a band pass characteristic. If we desire to have a very narrow band amplifier and need some rejection of nearby signal, then the bandpass characteristic may be desirable. If we hope to make a broader band amplifier then we may want to choose the solution of Figure 7-12c. As the value of complex impedance changes in a way to make the corresponding reflection coefficient larger, the bandwidth will tend to become more narrow [5]. Another reason for specific topology selection could be the convenience of physical circuit layout. Because the transistor must have a gate and drain bias feed, a shunt inductor adjacent to the complex impedance may be convenient as this component can be absorbed into the bias feed network design. For this reason the matching network of Figure 7-12b is chosen for the amplifier.

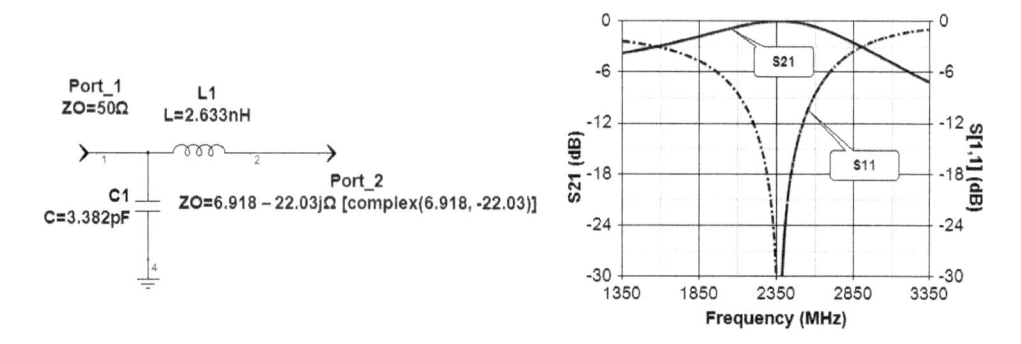

Figure 7-12a First L-network solution for the Input match

Single Stage Amplifier Design   333

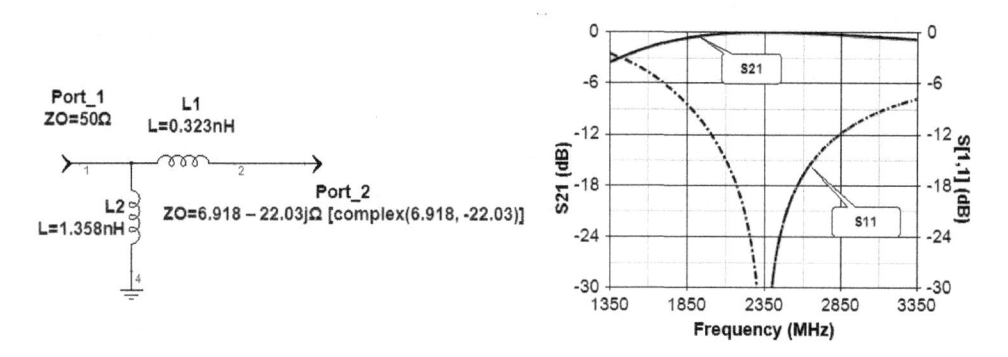

Figure 7-12b Second L-network solution for the input match

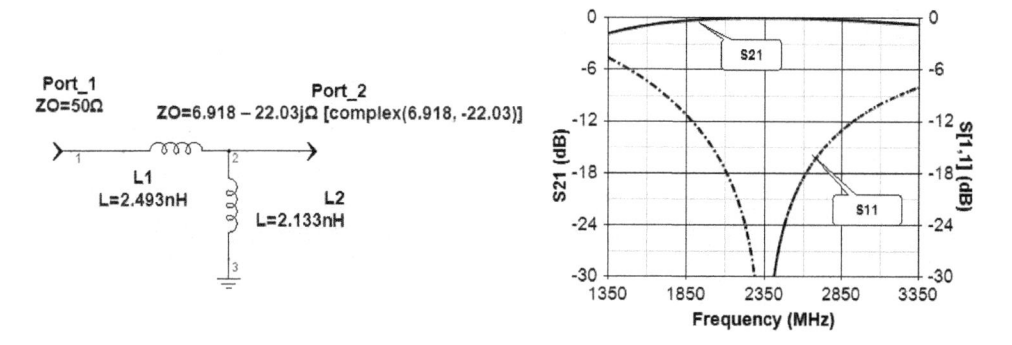

Figure 7-12c Third L-network solution for the Input match

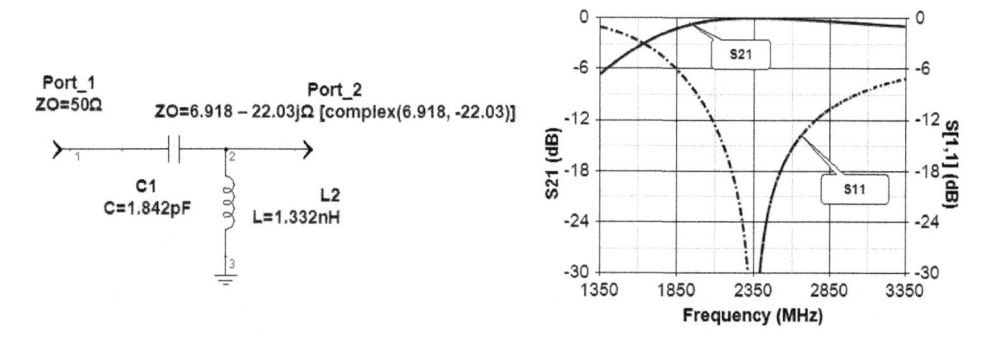

Figure 7-12d Fourth L-network solution for the input match

7.3.3 Synthesis Based Output Matching Networks

Example 7.3-2 Design the output matching network for the SHF-0189 at 2350 MHz.

Solution: Enter Z2 for the complex impedance at Port 2 in the L Network Impedance Matching Synthesis Tool. Scrolling through the four possible solutions we find that only two solutions give a good return loss and corresponding impedance match for the output network. The solution of Figure 7-13b is selected for the output matching network because of the broader band transmission characteristic.

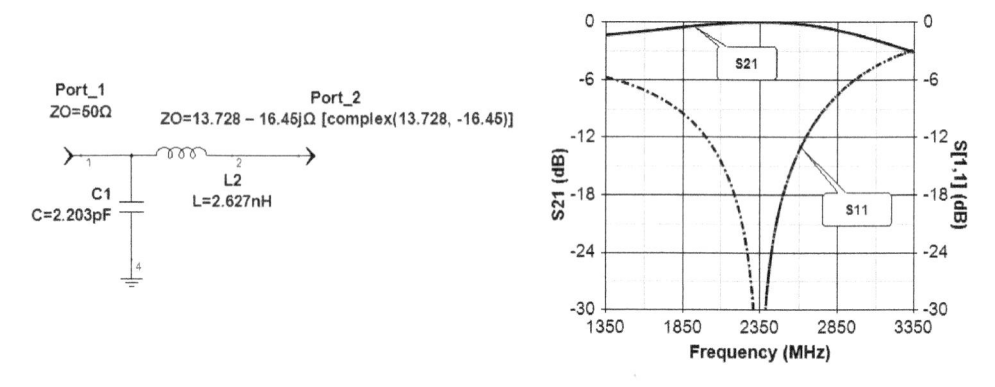

Figure 7-13a Output matching L-network for the first solution

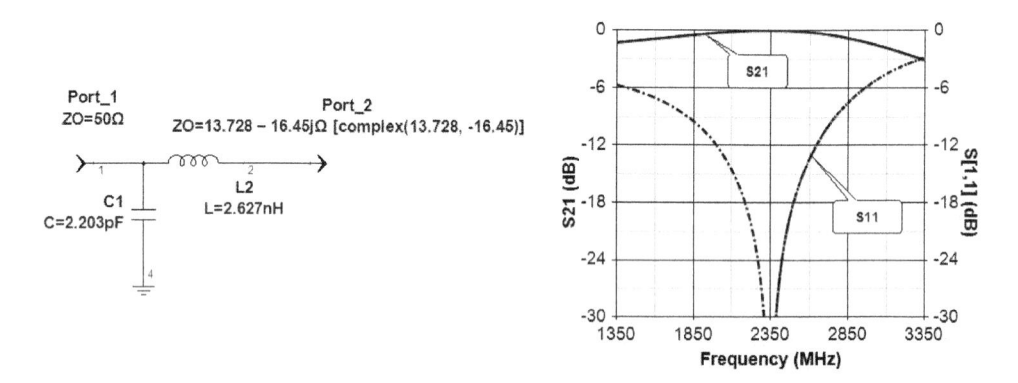

Figure 7-13b Output matching L-network for the second solution

7.3.4 Ideal Model of the Maximum Gain Amplifier

Example 7.3-3 Assemble and simulate the ideal maximum gain amplifier. Use a linear sweep from 100 MHz to 5000 MHz to analyze the response of the amplifier.

Single Stage Amplifier Design

Solution: Create the ideal model of the amplifier by attaching the synthesized input and output matching networks to the device's S parameter file based model as shown in Figure 7-14. Use a Linear Sweep from 100 MHz to 5000 MHz to analyze the response of the amplifier. The swept gain, input return loss, and output return loss is shown in Figure 7-15. As the return loss shows, an excellent match has been achieved at the input and output ports. Also the maximum gain (15.79 dB) of the amplifier at 2.35 GHz is exactly the originally calculated G_{MAX} indicating a successful simultaneous conjugate impedance match.

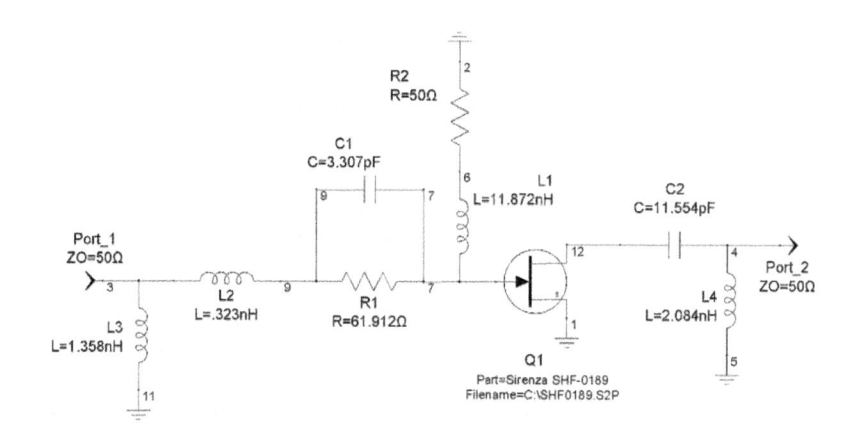

Figure 7-14 Complete schematic of the ideal amplifier circuit

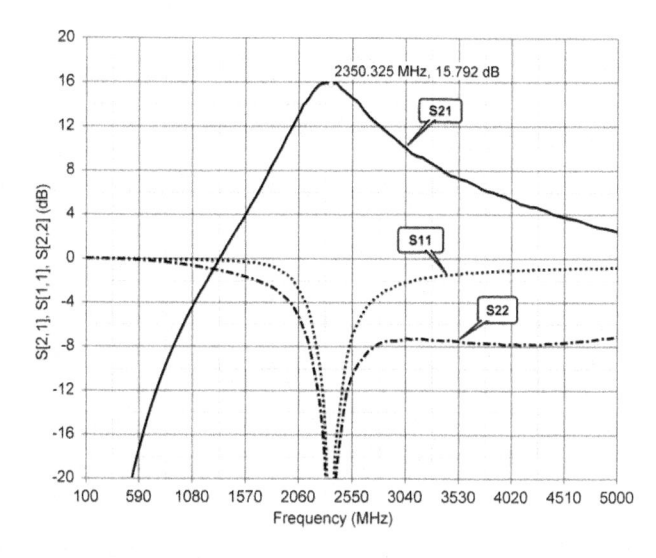

Figure 7-15 Swept response of the ideal amplifier design

The fractional bandwidth of the amplifier is calculated by sweeping the response over a narrower frequency range as shown in Figure 7-16. Markers are placed on the passband response at the 0.1 dB ripple bandwidth. As Figure 7-16 shows the 0.1 dB bandwidth is:

$$BW_{20dB} = 2396 - 2333 = 63 \text{ MHz}.$$

This corresponds to a fractional bandwidth of about 2.66 % consistent with a narrowband amplifier. The conjugate match is achieved at the single frequency of 2.35 GHz.

$$FBW_{20dB} = \frac{63}{\sqrt{(2396).(2333)}} = 2.66 \%$$

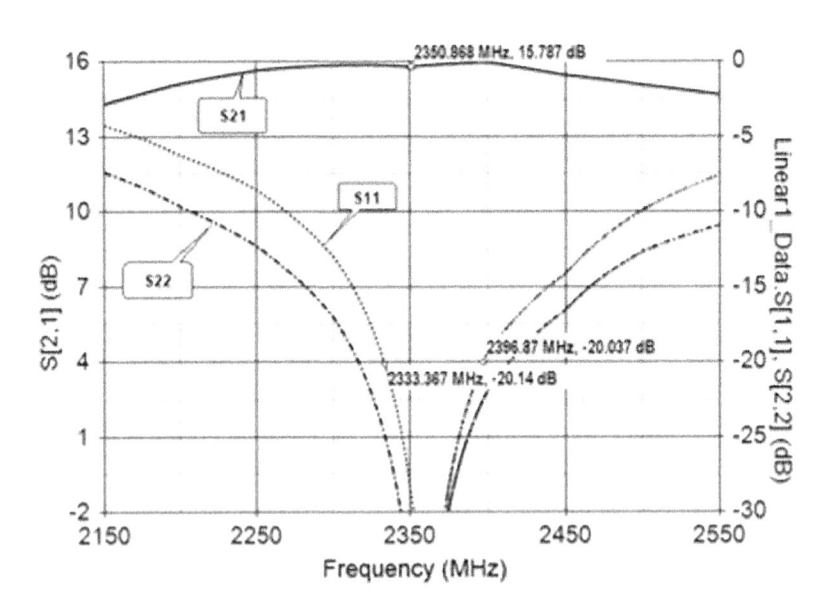

Figure 7-16 Ideal amplifier response with markers at 20 dB return loss

7.4 Physical Model of the Amplifier

Example 7.4-1: Use physical components to create the physical model.

Solution: The ideal amplifier circuit is a good starting point to confirm the matching network design and stability of the device but is not very useful

Single Stage Amplifier Design

for creating the actual physical amplifier. In Chapter 4 we learned that the parasitic effects of the real lumped element components can significantly alter the expected performance of a circuit. Additionally the microstrip PCB tracks will begin to alter the performance of the circuit. As the frequency of the circuit increases, each PCB trace represents a longer portion of a wavelength and will detune the circuit. To create the physical model of the amplifier we need to select a microstrip substrate to place the PCB circuit tracks. We will also need to design bias feed networks and replace the ideal lumped element components with a physical model equivalent. Select the Roger's RO3010 material with dielectric thickness of 0.025 inches and 1 oz. copper (.00134 inch) copper thickness.

7.4.1 Transistor Artwork Replacement

The Genesys artwork replacement library contains a variety of layout footprints for various transistor packages. The selection of transistor packages is so numerous that you may not always find your transistor's package outline in the Genesys library. The Genesys Footprint Editor makes it very easy to create custom artwork replacement elements for physical circuit layouts. Figure 7-17 shows the transistor package drawing of the SOT89 from the data sheet.

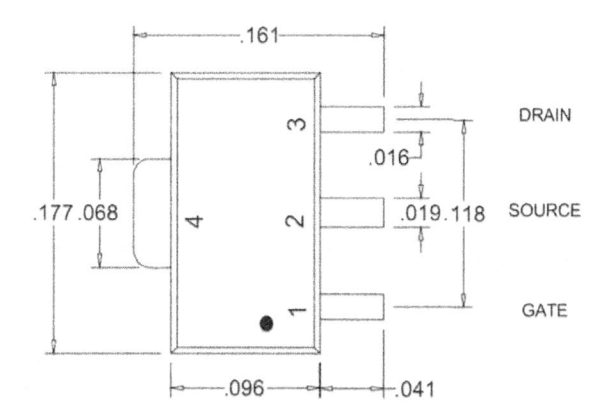

Figure 7-17 SOT89 package outline from data sheet

A drawing of the package can be created in any mechanical CAD program and imported into Genesys as a DWG or DXF file. Normally the file should be saved in the units of mils. Invoke the Genesys Footprint Editor by selecting on the toolbar:

- Tools
- Footprint Editor
- New Footprint

Once the Footprint Editor Window opens, return to the Toolbar and import the drawing by selecting:

- File
- Import
- DXF/DWG File

Browse to the location of the DWG file and select the file. At the Import window, assign the package drawing to the Top Assembly layer. The SOT89 package drawing imported into the Footprint Editor is shown in Figure 7-18.

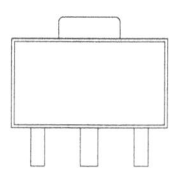

Figure 7-18 SOT89 package drawing imported into the Footprint Editor

Once the package drawing is loaded into the Footprint Editor, we can add PCB solder pads and port assignments. The source lead is to be connected directly to a ground pad. The device data sheet may give a suggested ground pad layout for optimum RF grounding and thermal conductivity.
Using the Draw Rect tool, on the Footprint Editor Toolbar, draw a set of PCB tracks as shown in Fig. 7-19. Make the pads for the gate and drain leads 25mil x 30mil rectangles. Make sure that these pads on assigned to the Metal layer. Then add the port assignments as shown in Figure 7-19. The port assignments should correspond to the order in which the ports are assigned on the schematic element. Because the transistor's gate and drain

Single Stage Amplifier Design 339

are the input and output respectively, these leads should be assigned port 1 and port 2. The source lead can then be assigned to port 3.

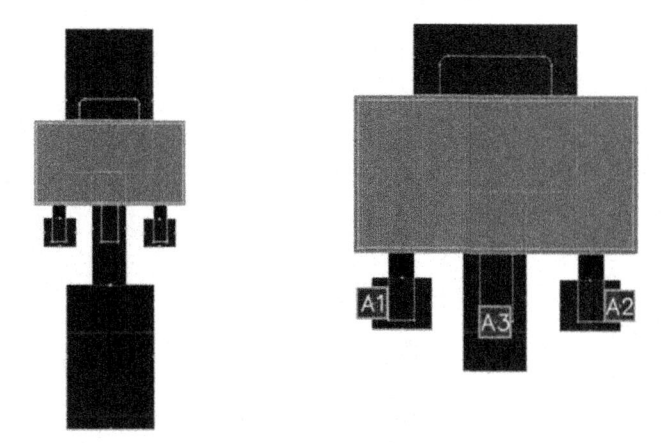

Figure 7-19 PCB tracks and ports added to the metal layer of the artwork file

Finally save the artwork replacement element. From the toolbar select:

- Tools
- Footprint Editor
- Save Footprint

Choose either an existing Genesys library or create a new library to save any custom artwork that you create.

7.4.2 Amplifier Physical Design and Layout

Start the physical design of the amplifier by placing the device S parameter file on the schematic. Use a microstrip Via Hole model for the connection of the transistor source lead to ground. Specify a hole radius of 12 mils. Then assign the artwork replacement element for the SHF-0189 by opening the Properties window of the S parameter file. Select the Change Footprint button and browse to the location of the artwork replacement element created in the previous section. The footprint will be displayed in the window as shown in Figure 7-20.

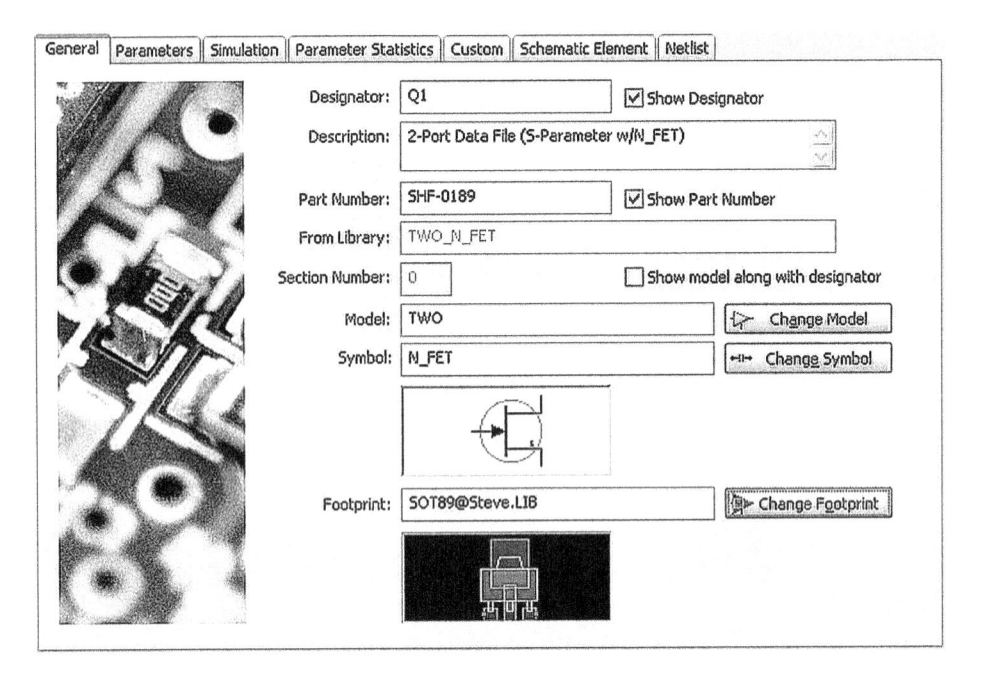

Figure 7-20 Assigning artwork replacement element

Proceed with the construction of the stabilization network portion of the design from Figure 7-14. This time add the Microstrip PCB tracks as shown in Figure 7-21. Use Modelithic surface mount components (30 mil x 60 mil) for the stabilization components as these will be easy to tune or optimize Use the S parameter files of the components that we do not intend to optimize. A 470 pF capacitor is added as a gate bypass capacitance. On the drain side of the transistor a 33 nH inductor is used as an RF choke to supply the drain voltage. Edit the footprint of the S parameter files and change the footprint to a 0603 surface mount component from the Genesys library. The Modelithic components will automatically have the correct footprint assigned to its properties.

Single Stage Amplifier Design

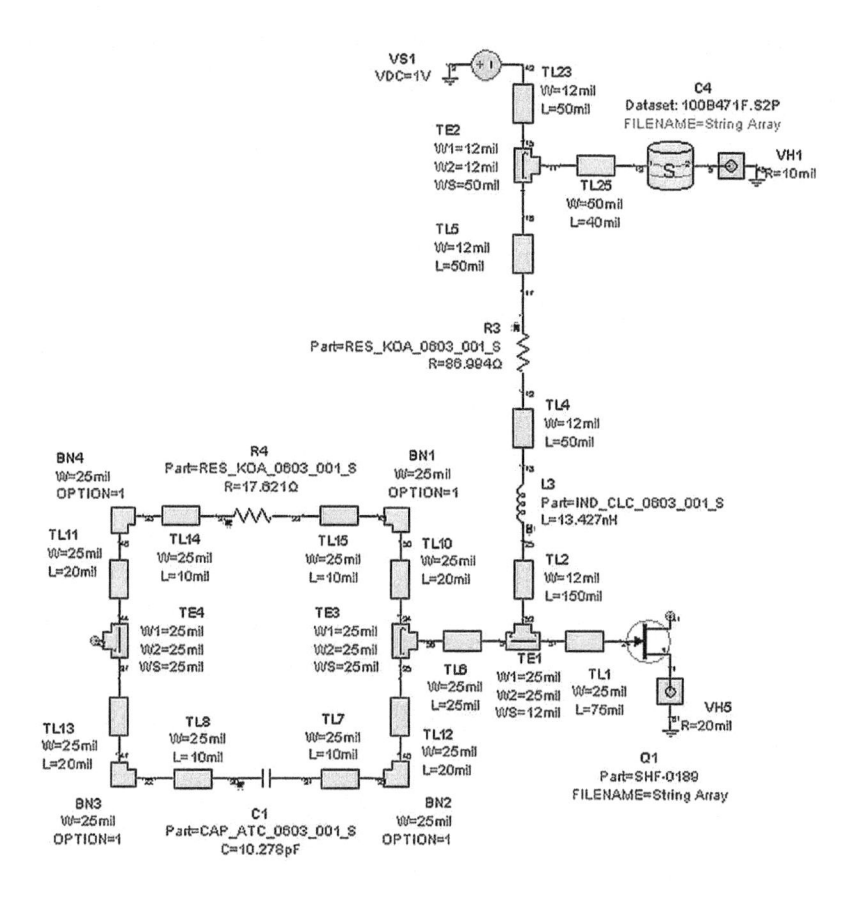

Figure 7-21 Construction of physical amplifier model

Continue with the physical model construction by adding the matching networks. Note that on the input side of the circuit the matching network has an inductor connected to ground. Because there will be a negative voltage applied to the gate of the transistor, a DC blocking capacitor must be added to keep the inductor from shorting out the gate bias. Add an S parameter file for a 1.5 pF chip capacitor to act as a DC block. The capacitance value should be chosen such that the capacitor is operating in series resonance for minimum insertion loss. Knowing that the chip capacitor typically has some series inductance, the small series inductance of 0.323 nH can be absorbed into the DC blocking capacitor. The output matching circuit includes a series capacitance so no additional DC blocking capacitor in needed. The completed schematic of the physical amplifier model is shown in Figure 7-22. The physical PCB layout is shown in Figure 7-23.

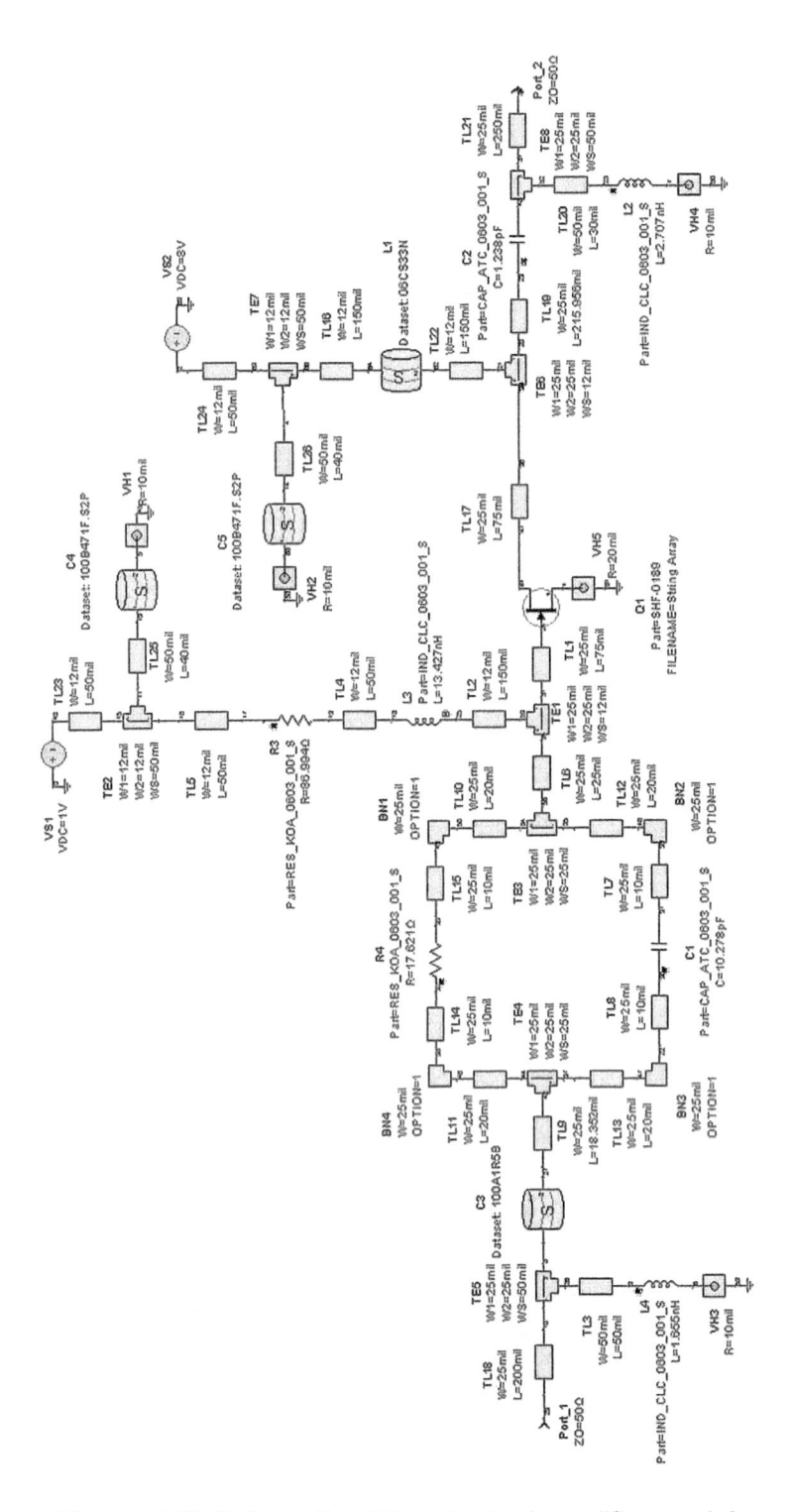

Figure 7-22 Schematic of the physical amplifier model

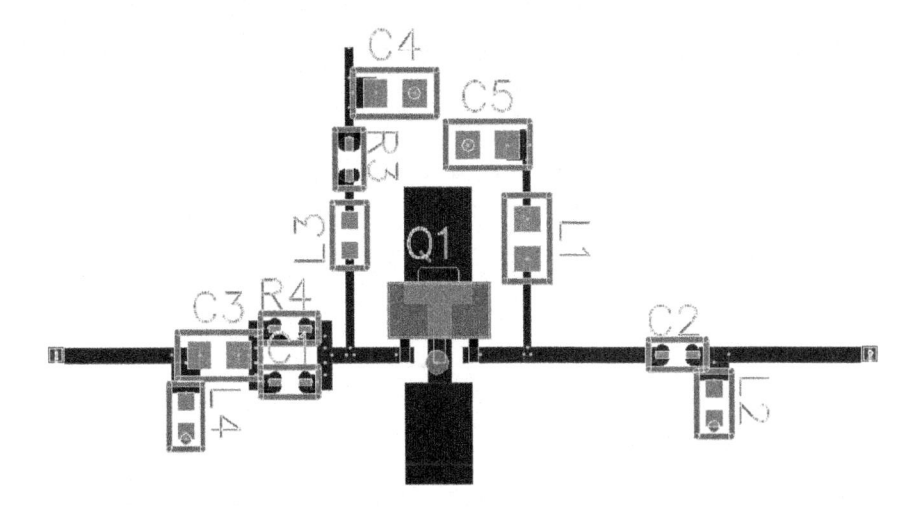

Figure 7-23 Physical layout of completed amplifier circuit

7.4.3 Optimization of the Amplifier Response

After completion of the physical circuit schematic of Figure 7-22 the circuit response is swept from 100 MHz to 5000 MHz, as shown in Figure 7-24. The response of the amplifier has significantly changed from the response of the ideal amplifier response of Figure 7-15. The gain and return loss have shifted in frequency. This is a typical result of replacing ideal elements with real physical models. This ability to deal with package parasitics and physical layout is a very powerful benefit of modeling the amplifier in Genesys. We know that it is possible to make the component element values tunable and manually tune each element while observing the change in response. This is quite useful and encouraged so that you can see the circuit sensitivity to particular components. As the number of components and tunable elements increase in a given design, the tuning of individual elements can be cumbersome. In this section we will introduce the use of Optimization to tune the amplifier's response to the desired characteristics. We can think of Optimization as an automatic circuit tuner. Selected components in the amplifier can be assigned as tunable variables. Multiple goals can then be set such as gain levels, return loss, etc. The optimization is then run and the Genesys software will execute a mathematical optimization algorithm to determine the values of the variables required to achieve the specified goals [3]. In this example set all seven of the

Modelithic model values as tunable. Also make the microstrip line lengths for the series sections TL9 and TL19 variable. We will let the software optimize these variables to achieve the desired response. Make sure to set response goals for out-of-band responses as well as in band response.

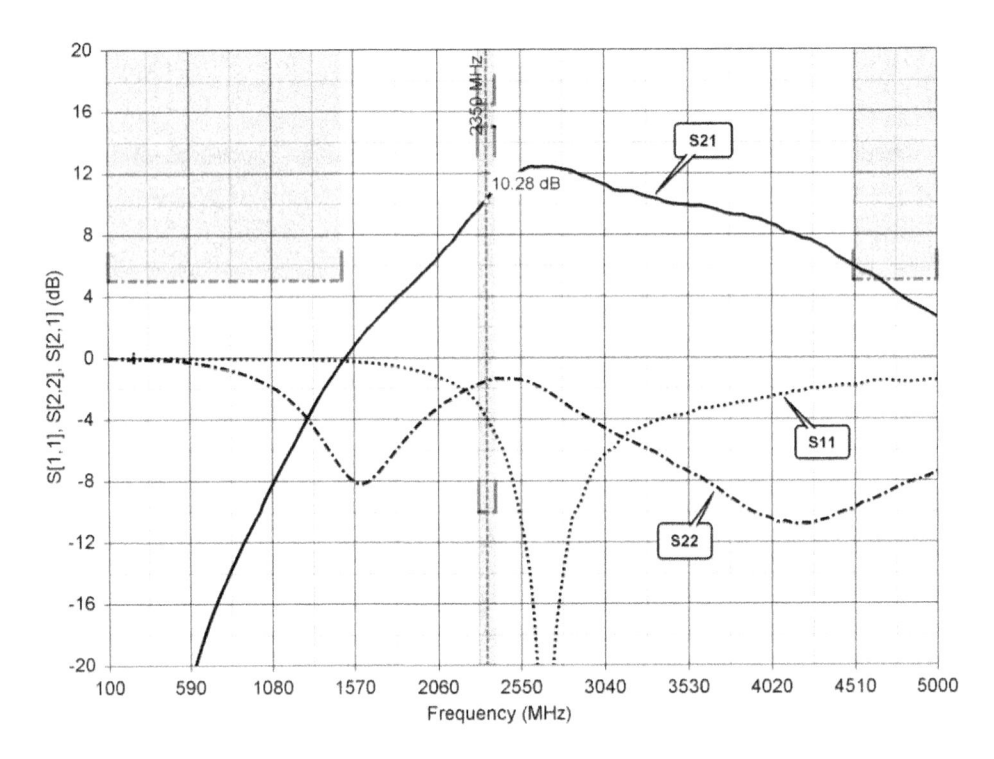

Figure 7-24 Initial response of the physical amplifier circuit of Figure 7-22

7.4.4 Optimization Setup Procedure

Add a Genesys Optimization through the Dsigns-Evaluations folder. The final specifications of the amplifier are given as:

- In-Band Frequency Range: 2.30 GHz – 2.40 GHz
    15.0 dB > Gain (S21) < 16.5 dB
    Input Return Loss (S11) < -10 dB
    Output Return Loss (S22) < -10 dB

- Out-of-Band Frequency Range: 100 MHz–1.5 GHz
    Gain (S21) < 5 dB

- Out-of-Band Frequency Range: 4.5GHz–5 GHz
    Gain (S21) < 5 dB

Select the proper Linear Analysis for the desired frequency range. On the Goals tab, select the measurements to be optimized as shown in Figure 7-25. Note that two line entries have been used to set the window for the gain flatness to lie between the target specification of 15 dB and 16.5 dB. Enter the frequency range in the Min and Max columns. The input and output return loss is simply set to be less than -10 dB. Then set the below-band gain to be less than 5 dB from 100 MHz to 1500 MHz. Finally set the above-band gain to be less than 5 dB from 4500 MHz to 5000 MHz. These goals are close to representing the response of the ideal amplifier response of Figure 7-15. Weighting can be assigned to the Goals to aid the optimizer in obtaining a solution. For this example the in-band response is determined to be much more important than the out-of-band response so the weighting is defined as 10 and 1 respectively.

Use	Measurement	Op	Target	Weight	Min	Max
✓	Linear2_Data.S[2,1]	<	16.5	10	2300	2400
✓	Linear2_Data.S[2,1]	>	15	10	2300	2400
✓	Linear2_Data.S[1,1]	<	-10	10	2300	2400
✓	Linear2_Data.S[1,1]	<	-10	10	2300	2400
✓	Linear2_Data.S[2,1]	<	5	1	100	1500
✓	Linear2_Data.S[2,1]	<	5	1	4500	5000
☐		<				

Figure 7-25 Setting up the goals of the amplifier optimization

On the Variables tab, select the Get Tuned Variables button to import all of the tunable elements to the Variable list. It is recommended to place a minimum and maximum component value on each element to constrain the Optimizer to stay within realistic values for the elements. The variable list and their minimum and maximum component values are shown in Figure 7-26.

Use	Variable	Min	Max
✓	Designs\PhysicalAmp\L4.L	1.6	10
✓	Designs\PhysicalAmp\L3.L	10	270
✓	Designs\PhysicalAmp\L2.L	1.6	20
✓	Designs\PhysicalAmp\C2.C	1	25
✓	Designs\PhysicalAmp\C1.C	.4	25
✓	Designs\PhysicalAmp\R4.R	10	100
✓	Designs\PhysicalAmp\R3.R	10	150
✓	Designs\PhysicalAmp\TL9.L	25	200
✓	Designs\PhysicalAmp\TL19.L	25	200

Figure 7-26 Importing the variables to be optimized and set their range

Execute the Optimization from the Workspace Tree by right clicking on the Optimization and select the type of optimization. Refer to the Genesys User's Guide for a thorough discussion on the various types of optimization algorithms [3]. For this optimization the Min-Max Gradient Search optimization was selected. The final optimized response is shown in Figure 7-27. The optimized component values are shown on the schematic of Figure 7-22. The amplifier's response has been retuned to be very close to the response of the ideal amplifier. Note that Genesys shows the boundary conditions for the optimization goals as shaded regions on the rectangular plot.

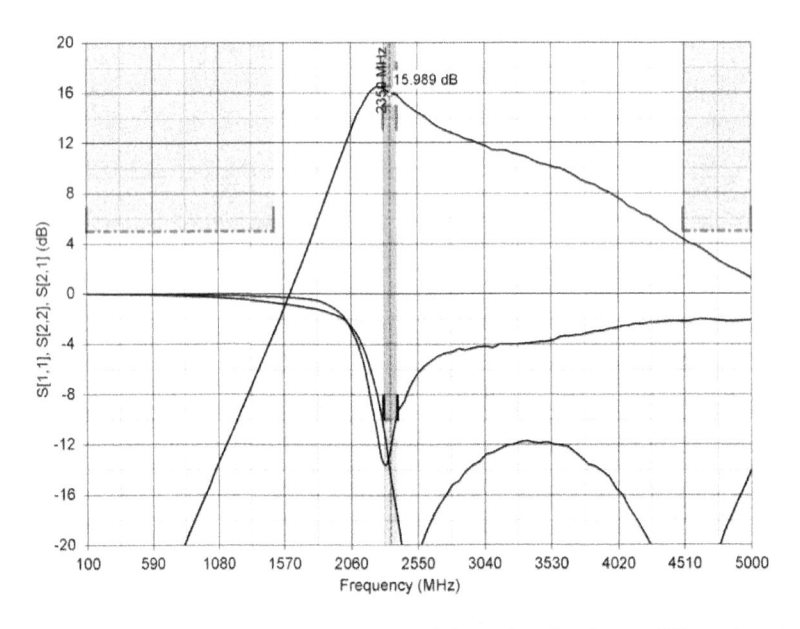

Figure 7-27 Final optimized response of the physical amplifier circuit

The measured response of the prototype amplifier was found to be very close to the optimized response of Figure 7-22. This exemplifies the importance of using the Genesys software to design and simulate a physical model of the amplifier. The photograph of the SHF-0189 prototype amplifier is shown in Figure 7-28.

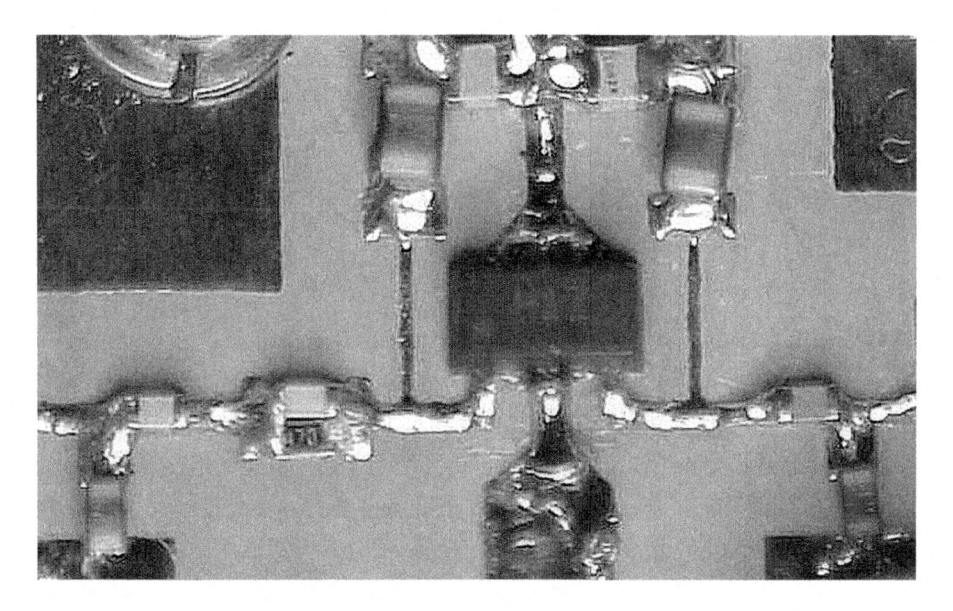

Figure 7-28 Photograph of SHF-0189 prototype amplifier

7.5 Specific Gain Amplifier Design

In the design of RF and microwave amplifiers, it is common to design with potentially unstable transistors. The designer must know how to deal with potentially unstable devices. This section outlines the design of a 2.3 GHz amplifier using the RT243 GaN HEMT device. The RT243 has a usable frequency range of 100 MHz to 5 GHz.

7.5.1 Specific Gain Match

There may be situations where the maximum gain is not desired from a given amplifier stage. This could be required to reduce the power and noise from one preceding amplifier stage to another thereby not over driving the succeeding stage. We will see in following sections that there are other reasons for matching to specific impedances such as matching for low noise or maximum output power. An amplifier can be matched to achieve any level of gain below its maximum conjugate gain, Gmax. A convenient means of designing for a specific gain level is through plotting the Constant Gain Circles on the Smith Chart. The Constant Gain circle is a locus of impedance values on the Smith Chart in which the selection of an impedance on the circumference of the circle will result in an amplifier that will achieve that specific gain. The two basic types of constant gain circle are:

1. Available Gain Circle, GA
2. Power Gain Circle, GP

The Available Gain Circle, GA, is a locus of impedance points that are referenced to the input of the active device. Selecting a point on the circumference of an Available Gain circle, GA, will selectively mismatch the input of the device to achieve the value of gain that corresponds to the Available Gain Circle. The GA value is therefore independent of the load impedance. The Available Gain can be defined as [4]:

$$GA = \frac{power\ available\ from\ the\ network}{power\ available\ from\ the\ source} \quad (7\text{-}10)$$

Once a given source impedance has been selected on the Available Gain Circle, an output impedance can be calculated to conjugately match the output of the device to the load. The Power Gain Circle, GP, is a locus of impedance that is referenced to the output of the device. Selecting a point on the circumference of a given Power Gain Circle, GP, will selectively mismatch the output of the device to achieve a specific value of gain. The Power Gain can be defined as [4]:

$$GP = \frac{power\ delivered\ to\ the\ load}{power\ input\ to\ the\ network} \quad (7\text{-}11)$$

Once the Power Gain Circle is plotted on the Smith Chart any location on the circumference of the circle can be chosen as the load impedance for

Single Stage Amplifier Design

the transistor. A Specific Gain at some value below the GMAX value of the device must be selected. The Constant Power Gain Circle for a specific value of gain can be derived from the following equations [4]. The center of the Power Gain Circle is defined as:

$$r_o = \frac{GC_2^*}{1+D_2G} \qquad (7\text{-}12)$$

where $G = \frac{SpecificGain}{|S_{21}|^2}$ note that the Specific Gain is the absolute gain value, not in dB

$$C_2 = S_{22} - D_S S_{11}^*$$

$$D_2 = |S_{22}|^2 - |D_S|^2$$

$$D_S = S_{11}S_{22} - S_{12}S_{21}$$

The radius of the Specific Gain Circle is then calculated from the following equation [4].

$$\rho_o = \frac{\sqrt{1 - 2K|S_{12}S_{21}|G + |S_{12}S_{21}|^2 G^2}}{1+D_2G} \qquad (7\text{-}13)$$

Once a given load impedance has been chosen on the circumference of a Power Gain Circle, an input impedance can be calculated that will conjugate match the input of the amplifier. The only caution in the selection of the load impedance on the circumference of the specific gain circle is to make sure that the impedance is not selected in a region in which the transistor could be unstable. In the Simultaneous Conjugate match of section 7.2.1, the device was forced to be unconditionally stable by a RC network placed at the input of the device. This is a useful technique when designing an amplifier so that the engineer can be confident that the device will not oscillate at frequencies inside or outside of the pass band regions. There may be cases, such as Low Noise Amplifier design, in which it would not be desirable to utilize an RC stabilization network because of the degradation of noise figure. In any event it is not strictly necessary to include a stabilization network if the designer selects source and load impedances such that they do not enter unstable regions. The stability circles represent the boundaries between those values of source and load impedances that are stable and unstable [3]. Selecting a load impedance in an unstable region may cause the device to oscillate. The perimeter of the circle represents the points at which K = 1. The following equations can be used to analytically define the stability circles on the Smith Chart. The center of the device input stability circle [4].

$$r_{s1} = \frac{C_1^*}{|S_{11}|^2 - |D_S|^2} \qquad (7\text{-}14)$$

where, $C_1 = S_{11} - D_S S_{22}^*$

The radius of the device input stability circle is defined by [4]:

$$\rho_{S1} = \left| \frac{S_{12} S_{21}}{|S_{11}|^2 - |D_S|^2} \right| \qquad (7\text{-}15)$$

The center of the device output stability circle is defined as [4]:

$$r_{s2} = \frac{C_2^*}{|S_{22}|^2 - |D_S|^2} \qquad (7\text{-}16)$$

The radius of the device output stability circle is defined by

$$\rho_{S2} = \left| \frac{S_{12} S_{21}}{|S_{22}|^2 - |D_S|^2} \right| \qquad (7\text{-}17)$$

Once the stability circles are plotted on the chart the stable and unstable sides of the circle must be determined. If the circle surrounds the center of the Smith Chart, then the inside of the circle is the stable region of impedances for that device. If the circle does not encircle the center of the chart, then the region outside the center of the circle represents stable

Single Stage Amplifier Design   351

impedances for the device. This relationship assumes that both S11 and S22 are less than one, when measured in a 50 Ω system.

7.5.2 Specific Gain Design Example

Example 7.5-2: Using the RT243 device, design a specific gain amplifier that achieves 16 dB gain at a frequency of 2.3 GHz. Examine the stability circles by creating a schematic with the small signal S parameter file.

Solution: Setup a Linear Analysis with both the start and stop frequencies set at 2.3 GHz with 2 points.

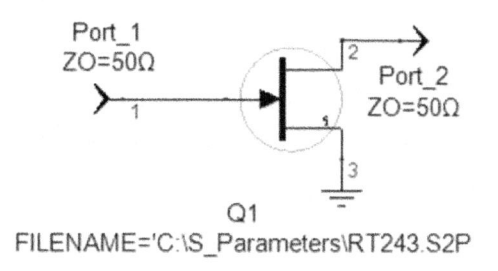

```
       Port_1
       ZO=50Ω                  2
                               →  Port_2
         →                        ZO=50Ω
          1
                               3
                      Q1       ─┬─
              FILENAME='C:\S_Parameters\RT243.S2P
```

Figure 7-29 Schematic with RT243 S parameter file

A sweep of the circuit will result in a plot of the center of each stability circle displayed. To display the entire circle, click on the center trace (dot) on the Smith Chart. As Figure 7-30 shows the circles are far away from the center of the chart therefore making the outside of the circles the stable region. There is only a small region of the output stability circle that represents an unstable region of impedance. These unstable regions are shaded as shown in Figure 7-30. If the device were unconditionally stable the stability circles would lie completely outside the circumference of the Smith Chart. Add another Linear Analysis from 100 MHz to 3000 MHz with 100 MHz steps. Then add a Table to display K, B1, and GMAX. Note that stability factor k is less than 1 indicating that the transistor is potentially unstable over its entire frequency range. G_{max} for the device without a stabilization network is 22.6 dB at 2.3 GHz. We want to design the amplifier to have a gain of 16 dB with a 50 Ω source and load impedance.

352   Microwave and RF Engineering

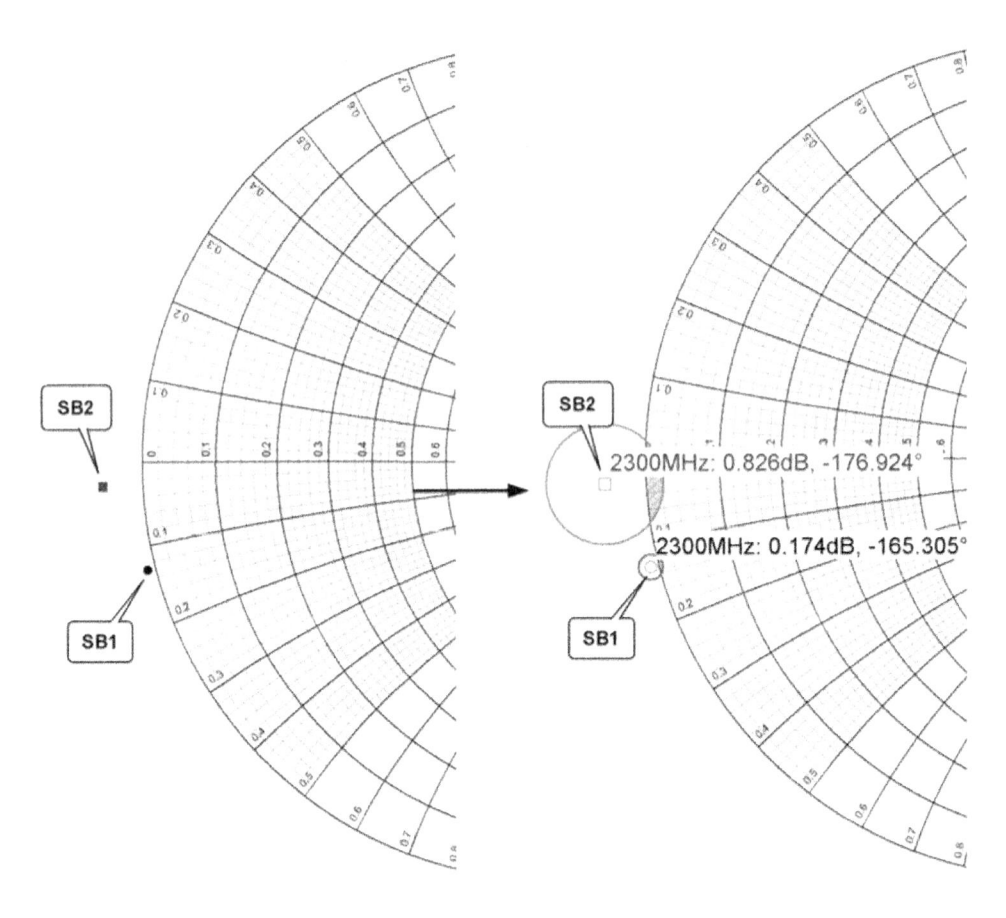

Figure 7-30 Stability circle centers and full circles at 2.3 GHz

F (MHz)	K ( )	B1 ( )	GMAX (dB10)
100	0.272	0.966	35.753
300	0.122	0.838	30.862
500	-0.236	0.904	28.363
700	0.266	0.853	27.462
900	0.316	0.845	26.4
1100	0.352	0.826	25.528
1300	0.379	0.8	24.778
1500	0.419	0.769	24.125
1700	0.43	0.739	23.546
1900	0.463	0.711	23.035
2100	0.537	0.69	22.75
2300	0.654	0.666	22.61
2500	0.802	0.641	22.448
2700	0.861	0.625	22.188
2900	0.931	0.609	21.911
3000	0.97	0.601	21.765

Table 7-1 RT243 device stability factors and GMAX

Fortunately Genesys has the ability to plot a series of constant gain circles. This enables the plot of the circles on the Smith Chart without the need to solve the Equations (7-5) and (7-6). Selecting the GP function (power gain circles) in the output graph plots a series of seven gain circles. When K < 1 the smallest circle represents the maximum gain, GMAX, and the inside of the circle is shaded [3]. This means that impedances inside this circle can be unstable. The successive Power Gain circles represent gain values of GMAX -1dB, GMAX-2dB, GMAX-3 dB, GMAX-4 dB, GMAX-5 dB, and GMAX-6 dB. Thus the circles represent the gain values of: 22.6 dB, 21.6 dB, 20.6 dB, 19.6 dB, 18.6 dB, 17.6 dB and 16.6 dB respectively. Figure 7-31 shows that the last or largest circle is the 16.6 dB power gain circle. An examination of the power gain and stability circles reveals another reason why we may not want to use the maximum gain from the device. The highest gain circle is very close to the region of instability on the output stability circle, SB2.

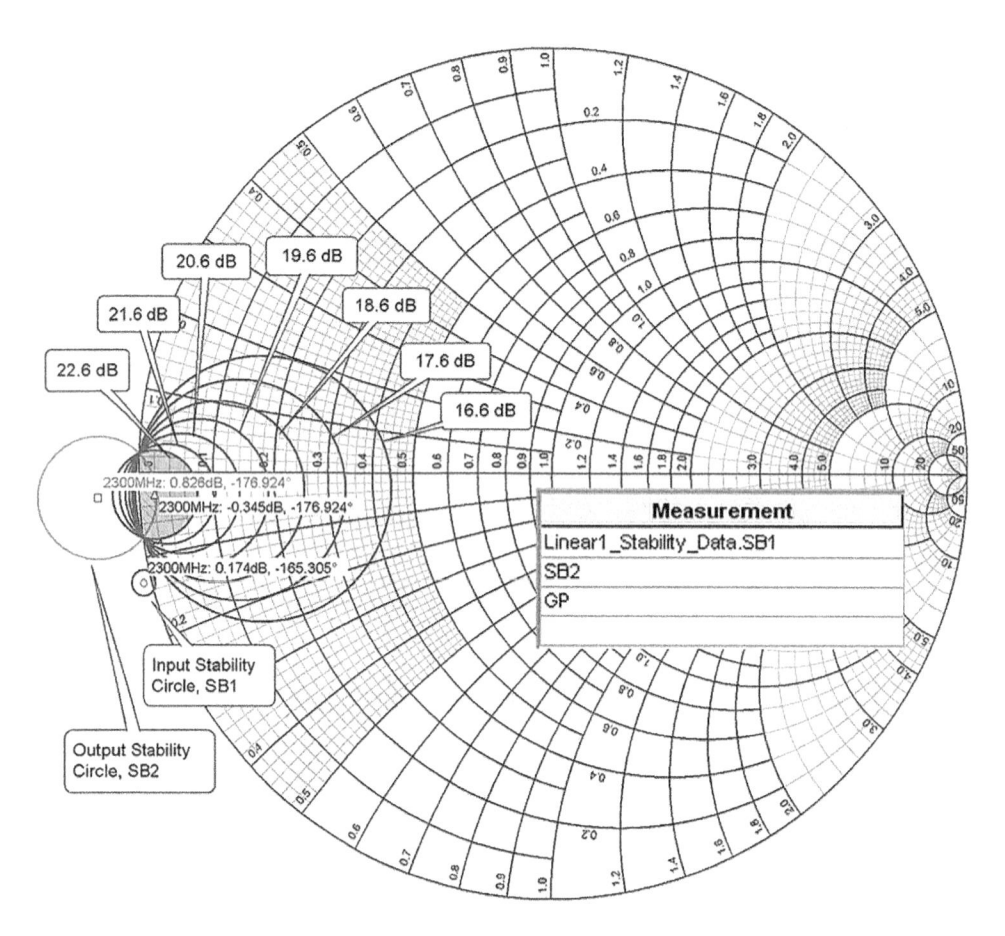

Figure 7-31 Constant power gain circles for the RT243 at 2.3 GHz

We'll choose the 16.6 dB circle because it is the closest to the design goal of 16 dB. The load impedance can be any point on the circumference of the 16.6 dB circle that does not enter the shaded, unstable, region. It is a good practice to choose locations that are not close to the unstable regions of the stability circles. The load impedance can be visually read from the Smith Chart. The load impedance selected on the 16.6 dB Power Gain Circle is shown in Figure 7-32. The impedance at this point is $22 + j0$ Ω. To precisely identify this impedance a new design schematic can then be created with an IMP element to define this impedance on the Smith Chart. The IMP element impedance can be plotted on the same Smith Chart as the Power Gain circles. The IMP element's resistance and reactance can be tuned to get the impedance to exactly overlay on the desired point on the circumference of the 16.6 dB Power Gain circle.

Single Stage Amplifier Design 355

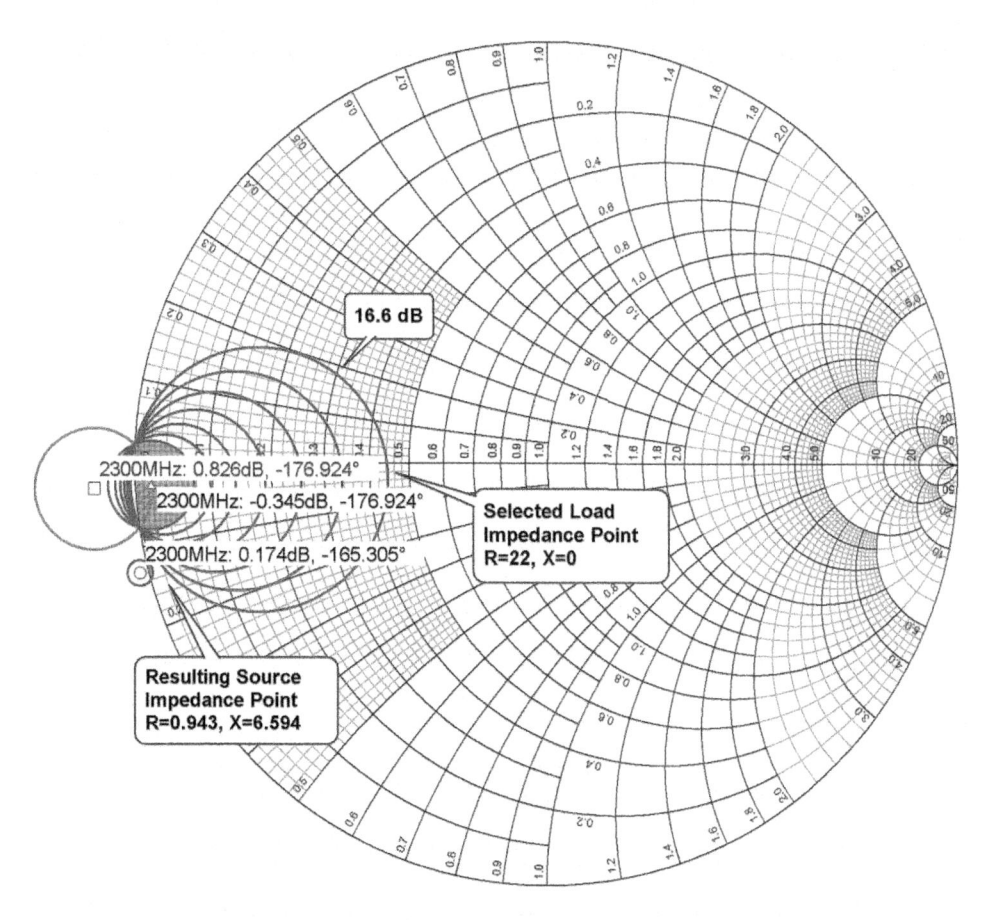

Figure 7-32 Location of source and load impedance for power gain match

Once the load impedance has been selected it is necessary to calculate the resulting source reflection coefficient that will achieve a conjugate match on the input. Equation 7-18 is used to calculate this source reflection coefficient [4]. The source reflection coefficient is therefore defined as a function of the chosen load reflection coefficient, Γ_L. Note that the input match is then a conjugate match of the calculated source reflection coefficient, Γ_S.

$$\Gamma_S = \left[S_{11} + \frac{S_{12}S_{21}\Gamma_L}{1-(\Gamma_L S_{22})} \right]^* \qquad (7\text{-}18)$$

The chosen load impedance must first be converted to a reflection coefficient, Γ_L. This is easily handled by the Equation Editor, as shown in

Figure 7-33. Simply define the Load variable as the reflection coefficient S[1,1]. Once the load impedance is defined as a polar formatted variable, Equation (7-18) can be solved within the same Equation Editor. Because two frequency step points were needed to display the circles, the equations are entered as swept array variables. The source impedance is calculated as 0.943-j6.594 as shown in Figure 7-33. Location of the source and load impedance for power gain match is shown in Figure 7-32.

```
1    'Enter Actual Load Impedance from Gain Cirle
2    RL=22
3    XL=0
4    'Create Complex number for the Load Impedance
5    'Define Load Imedance using the IMP element with
6    'Linear Analysis 2
7    Load=Linear2_Data.S[1,1]
8    'Calculate Source Reflection Coefficient
9    'Make sure to have same number of points in both
10   'Linear1 and Linear2 analysis
11   Num=Linear1_Data.S[2,1].*Linear1_Data.S[1,2].*Load
12   Denom=1-(Load.*Linear1_Data.S[2,2])
13   Source=Linear1_Data.S[1,1].+(Num/Denom)
14   'calculate the conjugate of the Source Impedance
15   Zsource=gammatoz(50,Source)    'polar to rectangular
16   Zsource_real=real(Zsource[1])
17   Zsource_imag=-imag(Zsource[1])    'make conjugate
18   'Enter the values of Zsource_real and Zsource_imag
19   'to an IMP element to perform the Input Matching
20   'Circuit Design
```

Figure 7-33 Solution of Equation (7-18) in Equation Editor

7.5.3 Graphical Impedance Matching Circuit Design

Example7.5-2: use the Smith Chart to graphically design the input and output matching networks.

Solution: The simple two-element LC matching circuit can be synthesized for both the input and output matching circuits. Setup a design schematic for the design of the output matching circuit. Use the generic impedance element, IMP, to define the load impedance. Then place a 50 Ω resistor to represent the 50 Ω output impedance in which the load impedance is to be matched. Place a capacitor, C1, in shunt with the 50 Ω resistor. Initially set the inductor value, L1, to 0 nH. Then tune the value of the capacitor, C1, until the resulting impedance is positioned on a constant resistance circle that is aligned with the load impedance. Then tune the inductance

Single Stage Amplifier Design

value to bring the resulting impedance to overlay the load impedance. The resulting LC values are shown in Figure 7-34.

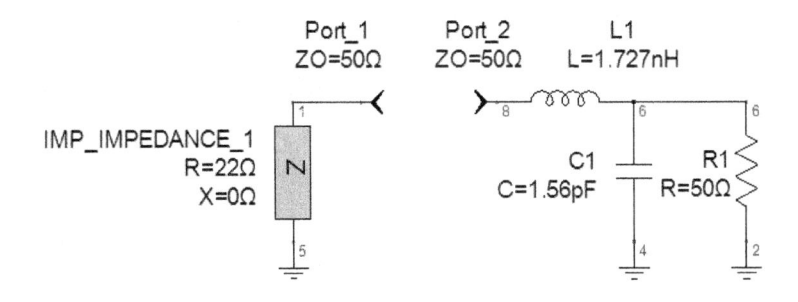

Figure 7-34 Schematic used to design the output matching circuit

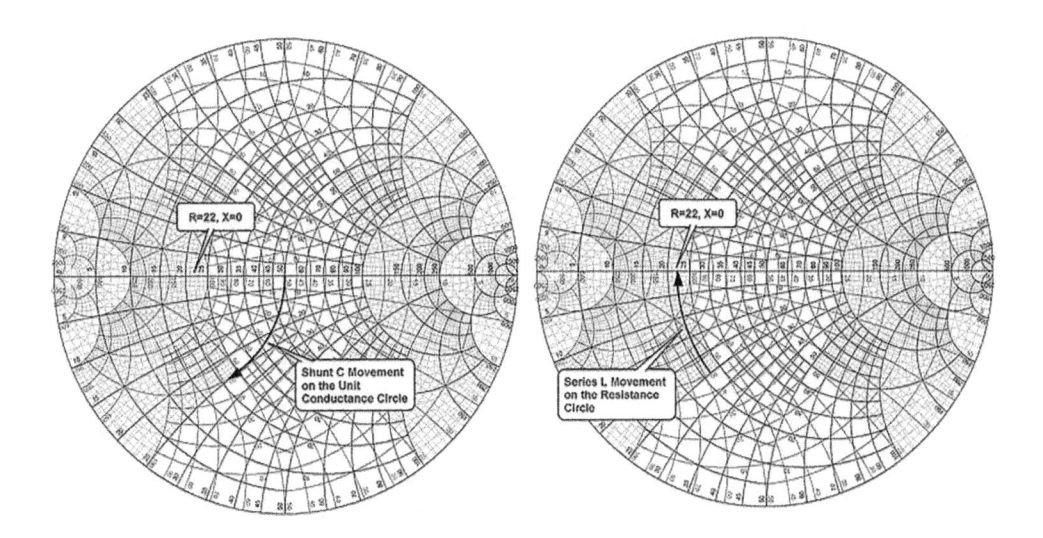

Figure 7-35 Shunt C and series L, output impedance match progression

In similar fashion the input matching circuit is designed by initially setting the inductor value of L1 to zero. Tune the capacitance value to a constant resistance circle near the source impedance. In this case the circuit is almost matched. Only a very small inductance, 0.014 nH, is required to complete the match. Here we see that the source impedance is located very near to the input stability circle. Very careful component selection would be required to make sure that the matching circuit would not push the match into the area of unstable performance. Any component tolerance or drift over temperature may result in an unstable amplifier circuit. It is this reason why an RC stability circuit, as included in section 7.2.1, is sometimes preferred.

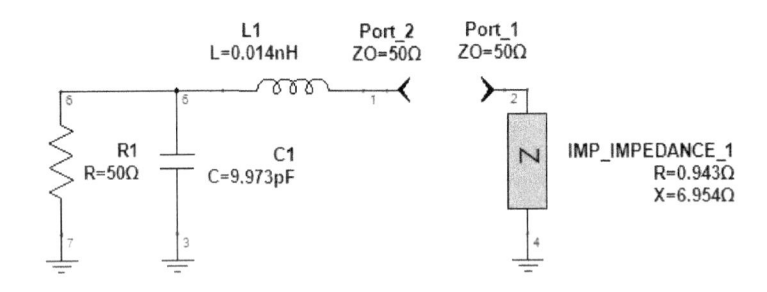

Figure 7-36 Schematic used to design the input matching circuit

The shunt C – series L, input impedance match progression is shown in Figure 7-37.

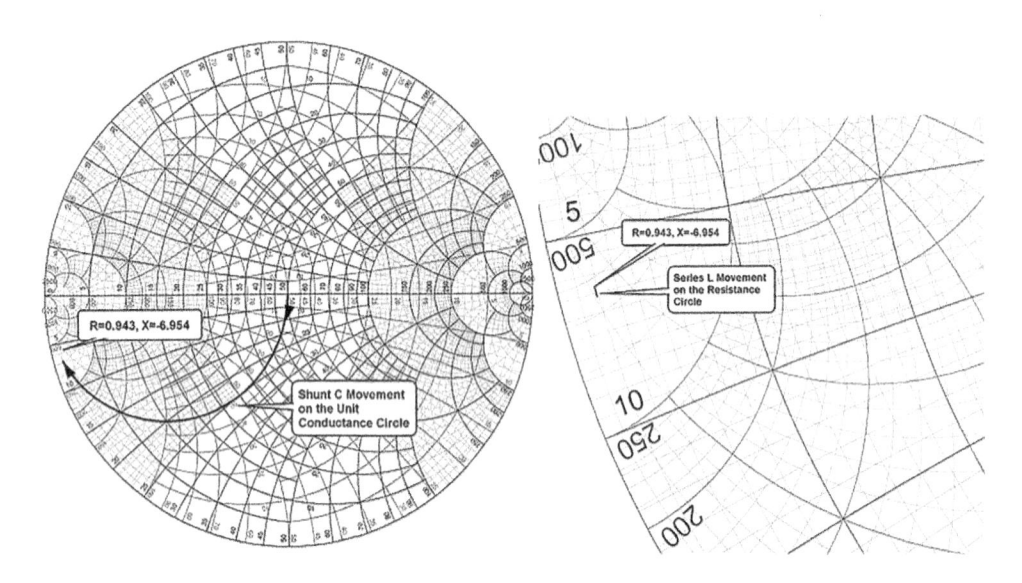

Figure 7-37 Shunt C–series L, input impedance match progression

7.5.4 Assembly and Simulation of the Specific Gain Amplifier

Example 7.5-3: Assembly and Simulation of the Specific Gain Amplifier.

Single Stage Amplifier Design 359

Solution: Create a new schematic and attach the input and output matching circuits to the device S parameter file. Create a Linear Analysis and sweep the amplifier from 1.3 GHz to 3.3 GHz. The resulting responses are shown in Figure 7-39. Note that the gain at 2300 MHz is 16.64 dB verifying that the Power Gain circle match has been successful. Also note a major difference in return loss compared to the simultaneous conjugate match case. Here we see that the output return loss is rather poor, -1.39 dB. This is a result of the selective mismatching process to achieve a much lower value of gain than the GMAX of the device. The input return loss is very good because a conjugate match of the resulting device and load impedance combination was performed.

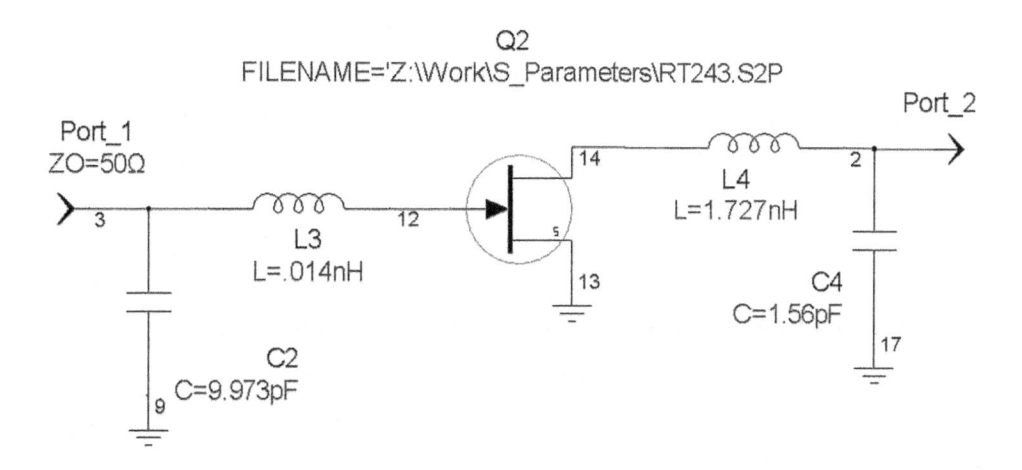

Figure 7-38 Ideal amplifier schematic for power gain matching example

Example 7.5-4: Construct the physical design for the power gain matched amplifier.

Solution: Begin with the package outline drawing of the transistor as was done in Section 7.3. Then progress with the microstrip PCB tracks and physical models for the inductors and capacitors. Clearly the 0.014 nH series inductor and possibly the 1.727 nH inductor will be absorbed into the physical design of the circuit. Finally optimize the amplifier to meet the original design goals.

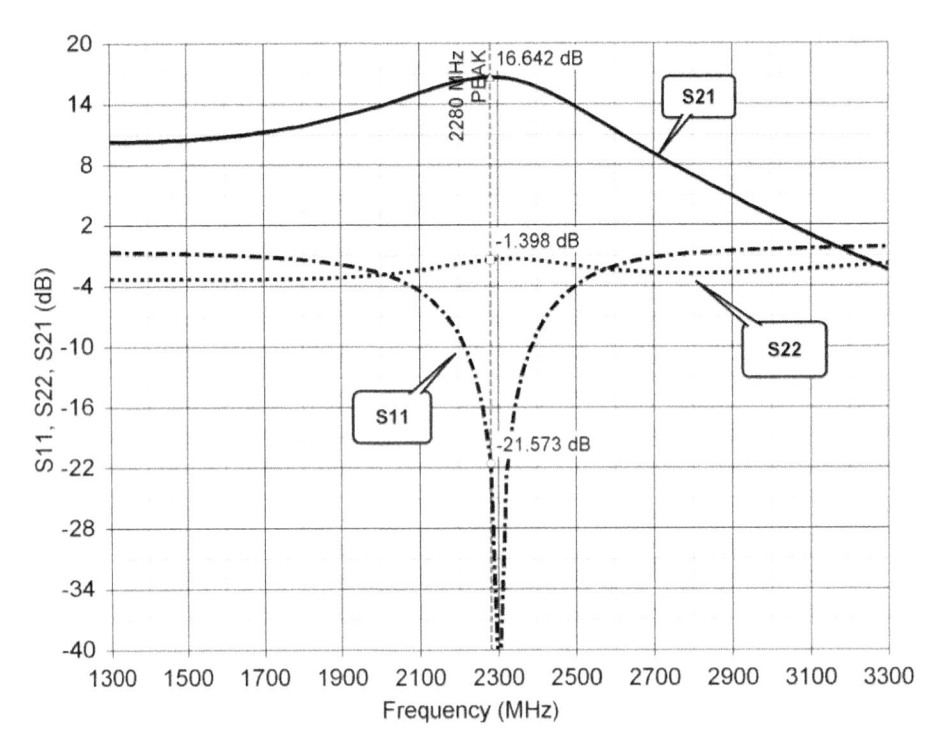

Figure 7-39 Response of the power gain matched amplifier

7.6 Low Noise Amplifier Design

One important case of selectively mismatched amplifier design is the Low Noise Amplifier. In LNA design the input is not matched to the reflection coefficient that results in maximum gain amplifier but rather to a reflection coefficient that gives the desired noise figure. Low noise amplifiers are frequently used as the input stage in a radio receiver or satellite down converter to minimize the noise that is added to the amplified signal. Figure 7-40 depicts a signal with noise at the input and output of an LNA.

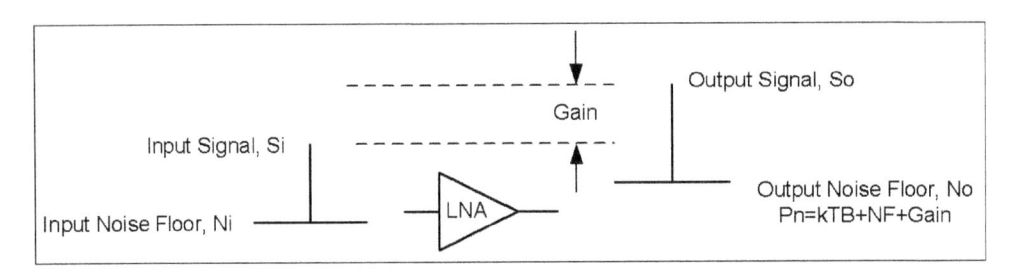

Figure 7-40 Signal and noise through the LNA

Single Stage Amplifier Design

All transistors will add some amount of noise to the input signal. Low Noise transistors are optimized so that they will add a minimum amount of noise to the signal. The signal at the output of a linear LNA is simply the signal level at the input plus the gain of the LNA in dB. However the noise at the output will be increased by the gain and the noise figure of the transistor. The noise figure of the LNA is the degradation of the signal to noise ratio of the input of the LNA to the signal to noise ratio of the output as given by Equation (7-19).

$$NF_{dB} = 10 \cdot \log \left[\frac{\left(S_i / N_i \right)}{\left(S_o / N_o \right)} \right] \quad (7\text{-}19)$$

The noise is considered to be a white noise distribution meaning that it is constant with frequency and thermal in nature. Therefore any measurement of noise must be referred to a specific bandwidth. The thermal noise power is also a function of temperature and is defined by Equation (7-20).

$$P_n = kTB \quad (7\text{-}20)$$

where,
    k = Boltzmann's constant, 1.374×10^{-23} J/°K
    T = Temperature of the input noise source (290 °K = room temperature)
    B = Bandwidth of Measurement in Hz

If we normalize the measurement bandwidth to 1 Hz i.e., B=1 the thermal noise can be calculated using Equation (7-20):

$$P_n = \left(1.374 \cdot 10^{-23}\right)(290)(1) = 3.984 \cdot 10^{-21} \; Watts/Hz \; or \; 3.984 \cdot 10^{-18} \; mW/Hz$$

Converting mW to dBm, we can express the thermal noise floor as:

$$P_{no} = 10 \cdot \log\left(3.984 \cdot 10^{-18}\right) = -173.9 \; dBm/Hz$$

Working with the thermal noise floor normalized to a 1 Hz bandwidth it is straight forward to calculate the output noise from the amplifier using the amplifier's noise figure and gain as given by Equation (7-21).

$$P_n(1Hz) = kTB_{1Hz} + Gain_{dB} + NoiseFigure_{dB} \qquad (7\text{-}21)$$

Using Equation (7-22) we can normalize the noise power in a one Hertz bandwidth to any measurement bandwidth.

$$P_n(B) = P_n(1Hz) + 10 \cdot \log(B) \qquad (7\text{-}22)$$

where B = the measurement bandwidth in Hz

7.6.1 Noise Circles

In order to perform noise figure simulation in Genesys the device S parameter file must include noise parameters. Figure 7-41 shows an S parameter file with noise parameters appended to the end of the file. The noise parameters include the optimum noise figure, NF_{opt} at a corresponding frequency in the left most column. The NF_{opt} is the lowest possible noise figure that can be achieved with this device when it is matched to Γ_{opt}. The third and fourth column gives the magnitude and angle of the Γ_{opt}. The final column gives the normalized noise resistance for the device. These noise parameters are usually provided by the device manufacturer on the data sheet.

Single Stage Amplifier Design

```
!S and NOISE PARAMETERS at Vce=2.7V  Ic=5mA.
# ghz s ma r 50
0.1    0.88   -11    13.32  169   0.01   83   0.97   -5
0.5    0.70   -48    10.49  134   0.04   64   0.82   -22
0.9    0.49   -72     7.73  111   0.05   57   0.70   -30
1.0    0.46   -77     7.15  107   0.06   55   0.69   -31
1.5    0.30  -100     5.27   89   0.07   53   0.62   -35
1.8    0.24  -112     4.52   81   0.08   52   0.59   -37
2.0    0.21  -120     4.13   76   0.09   52   0.58   -39
2.4    0.16  -140     3.52   67   0.10   50   0.57   -42
3.0    0.13  -172     2.88   55   0.12   48   0.55   -46
4.0    0.15   137     2.25   38   0.15   44   0.53   -55
5.0    0.21   106     1.87   22   0.18   39   0.52   -64
! Noise Parameters
!Freq  NFopt   Γopt        rn=Rn/50
0.5    1.1    0.77   9      1.10
0.9    1.2    0.71  18      0.96
1.8    1.5    0.60  45      0.66
2.4    1.8    0.51  65      0.47
```

Figure 7-41 AT30511 S parameter file with noise parameters included

Matching the device to Γ_{opt} can potentially lead to very poor input return loss and instability. Just as we seen in the example of section 7.4.2 the designer must be aware of the location of Γ_{opt} with respect to the input stability circle. In LNA design we usually do not want to add resistive loading to the input of the device to improve the stability because this would increase the thermal noise power and increase the noise figure. In certain critical designs the poor input return loss is accepted and an isolator is added to the input of the LNA to provide a good input return loss. This also involves tradeoffs as the losses in the isolator add directly to the noise figure. A trade off is required among input return loss, stability, and accepting a higher noise figure than that which is provided by Γ_{opt}. A reflection coefficient can be chosen that corresponds to a noise figure that is greater than NF$_{opt}$. It is sometimes helpful to examine the constant noise figure contours on the Smith Chart. The noise contours are a family of noise figure circles on the Smith Chart that represent the reflection coefficients that correspond to a variety of resulting noise figures for a given device. To plot the noise figure circles in Genesys, create a Linear Analysis with a two-point frequency step. Create a Smith Chart graph and plot the NCI, noise circle contours. A family of eight circles is plotted as shown in Figure 7-42. These circles represent noise figure degradation from NF_{opt} in the increments of: NF_{opt} + 0.25, 0.5, 1, 1.5, 2, 2.5, 3 and 6 dB. Therefore the inner most circle represents (NF_{opt} + 0.25) dB, the second circle represents (NF_{opt} + 0.5) dB and so on. The circles can also be helpful when designing a wide band LNA. It is sometimes helpful to plot the stability circles along with the noise circles so that a given noise match reflection coefficient can be selected that is not too close to the stability circles. The resulting noise figure can be calculated for any source reflection coefficient, Γs, from Equation (7-23) [9].

$$NF = NF_{min} + \frac{4R_n}{Z_0\left|1+\Gamma_{opt}\right|^2} \cdot \frac{\left|\Gamma_s - \Gamma_{opt}\right|^2}{\left(1-\left|\Gamma_s\right|^2\right)} \tag{7-23}$$

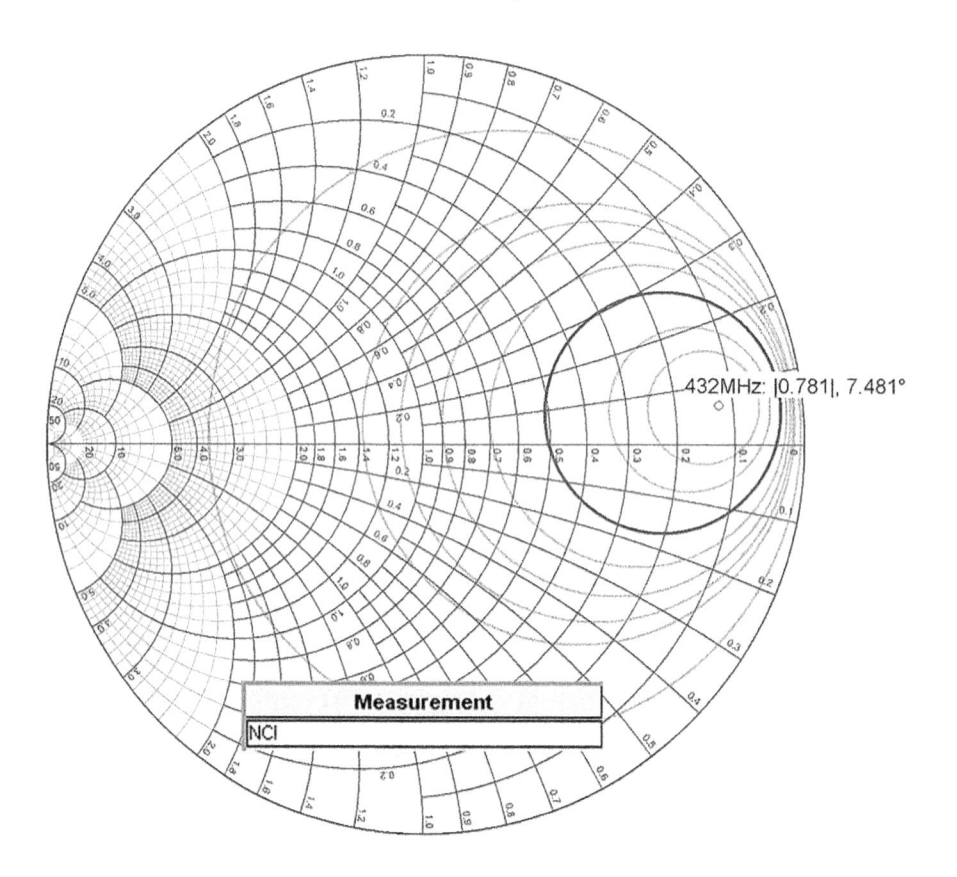

Figure 7-42 Noise Figure circles and Γ_{opt}

Single Stage Amplifier Design 365

7.6.2 LNA Design Example

Example 7.6-1: Design a single stage Low Noise Amplifier for a UHF satellite downlink. The amplifier is intended to operate with a source and load impedance of 50 Ω. The design specifications are given as:

Center Frequency:	432 MHz
Gain:	20 dB minimum
Noise Figure:	1.2 dB maximum
Output Return Loss:	Less than -10 dB

Solution: The Avago AT30511 low noise transistor will be used for this design. The S parameter and noise parameter file for this device is shown in Figure 7-41. Create a Genesys schematic with the device S parameter file as shown in Fig 7-43. Add a Table to display the stability parameters K and B1 and the noise parameters Γ_{opt} and Z_{OPT}. Also add a Smith Chart with Γ_{opt} and the stability circles. Then create a Smith Chart and display the stability circles along with the noise circles.

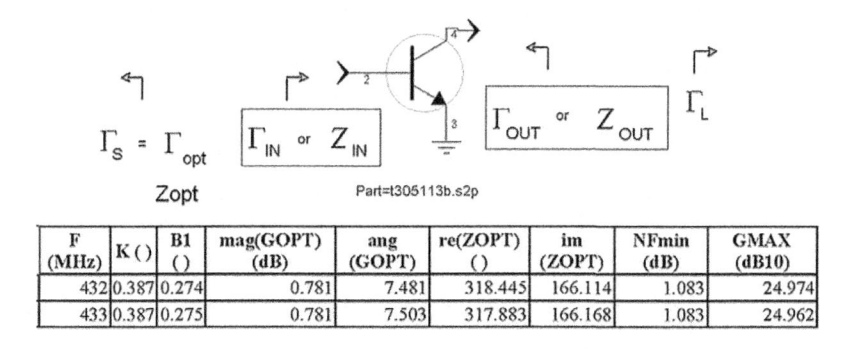

F (MHz)	K ()	B1 ()	mag(GOPT) (dB)	ang (GOPT)	re(ZOPT) ()	im (ZOPT)	NFmin (dB)	GMAX (dB10)
432	0.387	0.274	0.781	7.481	318.445	166.114	1.083	24.974
433	0.387	0.275	0.781	7.503	317.883	166.168	1.083	24.962

Figure 7-43 AT30511 S parameter, K, B1, Γ_{opt}, Z_{opt}, NFmin, and GMAX

Figure 7-43 shows that K < 1 meaning that the device is potentially unstable. This means that there are source and load reflection coefficients that can result in instability and oscillation. Figure 7-44 shows a plot of the input stability circle along with the noise circles and Γ_{opt}. In this example, the inside of the input stability circle, SB1, represents the region of stable source reflection coefficients. Genesys shades the region of the stability circle that represents unstable reflection coefficients. Because the input reflection coefficient, Γ_{opt}, is well inside of the input stability circle, SB1, and we have no specification for input return loss, we can proceed with matching the input to Γ_{opt}. From the discussion of Specific Gain matching of Section 7.4.1 we can deduce that because a specific reflection coefficient is selected at the input of the device, the available gain circles can be used to determine the expected gain from the device when matched to Γ_{opt}. Figure 7-45 shows the available gain circles plotted along with Γ_{opt}. Gmax was defined as 24.9 dB in Figure. 7-43. The available gain circles are drawn for gains of 0, 1, 2, 3, 4, 5 and 6 dB less than maximum gain. Figure 7-45 shows the available gain circles for the AT30511 at 432 MHz. The 21.9 dB circle intersects Γ_{opt} therefore the expected gain of the device when matched to Γ_{opt} is 21.9 dB.

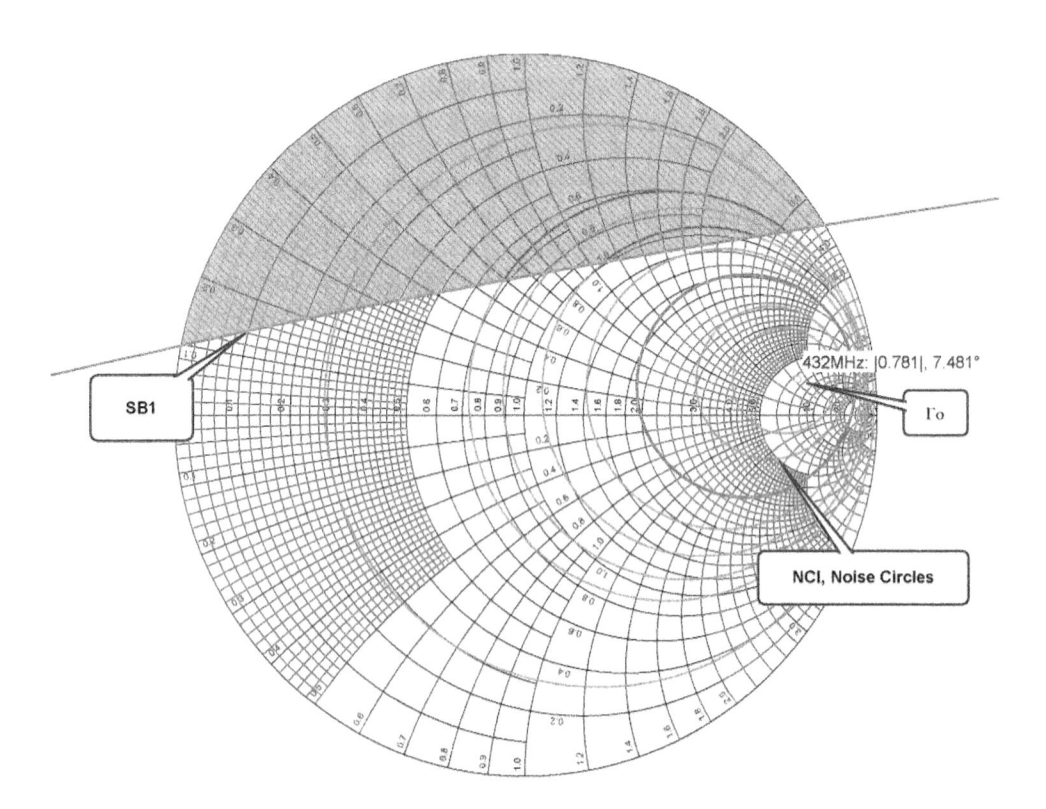

Figure 7-44 Input stability circle with noise circles

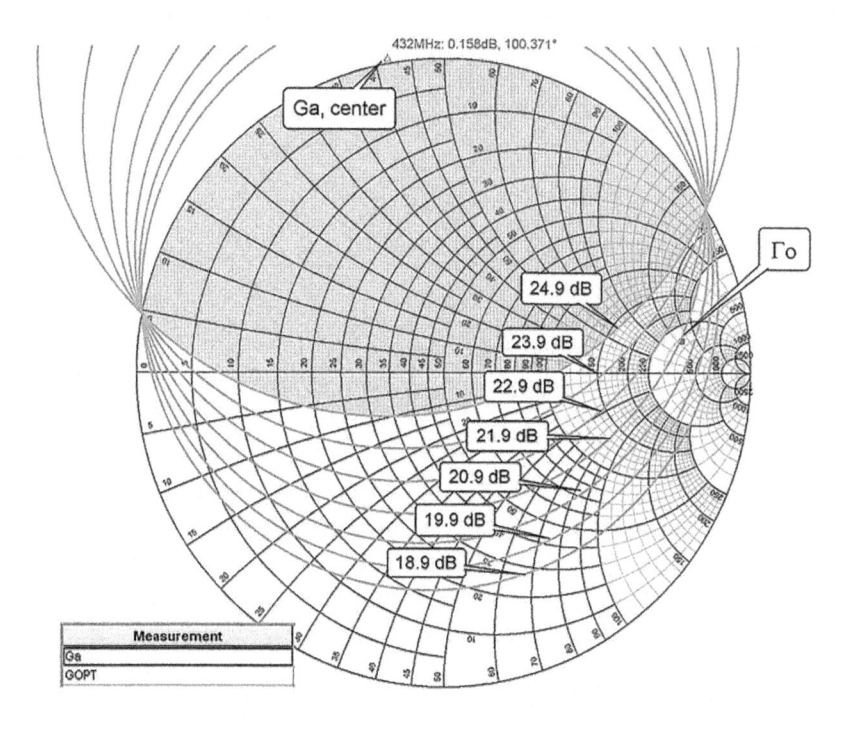

Figure 7-45 Available gain circles and Γ_{opt}

7.6.3 Analytical Design of the LNA Input Matching Network

Example7.6-2: Design the LNA Input Matching Network

Solution: The input and output matching circuits will be analytically designed using the single-stub matching networks. Because we are designing essentially a fixed frequency LNA, a simple single-stub matching network can be used. The equations developed in chapter 6 section 6.4 will be used to determine the input and output matching networks. Referring to Figure 7-43, we match the 50 Ω source impedance to the input impedance Z_{in} of the device. Z_{opt} is the impedance looking from the device back toward the 50 Ω source. Therefore, in the design of the input matching network, we must use the conjugate of Z_{opt} for the complex load impedance. Figure 7-43 gives $Z_{opt} = 318.4 + j166.1$ Ω, therefore, Z_{in} is defined as $Z_{opt}^* = 318.4 - j166.1$ Ω. As Figure 7-46 shows there are two single-stub matching networks that match the 50 Ω source impedance to the load impedance.

```
1   %Enter input souce and Zopt impedances
2   RS=50
3   RL=318.6
4   XL=-166.1
5   f=432e6
6   %Normalize the load impedance
7   r=RL/RS
8   x=XL/RS
9   %Calculte t1 and t2
10  t1=(x+sqrt(r*(r^2+x^2-2*r+1)))/(r-1)
11  t2=(x-sqrt(r*(r^2+x^2-2*r+1)))/(r-1)
12  %Calculate the electrical lenghts of the lines d1 and d2
13  d1=360*(pi+atan(t1))/(2*pi)
14  d2=360*(pi+atan(t2))/(2*pi)
15  %Calculate B1 and B2
16  B1=(x*t1^2+(r^2+x^2-1)*t1-x)/(RS*(r^2+x^2+t1^2+2*x*t1))
17  B2=(x*t2^2+(r^2+x^2-1)*t2-x)/(RS*(r^2+x^2+t2^2+2*x*t2))
18  %Calculate the electrical lenghts of the open-ended stubs
19  so1=360*(pi-atan(RS*B1))/(2*pi)
20  so2=-360*atan(RS*B2)/(2*pi)
```

Figure 7-46 Calculation of the single-stub input matching network

For the input matching network, we select the shorter open-stub length so2 = 68.216 degrees and the corresponding line d2 = 105.58 degrees, as shown in Figure 7-47.

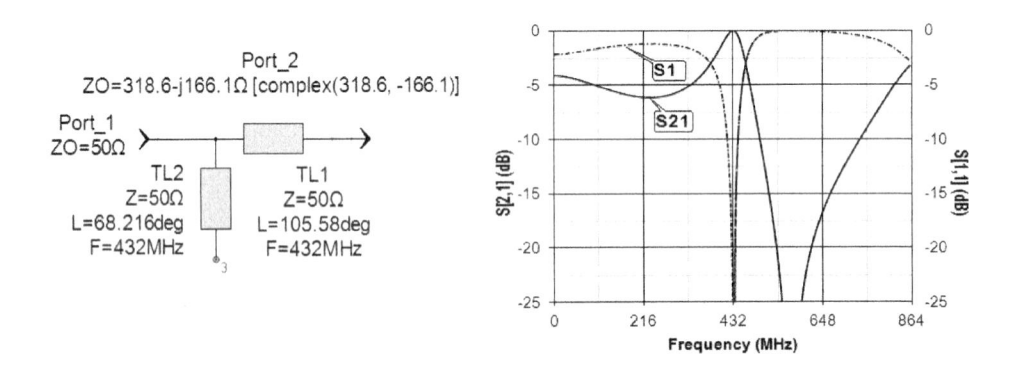

Figure 7-47 Single-stub input matching network and simulated response

7.6.4 Analytical Design of the LNA Output Matching Network

Example 7.6-3: Design the LNA Output Matching Network

Single Stage Amplifier Design

The output load reflection coefficient, Γ_L, is defined in terms of the selected source reflection coefficient, Γ_S, as given by Equation (7-24) [4]. The Genesys Equation Editor is used to calculate the load reflection coefficient, Γ_L.

$$\Gamma_L = \left(S_{22} + \frac{S_{12}S_{21}\Gamma_S}{1 - S_{11}\Gamma_S} \right)^* \qquad (7\text{-}24)$$

where, $\Gamma_S = \Gamma_{opt}$.

```
1   'Obtain Device S-Parameters from Linear1 Dataset
2   Num=Linear1_Data.S(1,2).*Linear1_Data.S(2,1).*Linear1_Data.GOPT
3   Denom=1-(Linear1_Data).S(1.1).*Linear1_Data.GOPT)
4   GL=Linear1_Data.S(2,2).+(Num./Denom)
5   GL=conj(GL)
6   GL_mag=mag(GL(1))
7   GL_ang=ang(GL(1))*(180/PI)
8   'Calculate Load Impedance
9   ZL=gammatoz(50,ZL)
10  Zload_real=real(ZL(1))
11  Zload_imag=imag(ZL(1))
```

Figure 7-48 Equation Editor solution of Γ_L and Z_L

To use the impedance matching utility we need the conjugate of Z_L or Z_{OUT} of the device or in this case: $Z_{OUT} = 71.9 - j54.4 \, \Omega$. Check the location of Γ_L with respect to the output stability circle to make sure that the impedance does not lie in the region of instability. Figure 7-49 shows the location of Γ_L with respect to the output stability circle, SB2. The impedance corresponding to Γ_L can be entered into an impedance element and plotted on the same graph as the output stability circle. The inside of the output stability circle is shaded which means that this area represents unstable output load reflection coefficients. Because Γ_L is safely outside of the SB2 circle we can use Γ_L as an acceptable output reflection coefficient.

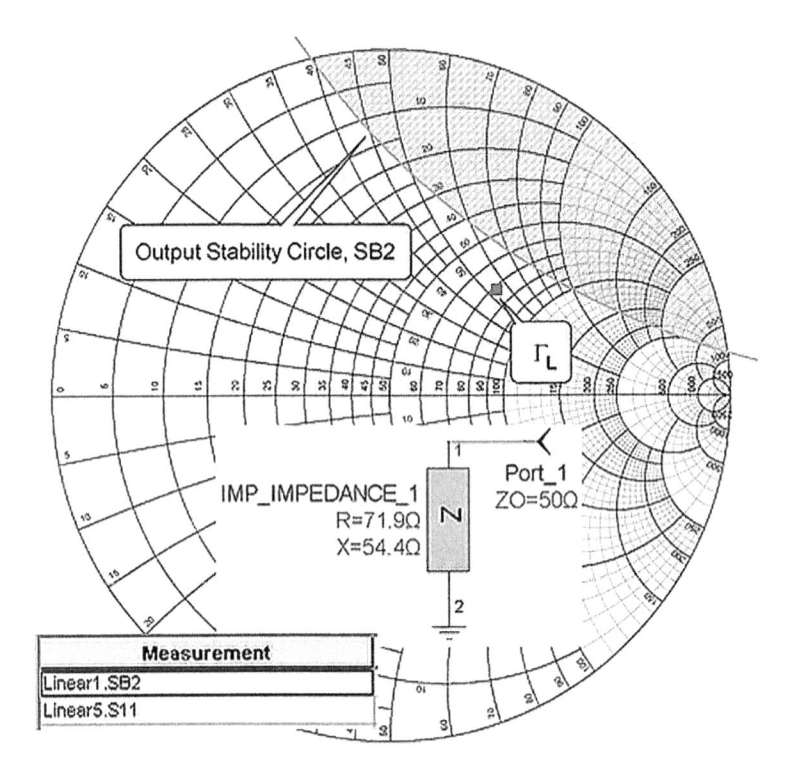

Figure 7-49 Output stability circle and Γ_L

```
1   %Enter output and load impedances
2   RS=50
3   RL=71.9
4   XL=-54.4
5   f=432e6
6   %Normalize the load impedance
7   r=RL/RS
8   x=XL/RS
9   %Calculte t1 and t2
10  t1=(x+sqrt(r*(r^2+x^2-2*r+1)))/(r-1)
11  t2=(x-sqrt(r*(r^2+x^2-2*r+1)))/(r-1)
12  %Calculate the electrical lenghts of the lines d1 and d2
13  d1=360*(atan(t1))/(2*pi)
14  d2=360*(pi+atan(t2))/(2*pi)
15  %Calculate B1 and B2
16  B1=(x*t1^2+(r^2+x^2-1)*t1-x)/(RS*(r^2+x^2+t1^2+2*x*t1))
17  B2=(x*t2^2+(r^2+x^2-1)*t2-x)/(RS*(r^2+x^2+t2^2+2*x*t2))
18  %Calculate the electrical lenghts of the open-ended stubs
19  so1=360*(pi-atan(RS*B1))/(2*pi)
20  so2=-360*atan(RS*B2)/(2*pi)
```

Figure 7-50 Calculation of the single-stub output matching network

The calculations in Figure 7-50 show that there are two single-stub matching networks that match the 50 Ω load impedance to the LNA output

Single Stage Amplifier Design

impedance. For the output matching network, we select the shorter open-stub length so2 = 44.364 degrees and the corresponding line d2 = 99.959 degrees, as shown in Figure 7-51.

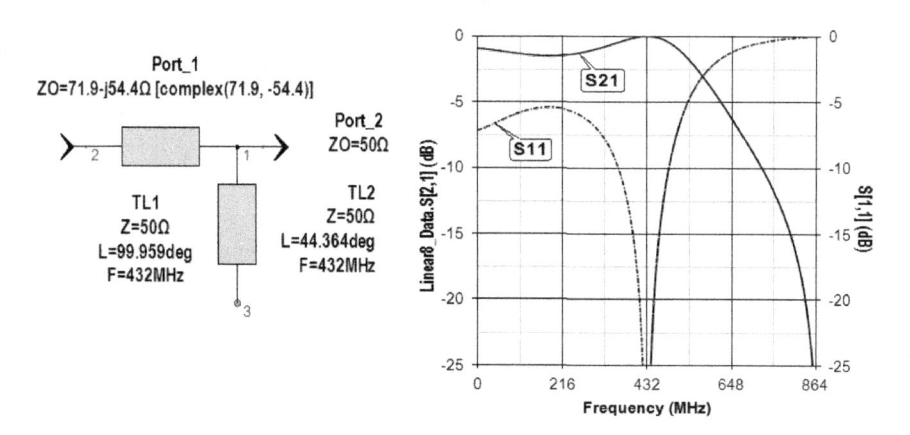

Figure 7-51 Output matching network and simulated response

7.6.5 Linear Simulation of the Low Noise Amplifier

Example 7.6-4: Simulation of the Low Noise Amplifier

Solution: Create a new schematic in Genesys with the input and output matching networks attached to the device. Add a new Linear Analysis that sweeps the amplifier from 400 MHz to 500 MHz. Create a graph and plot the amplifier noise figure NF and gain S21, as shown in Figure 7-53.

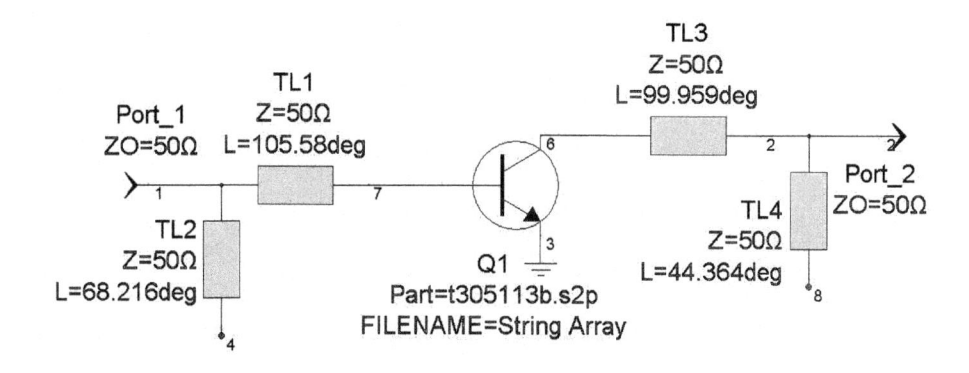

Figure 7-52 LNA schematic using single-stub matching networks

The amplifier gain at 432 MHz is 21.823 dB which is well within the acceptable range of the expected 21.9 dB. The simulated noise figure is 1.083 dB which is exactly the NFmin that was predicted in Figure 7-43. Note that this is the ideal schematic of the LNA. The remaining design sequence should include the microstrip interconnecting lines and bias feeds. The circuit should then be optimized to bring the final response as close as possible to the ideal circuit results as was done in Section 7.2.

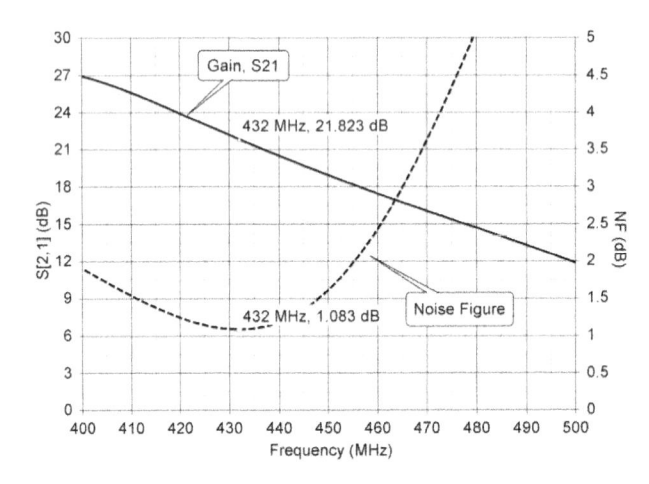

Figure 7-53 LNA noise figure and gain

Figure 7-54 shows that the LNA input is selectively mismatched to achieve the minimum noise figure while the output is perfectively matched to achieve maximum gain.

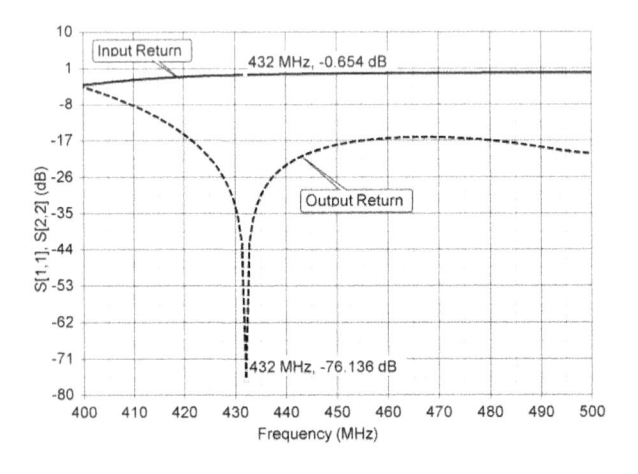

Figure 7-54 LNA input and output return loss

7.6.6 Amplifier Noise Temperature

Low Noise Amplifiers used at higher frequencies in the microwave region often have noise figure referred to in terms of noise temperature. Satellite engineers calculate the ratio of the antenna gain to the equivalent system noise temperature, G/T, as an important performance parameter. A higher system G/T indicates better signal to noise ratio and a better chance of receiving signals from space. Because the LNA is the first active component connected to the antenna it is convenient to also refer to its noise performance in terms of equivalent noise temperature. Noise Figure can be converted to noise temperature by first calculating the noise factor. The noise factor is simply the noise figure expressed in absolute units as calculated by equation (7-25).

$$NoiseFactor, \ F = 10^{\left(\frac{NF_{dB}}{10}\right)} \quad (7\text{-}25)$$

The equivalent noise temperature is then defined by equation (7-26).

$$NoiseTemperature, \ K = (F-1)T \quad (7\text{-}26)$$

where,

T = Temperature of the input noise source (290 °K = room temperature)

Equation (7-26) gives us a good appreciation for the fact that Noise Temperature, or Noise Figure, is always a function of the temperature of the source of noise. It is not necessarily the temperature of the amplifier itself but the temperature of the source of the input noise. When measuring extremely low noise amplifiers it is important to account for the temperature of the noise source. Equation (7-26) can easily be solved directly in the properties window of the output graph. Note that when absolute units (Abs) are selected, the noise figure (NF) is automatically converted to noise factor. As Figure 7-55 shows the minimum noise temperature for our LNA is 82.1 degrees Kelvin.

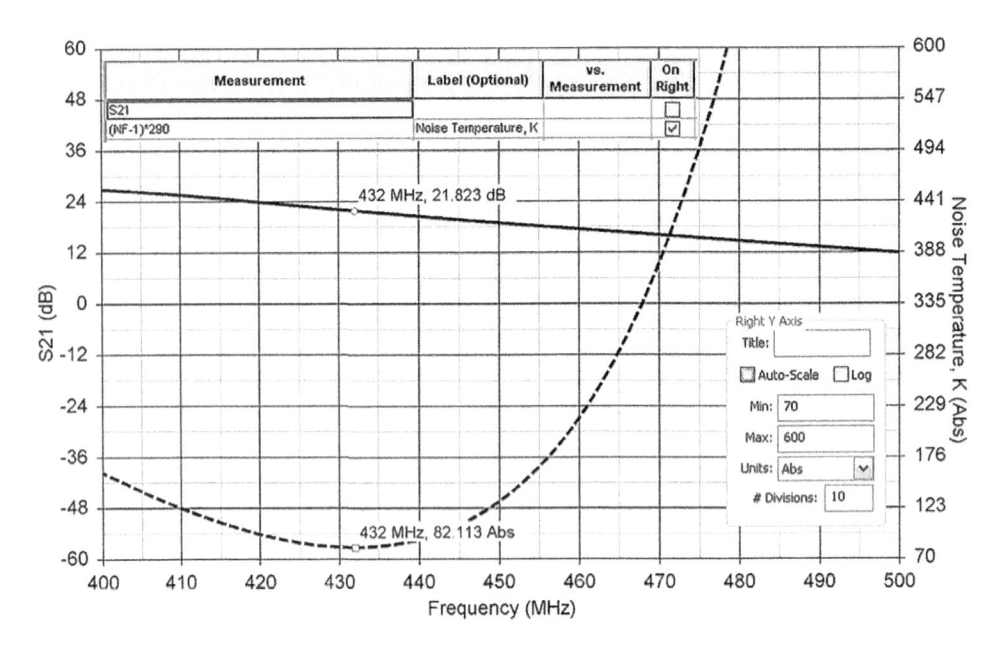

Figure 7-55 LNA noise temperature and gain

It is often desirable to convert between noise temperature and noise figure when comparing transistors used in LNA designs. The Genesys Equation Editor can be used to generate a reference table for noise figure to noise temperature conversions. Figure 7-56 shows a simple routine to convert noise figure to noise temperature. Three column arrays are created that contain the equivalent noise parameters. Equations (7-25) and (7-26) are then used to convert the noise figure to noise factor and temperature. The results of the Equation Editor are then used to create Table 7-2.

```
1   NF_dB=0.1:0.05:2.05
2   NoiseFactor=0.1:0.05:2.05
3   NoiseTemp_K=0.1:0.05:2.05
4
5   For i=1 to 39
6     NoisFactor[i]=10^(NF_dB[i]/10)
7     NoiseTemp_K[i]=(10^(NF_dB[i]/10)-1)*290
8   next
```

Figure 7-56 Noise figure to noise temperature conversion

NF_dB ()	NoiseFactor ()	NoiseTemp_K ()
0.1	1.023	6.755
0.15	1.035	10.191
0.2	1.047	13.667
0.25	1.059	17.184
0.3	1.072	20.741
0.35	1.084	24.339
0.4	1.096	27.979
0.45	1.109	31.661
0.5	1.122	35.385
0.55	1.135	39.153
0.6	1.148	42.965
0.65	1.161	46.82
0.7	1.175	50.72
0.75	1.189	54.666
0.8	1.202	58.657
0.85	1.216	62.694
0.9	1.23	66.778
0.95	1.245	70.909
1	1.259	75.088
1.05	1.274	79.316
1.1	1.288	83.592
1.15	1.303	87.918
1.2	1.318	92.294
1.25	1.334	96.721
1.3	1.349	101.199
1.35	1.365	105.729
1.4	1.38	110.311
1.45	1.396	114.947
1.5	1.413	119.636
1.55	1.429	124.379
1.6	1.445	129.178
1.65	1.462	134.031
1.7	1.479	138.941
1.75	1.496	143.908
1.8	1.514	148.933
1.85	1.531	154.015
1.9	1.549	159.157
1.95	1.567	164.358
2	1.585	169.619

Table 7-2 Noise Figure to Noise Temperature conversion

7.7 Power Amplifier Design

The amplifier circuits described thus far have been designed using the measured S parameters that represent the active device. The S parameters are based on the small signal characteristics of the device. That is to say that the device is operating well below its maximum output power capability. This would suggest that the S parameters are defined for class A amplification, operating well within the linear portion of their power transfer characteristic. The S parameters can therefore be used to design an amplifier for specific values of gain but not output power. As we drive the transistor closer to its maximum output power the signal excursion is occurring over a much wider range of the transistor's load line. This is to say that the device is now operating under large signal conditions. There exists a specific source and load impedance into which the transistor can produce its maximum output power. Because the device S parameters are no longer defined under large signal conditions, they cannot be used to determine the optimum load impedance for maximum output power. These impedances are normally determined by performing a load pull measurement on the device. A typical load pull test setup is shown in Figure 7-57. This test setup empirically tunes the device for maximum output power using the input and output tuners. The load pull test creates a set of contours on the Smith Chart that correspond to a given output power level. Selecting an impedance on the constant power contour will result in an impedance match that results in a power match. This is similar to the power gain circles that we seen in section 7.5 however power contours never form a circle. Power contours tend to have an elliptical shape when plotted on the Smith Chart. Cripps has introduced a technique to approximate the optimum load impedance using linear design techniques [6]. This concept is based on determining an optimum load resistance from the classical load line theory.

Once the load and source impedance are known, we can use the linear techniques developed earlier in this chapter to design the matching networks.

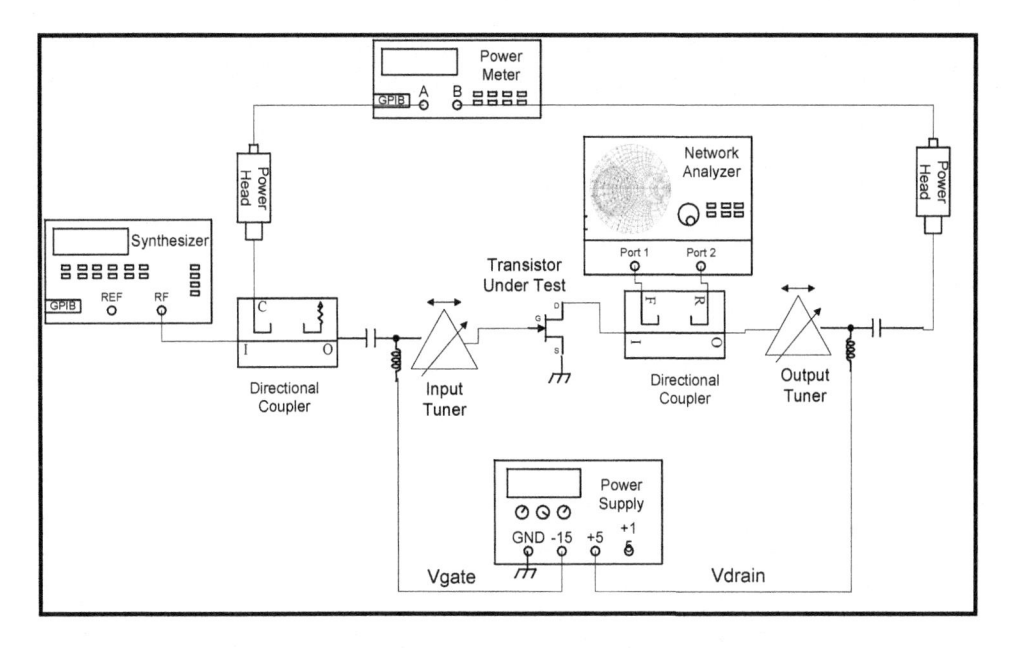

Figure 7-57 Load pull test for source and load impedance measurement

7.7.1 Data Sheet Large Signal Impedance

In many cases high power devices are intended for specific applications such as WiMAX, cellular base station, or mobile radio. In these applications the device manufacturer may perform the load pull analysis at specific frequencies and present the source and load impedance as an equivalent series circuit on the data sheet. This allows the designer to treat the power matching process as a simple impedance matching exercise. Figure 7-58 shows an excerpt of a data sheet for the Nitronex NPT25100 GaN HEMT device. As the data sheet shows the optimum source and load impedances are given in tabular and Smith Chart forms.

Frequency (MHz)	Z_S (Ω)	Z_L (Ω)
2140	12.1 - j20.0	2.6 - j2.6
2300	10.0 - j3.0	2.5 - j2.3
2400	9.5 - j3.0	2.5 - j2.5
2500	9.0 - j3.0	2.5 - j2.7
2600	8.5 - j3.0	2.5 - j3.1
2700	8.0 - j3.0	2.5 - j3.3

Z_S is the source impedance presented to the device.
Z_L is the load impedance presented to the device.

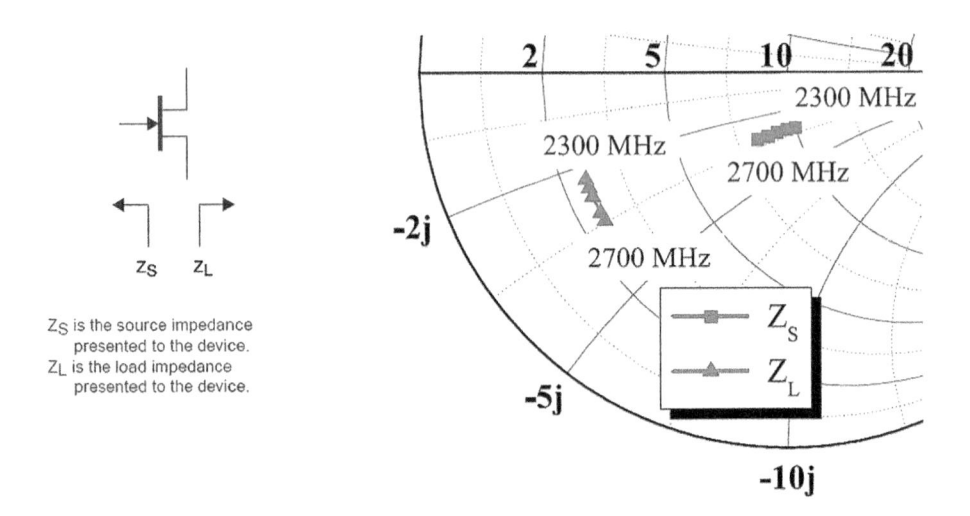

Figure 7-58 Optimum source and load impedance (*courtesy of Nitronex*)

As Figure 7-58 shows the listed Z_S and Z_L is not the device impedance but rather the impedance that is presented to the device. This is analogous to the Γ_S and Γ_L that was calculated in section 7.2 for the simultaneous conjugate match. The actual device impedance is the conjugate of the given optimum source and load impedance. The designer must be cautious when interpreting the optimum source and load impedances from various vendor data sheets. Some manufacturers may list the actual device impedance as shown on the Freescale Semiconductor data sheet of Figure 7-59. We also need to read the impedance data from the tables rather than directly from the Smith Chart. Because of the very low input and output impedance of power transistors, it is common to normalize the Smith Chart to 10 Ω so that the impedance locus is not compressed on the left hand side of the Smith Chart. The impedances given in the table are the actual impedance rather than the normalized impedance.

Single Stage Amplifier Design

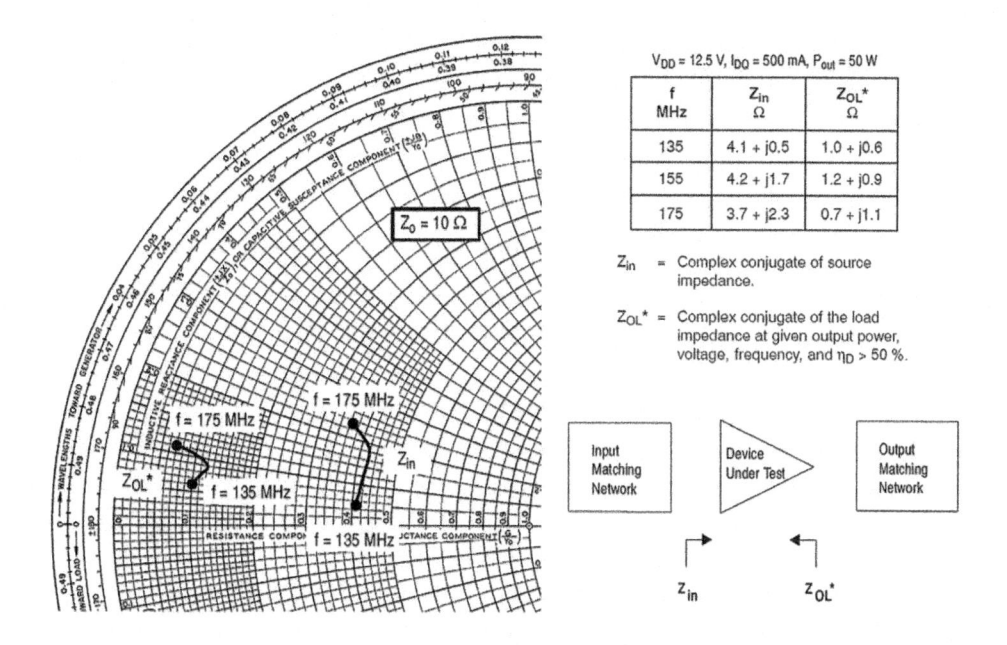

f MHz	Z_{in} Ω	Z_{OL}^* Ω
135	4.1 + j0.5	1.0 + j0.6
155	4.2 + j1.7	1.2 + j0.9
175	3.7 + j2.3	0.7 + j1.1

V_{DD} = 12.5 V, I_{DQ} = 500 mA, P_{out} = 50 W

Z_{in} = Complex conjugate of source impedance.

Z_{OL}^* = Complex conjugate of the load impedance at given output power, voltage, frequency, and η_D > 50 %.

Figure 7-59 Large signal series equivalent impedance (*courtesy of Freescale*)

7.7.2 Power Amplifier Matching Network Design

In this section we will design the matching networks for the NPT25100 GaN power transistor at a frequency of 2.140 GHz. The matching networks of previous examples have all been based on two-element L-networks. In this example we will design three-element Pi network matching circuits. The Pi network is often preferred over a two element network because the component values are less sensitive when physically realizing the impedance match. It can also help to keep the circuit Q lower for improved bandwidth. We will illustrate both graphical techniques and synthesis techniques. The input matching circuit will be designed using graphical techniques while the output matching network will be designed using network synthesis techniques.

7.7.3 Input Matching Network Design

Example 7.7-1: Design the power amplifier input matching network.

Solution: Graphical matching techniques will be employed for the input matching network using the Smith Chart in Genesys. We will design the matching circuit moving from 50 Ω at the center of the Smith Chart to the Z_S as given on the data sheet. A one port S parameter file can be created with impedance data as shown in Figure 7-60. Note the differences in line number 2 of the file. This line defines the format of the data contained within the file. The (Z) defines the file as containing impedance data. The number (1) means that the data is normalized to one, the actual device impedance.

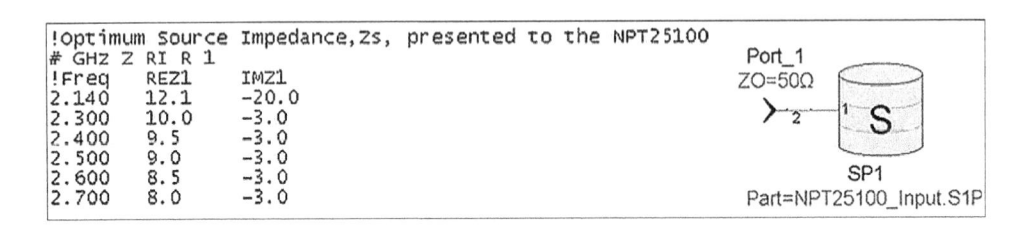

```
!Optimum Source Impedance,Zs, presented to the NPT25100
# GHZ Z RI R 1
!Freq    REZ1    IMZ1
2.140    12.1    -20.0
2.300    10.0    -3.0
2.400    9.5     -3.0
2.500    9.0     -3.0
2.600    8.5     -3.0
2.700    8.0     -3.0
```

Port_1
ZO=50Ω

SP1
Part=NPT25100_Input.S1P

Figure 7-60 One port S parameter data file containing the Z_S data

Setup a schematic in Genesys with the one-port Z_S impedance data and sweep it with a Linear Analysis at a fixed frequency of 2140 MHz. Then setup a second schematic with a 50 Ω resistor to represent the center of the Smith Chart and begin the Pi network design. Plot the impedance of both networks on the same Smith Chart. Figure 7-61 through 7-63 show the progression of the three-element Pi matching network design. We know that the shunt inductors of a Pi network will travel clockwise on the conductance circles while the series inductor will travel clockwise on a resistance circle. Note the location of the conductance circle in which Z_S is located. The first capacitor and series inductor L section must move to the location of the Z_S conductance circle.

Single Stage Amplifier Design

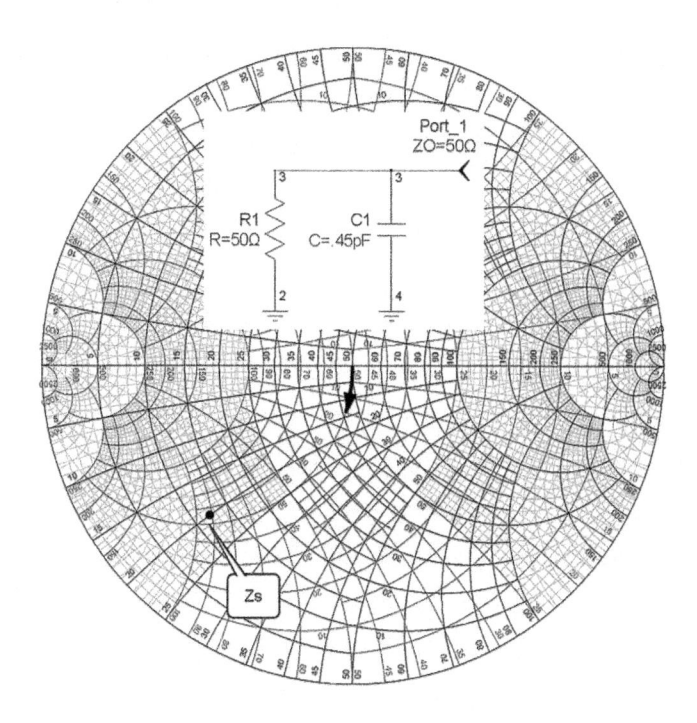

Figure 7-61 Tuning the First Shunt Capacitor

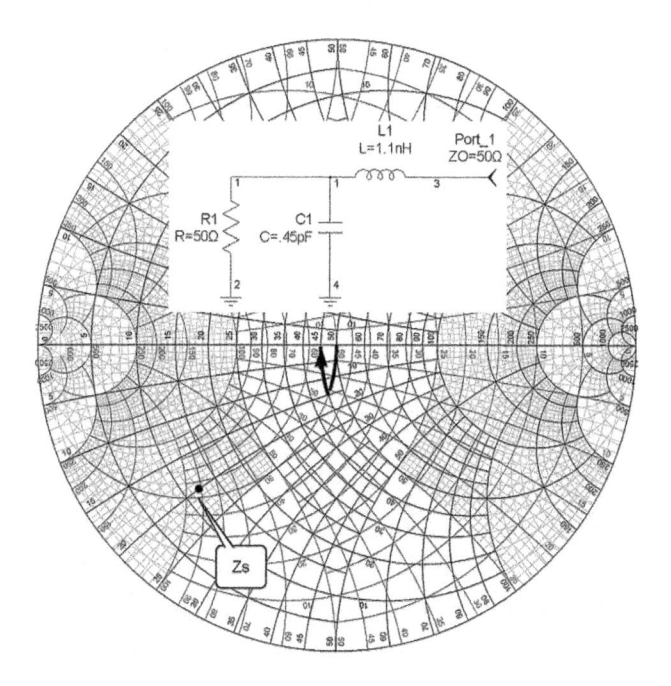

Figure 7-62 Tuning the series inductor to intercept the conductance circle

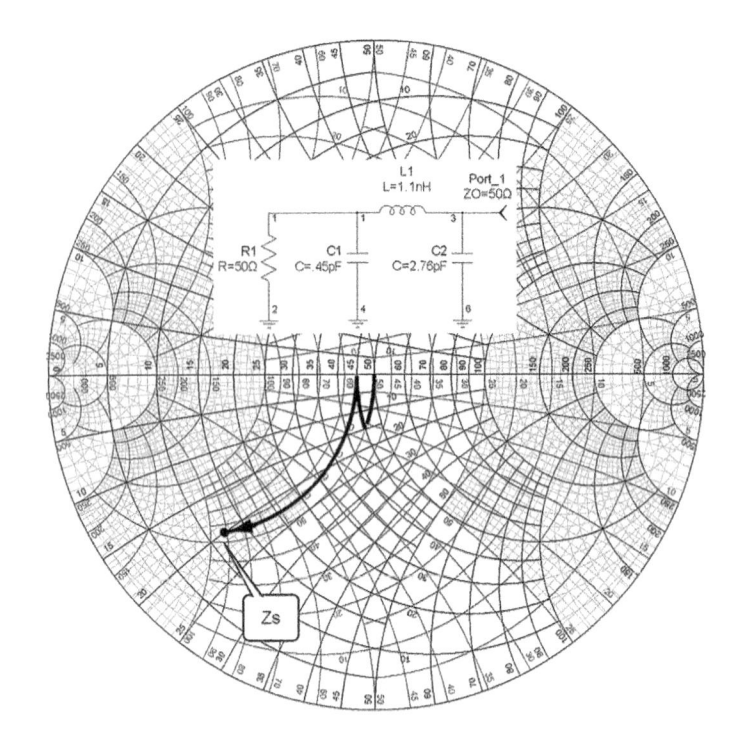

Figure 7-63 Tuning the second shunt capacitor to intersect Z_S

7.7.4 Output Matching Network Design

Example 7.7-2: Design the power amplifier output matching network.

Solution: The output matching network will be designed using the Impedance Matching Synthesis utility. When using the synthesis utility we need to have the device impedance rather than the impedance looking into the matching network toward the load. Therefore a one port S parameter file is created with the conjugate of the Z_L given on the data sheet as shown in Figure 7-64. Use caution when editing S parameter files to be read into a Genesys simulation. The first time that an S parameter file is read from disk it is loaded into the Genesys Workspace. Subsequent simulations will not go back to disk to read the file but rather read the file from the Workspace for faster simulation speed. The file will not be read from disk again until the copy loaded into the Workspace has been deleted. Create an Impedance Match Synthesis for the output matching network. Set the properties as shown in Figure 7-65. Enter a fixed frequency of 2140 MHz.

Single Stage Amplifier Design 383

On the Sections tab, read the one-port S parameter file that contains the conjugate of the Z_L from the data sheet. Next select an LC Pi Network for the matching circuit with an inductive tendency. Select Calculate to synthesize the matching network shown in Figure 7-66.

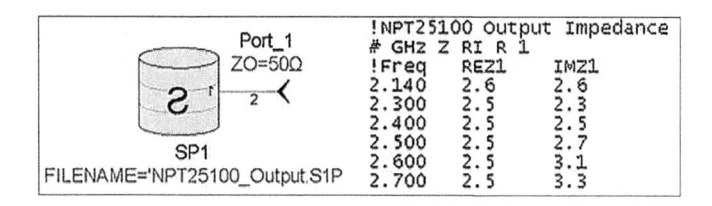

```
                Port_1          !NPT25100 Output Impedance
                ZO=50Ω          # GHZ Z RI R 1
                                !Freq     REZ1      IMZ1
                    2           2.140     2.6       2.6
                                2.300     2.5       2.3
                                2.400     2.5       2.5
                SP1             2.500     2.5       2.7
FILENAME='NPT25100_Output.S1P   2.600     2.5       3.1
                                2.700     2.5       3.3
```

Figure 7-64 S parameter data file containing the conjugate of Z_L

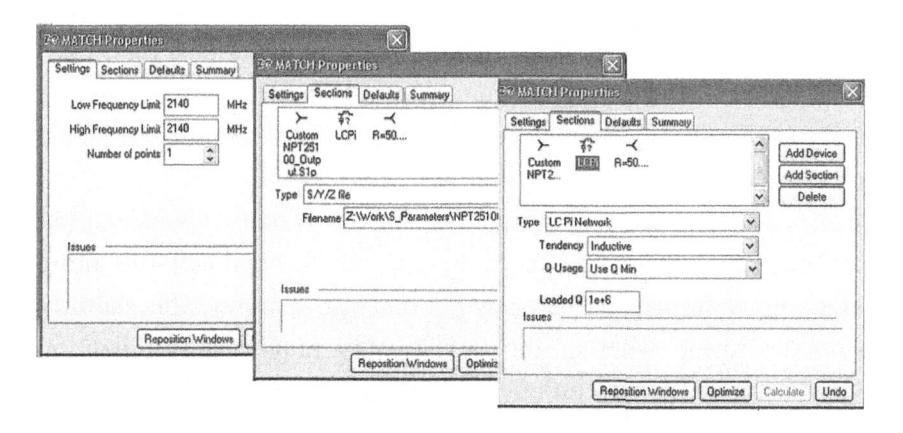

Figure 7-65 Setup of the output matching network synthesis parameters

The impedance matching synthesis program is an excellent tool for quickly examining various matching networks and their response.

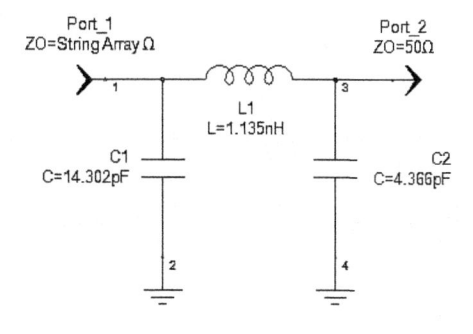

```
      Port_1                                    Port_2
      ZO=StringArray Ω                          ZO=50Ω
       ─1─────────────⌒⌒⌒─────────────3─
                        L1
                     L=1.135nH
              C1                         C2
              C=14.302pF                 C=4.366pF
               │                          │
               2                          4
               ─                          ─
```

Figure 7-66 Output matching network

Finally a two-port S parameter file can be created that contains the impedance data for the conjugate of Z_S and Z_L. This would be representative of the large signal input and output impedance of the device.

```
!Large Signal Impedances of the NPT25100
# GHz Z RI R 1
!Freq      REZ1   IMZ1     M(S21)    A(S21)    M(S12)    A(S12)    REZ2   IMZ2
 2.140     12.1   20.0     0         0         0         0         2.6    2.5
 2.300     10.0   3.0      0         0         0         0         2.5    2.3
 2.400      9.5   3.0      0         0         0         0         2.5    2.5
 2.500      9.0   3.0      0         0         0         0         2.5    2.7
 2.600      8.5   3.0      0         0         0         0         2.5    3.1
 2.700      8.0   3.0      0         0         0         0         2.5    3.3
```

Figure 7-67 S parameter data file containing the conjugate of Z_S and Z_L

Figure 7-68 shows the resulting matching networks attached to the large signal impedance data for the device. This allows examination of the return loss of the amplifier at the design frequency of 2140 MHz and the usable bandwidth. From Figure 7-67 note that the forward (S21) and reverse (S12) transmission parameters have been set to zero. Because we have no definition of the transmission parameters we cannot evaluate the gain of the circuit. If the manufacturer's data sheet includes an S parameter file along with the large signal impedance data, we could also examine the stability parameters to determine whether the device may require a stabilization network. The Linear Design techniques do provide a means of performing the matching network design from which the amplifier physical design can be realized. The circuit can then be built and empirically tuned and optimized on the bench for the desired performance. A thorough CAD design of the power amplifier requires a nonlinear physical model for the transistor. Under large signal conditions the strong nonlinearities will cause the gain to change as the drive (input) power changes. The gain of the device will go into compression and decrease as the input power is further increased. Harmonic energy is created by these nonlinearities that will further influence the behavior of the amplifier. The nonlinear design and characterization of power amplifiers is thoroughly treated in Microwave and RF Engineering Fundamentals, Volume 2, Nonlinear Design Concepts.

Single Stage Amplifier Design

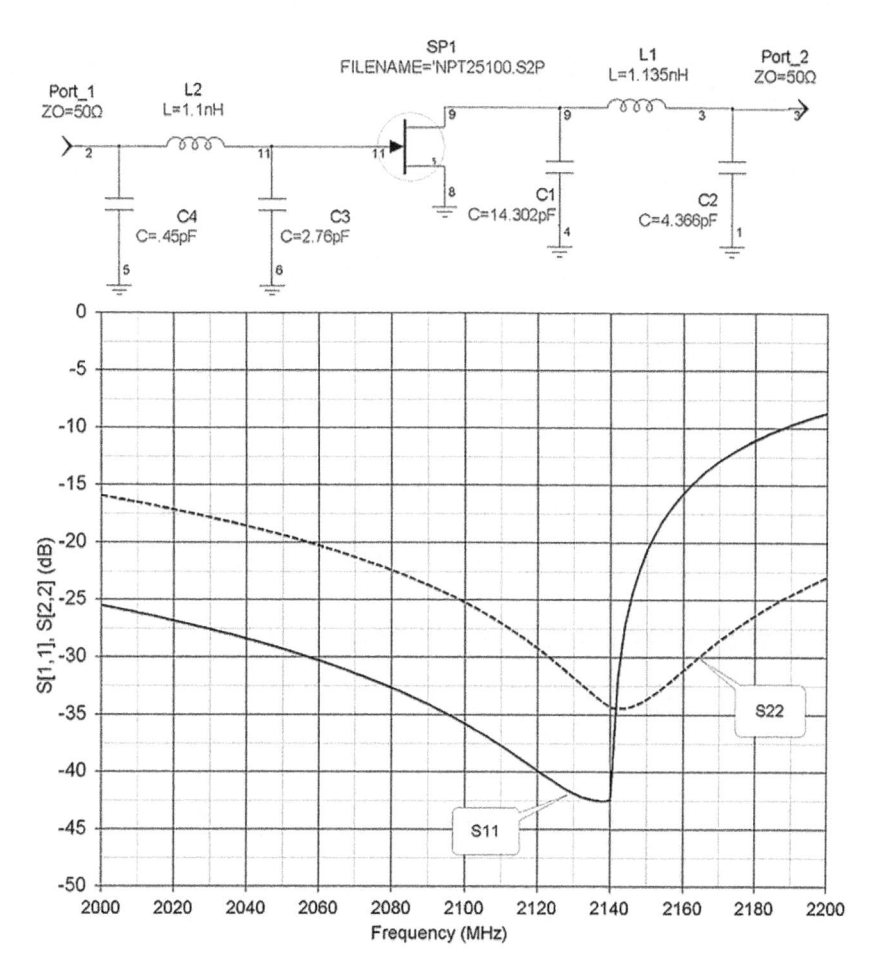

Figure 7-68 Resulting matching networks and return loss

The impedance matching techniques learned in Chapters five and six have been utilized to perform basic linear amplifier matching. These matching techniques have been solved analytically using the equations written in the Genesys Equation Editor as well as synthesis routines created in VBScript. Additionally the powerful matching synthesis tools that exist within Genesys have also been applied. Graphical techniques have also been applied to the design of multi-element impedance matching using the Smith Chart. These techniques provide the engineer with a comprehensive set of tools to apply to amplifier impedance matching. These impedance matching techniques have been used to the introduce four of the primary RF and microwave amplifier circuits including: maximum gain, specific gain, low noise, and power amplifier circuits.

References and Further Reading

[1] Dale D. Henkes, FAST: *Fast Amplifier Synthesis Tool*, Artech House Publishers, Norwood, MA. 2004

[2] Chris Bowick, *RF Circuit Design*, Butterworth-Heinemann, Newton, MA, 1982

[3] Keysight Technologies, *Genesys 2014.03, Users Guide*, 2014
www.keysight.com

[4] Guillermo Gonzales, *Microwave Transistor Amplifiers – Analysis and Design*, Second Edition, Prentice Hall Inc., New Jersey

[5] Randy Rhea, *The Yin-Yang of Matching: Part 1 – Basic Matching Concepts*, High Frequency Electronics, March 2006

[6] Steve C. Cripps, *RF Power Amplifiers for Wireless Communications*, Artech House Publishers, Norwood, MA. 1999

[7] David M. Pozar, *Microwave Engineering*, Fourth Edition, John Wiley & Sons, New York, 2012

[8] R. Ludwig, P. Bretchko, *RF Circuit Design*, Theory and Applications, Prentice Hall, Upper Saddle River, NJ, 2000

[9] Ali A. Behagi and Stephen D. Turner, *Microwave and RF Engineering,* BT Microwave LLC, State College, PA, 2011

[10] Keysight Technologies, *Genesys 2014.03, Users Guide*, 2014
www.keysight.com

Single Stage Amplifier Design 387

Problems

7-1. Design a Maximum Gain Amplifier

(a) For the Agilent HBFP0405 Transistor, calculate the stability factor, K, and the stability measure, B1 as well as Γ_{MS} and Γ_{ML} and conjugate match impedances at 2 GHz.

(b) Use Genesys built in functions to determine the same parameters at 2 GHz.

(c) Sweep the frequency range of the device over the entire range of frequencies contained in the S parameter file.

(d) Plot of the stability parameters and GMAX for the Agilent HBFP0405 device from 100 to 4000 MHz.

(e) Employ a parallel RC circuit on the input of the device to stabilize the transistor. Make both the R and C values tunable. Tune the resistance and capacitor values until K > 1 for frequencies above 500 MHz.

(f) Add an RL network in shunt with the input of the transistor for additional low frequency stability. Tune the resistance and inductor values until K > 1 for frequencies above 500 MHz.

(g) Use the analytical impedance matching techniques developed in Chapter 5 to design the input and output matching L-networks. When performing the analytical match we are; matching 50 Ω source impedance to the impedance looking into the device; and 50 Ω load impedance to the impedance looking into output of the device.

(h) Complete the amplifier design in Genesys by attaching the input and output matching circuits to the stabilized transistor. Use a Linear Sweep from 100 MHz to 4000 MHz to analyze the response of the amplifier, S11, S22, and S21 all in dB.

(i) Place a marker on S12 to measure the maximum gain of the amplifier at 2 GHz. Compare the maximum gain with the value originally calculated as G_{MAX}.

7-2. Design a Constant Gain Amplifier

(a) Use the Mitsubishi MGF0911a device to design a specific gain amplifier that achieves 10 dB gain at the frequency of 2 GHz. Examine the stability circles for the device in Genesys by creating a schematic with the small signal S parameter file.

(b) Setup a Linear Analysis with both the start and stop frequencies set at 2 GHz. Add a Linear Analysis from 100 MHz to 3000 MHz with 100 MHz steps. Then add a Table to display K, B1, and GMAX.

(c) Plot a series of constant power gain circles on the Smith Chart and choose the circle which is the closest to the design goal of 10 dB. The load reflection coefficient, defined as a reflection coefficient, Γ_L, is any point on the circumference of the circle that does not enter the unstable region. Choose a point that is not close to the unstable regions of the stability circles. Read the load impedance visually from the Smith Chart at this point.

(d) Calculate the source impedance from Equation 7-18. The source impedance is therefore defined as a function of the chosen load reflection coefficient.

(e) Use the Smith Chart to graphically design the simple two-element LC input and output matching networks.

(f) Create a new schematic and attach the input and output matching circuits to the device. Create a Linear Analysis that sweeps the amplifier from 1 GHz to 3 GHz. Display S11 and S21 in a rectangular plot.

(g) Place a marker on S12 to measure the gain of the amplifier at 2 GHz. Compare the gain with the value originally selected as specific gain.

7-3. Design a Low Noise Amplifier

Design a single stage Low Noise Amplifier using Agilent AT-32011 device. The amplifier is intended to operate with a source and load impedance of 50 Ω. The design specifications are given as:

Center Frequency:	500 MHz
Gain:	18 dB minimum
Noise Figure:	1.2 dB maximum
Output Return Loss:	Less than -10 dB

(a) Create a Genesys schematic with the device S parameter file. Setup a Linear Analysis and add a Table to display the stability parameters K and B1 and the noise parameters Γ_{opt} and Z_{OPT}. Also add a Smith Chart with Γ_{opt} and the stability circles. Then create a Smith Chart and display the stability circles along with the noise circles.

(b) Drawn the available gain circles are for gains of 0, 1, 2, 3, 4, 5 and 6 dB less than maximum gain.

(c) Utilize the "L" Network Impedance Matching Synthesis Tool to match the 50 Ω source impedance to Z_{opt} of the device at 500 MHz. In the impedance matching utility you must enter the conjugate of Z_{opt} for the complex load impedance.

(d) Use Genesys Equation Editor to calculate the load reflection coefficient, Γ_L, from Equation (7-24). Convert Γ_L to output impedance Z_L. Z_{OUT} is the conjugate of Z_L. Check the location of Γ_L with respect to the output stability circle to make sure that the impedance does not lie in the region of instability. If Γ_L is safely outside of the SB2 circle, use Γ_L as an acceptable output reflection coefficient.

Chapter 8

Multi-Stage Amplifier Design and Yield Analysis

8.1 Introduction

In the amplifier designs of Chapter 7, impedance matching was defined between the active device and a source or load impedance. The complex impedance of the device was dependent on the type of amplifier that was being designed i.e. Maximum Gain, Specific Gain, Low Noise, or Power Amplifier. Practical amplifiers often involve a cascade of amplifier stages to provide the required gain for a specific application. In an amplifier cascade, the output impedance of one device is typically matched to the input impedance of the succeeding stage rather than a 50 Ω resistance. This chapter provides an overview of amplifier inter-stage matching. Monte Carlo and Yield Analysis is also introduced in this chapter. Monte Carlo and Yield analysis are important to the successful manufacture of microwave circuit designs as variation to component tolerance is evaluated. For simplicity much of this chapter will focus on ideal schematic elements for both inter-stage matching and the Yield Analysis. Finally a brief introduction to cascade analysis is presented as it can be applied to linear simulation in Genesys.

8.2 Two-Stage Amplifier Design

Figure 8-1 shows a block diagram of a two stage amplifier with the matching networks represented by a box. The input impedance of each device is designated as ZM1 while the output is designated as ZM2. We will revisit two of the transistors used in Chapter 7 to design a two stage amplifier cascade. The first stage is the SHF0189 and the second stage is the RT243. The design specifications are given as:

>Center Frequency = 2350 MHz
>Bandwidth: 2260 MHz to 2380 MHz
>Gain: $\geq$ 34 dB
>Input & Output Return Loss $\leq$ -10 dB

The cascade is designed as a maximum gain-conjugate matched amplifier. Therefore three conjugate matching networks must be designed. The first conjugate matching network is designed between the 50 Ω source and the first stage input impedance, ZM1. The second matching network is the inter-stage matching network. This network is designed to match the first stage output impedance, ZM2, to the second stage input impedance, ZM1. The third conjugate matching network is designed between the second stage output impedance, ZM2, and the 50 Ω load impedance.

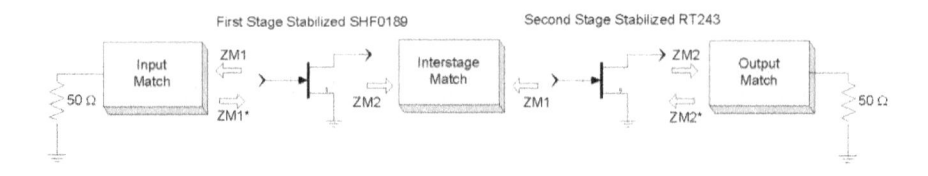

Figure 8-1 Two-stage amplifier with impedance matching networks

8.2.1 First Stage Matching Network Design

The first stage of the amplifier utilizes the SHF-0189 device. We know that the device must be unconditionally stable for a maximum gain conjugate matched amplifier [1]. Using the techniques of Chapter 7.2 a stabilizing network is designed around the SHF-0189.

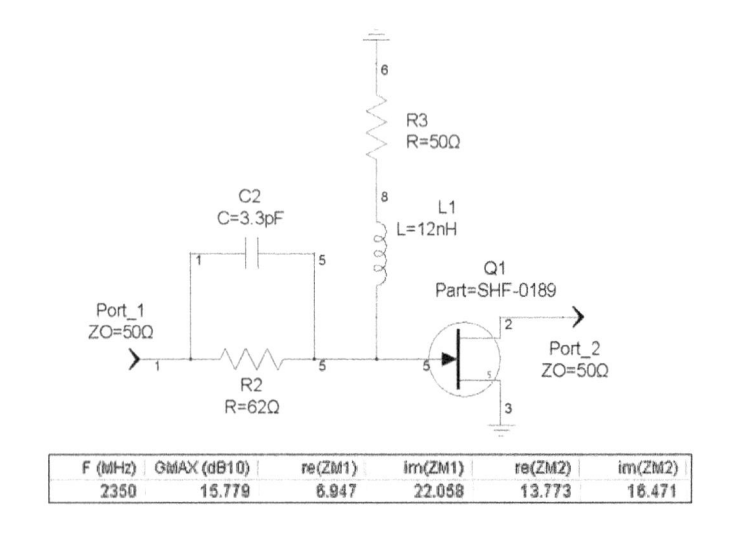

F (MHz)	GMAX (dB10)	re(ZM1)	im(ZM1)	re(ZM2)	im(ZM2)
2350	15.779	6.947	22.058	13.773	16.471

Figure 8-2 Stabilized first stage with conjugate match impedance

Once the device is stabilized the maximum stable gain and the required input and output impedance can be determined using the built-in functions in Genesys. The Genesys output is shown in the table inset of Figure 8-2. The equivalence between the Genesys function notation and the respective impedances is given below.

First Stage:
$$ZM1 = Z_{MS} = 6.947 + j22.058 \ \Omega$$
$$ZM2 = Z_{ML} = 13.773 + j16.471 \ \Omega$$
$$G_{max} = 15.779 \ dB$$

8.2.2 Design of the Amplifier Input Matching Network

Example 8.2-1: Design the Amplifier Input Matching Network

Solution: When designing the amplifier input matching network we are matching the 50 Ω source impedance to the conjugate impedance looking into the SHF-0189 device, ZM1, as given in Figure 8-2. Therefore the matching of the complex impedance is the conjugate of ZM1 as written in the following Equation editor. Normalize the load impedance and use Equations (5-32) and (5-33) to calculate the matching element values.

```
1    %Enter Design Parameters
2    Z0=50
3    RL=6.947
4    XL=-22.058
5    f=2.35e9
6    %Normalize Load Impedance
7    r=RL/Z0
8    x=XL/Z0
9    %Write Matching Equations
10   B1=sqrt((1-r)/r)/Z0
11   X1=Z0*(sqrt(r*(1-r))-x)
12   %Calculate Matching Elements
13   L1=X1/(2*pi*f)
14   C2=B1/(2*pi*f)
```

Figure 8-3 Calculation of the element values of the input matching network

The calculations in the Equation Editor show that the series element is an inductor L1= 2.665 nH and the shunt element is a capacitor, C2 = 3.372 pF. To assemble the schematic, display the response and measure the bandwidth, set up a new Schematic and select the element values given above. Simulate the schematic from 1350 MHz to 3350 MHz and display the insertion loss, S21, and input return loss, S11, in dB. The schematic of the matching network and the simulated response are shown in Figure 8-4. Put two markers at 20 dB points to measure the bandwidth as shown in Figure 8-4. The SHF-0189 output matching network is not designed here because we want to cascade the SHF-0189 with the RT243 device and design an inter-stage matching network between the two stages.

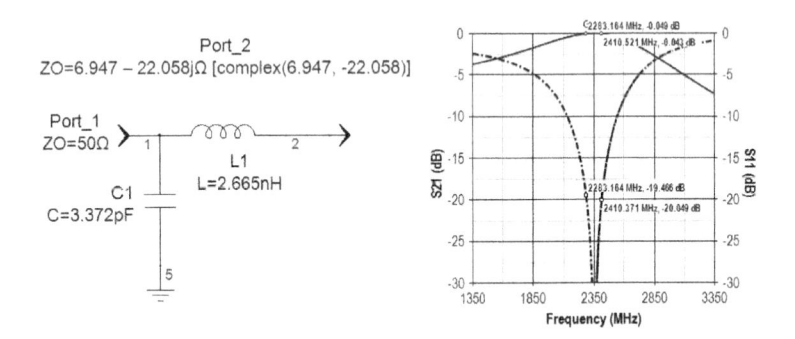

Figure 8-4 Schematic of the input matching network and response

8.2.3 Second Stage Matching Network Design

The second stage device along with its stabilization network is shown in Figure 8-5. Once the device is stabilized the maximum stable gain and the required input and output impedance are given by the built-in functions in Genesys. As the table inset of Figure 8-5 shows, the stabilized RT243 device has a GMAX = 18.973 dB at 2.35 GHz.

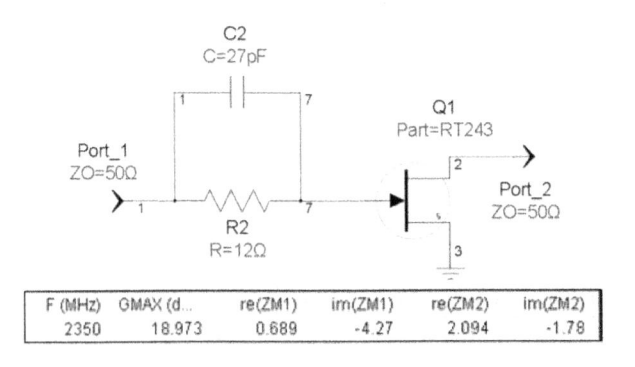

F (MHz)	GMAX (d...	re(ZM1)	im(ZM1)	re(ZM2)	im(ZM2)
2350	18.973	0.689	-4.27	2.094	-1.78

Figure 8-5 Stabilized second stage with conjugate match impedance

The equivalence between the Genesys function notation are.

$$ZM1 = Z_{MS} = 0.68 - j4.27 \ \Omega$$
$$ZM2 = Z_{ML} = 2.09 - j1.78 \ \Omega$$
$$G_{max} = 18.97 \ dB$$

8.2.4 Inter-Stage Matching Network Design

Example 8.2-2: Design the Amplifier Input Matching Network

Solution: One common method of inter-stage matching is to design each stage separately into a 50 Ω system and then cascade them directly with no additional matching. Many RF transistors have very low input and output impedance, much less than 50 Ω, making a low impedance interstage network desirable. In cascaded narrowband amplifier design a more efficient method of interstage matching is to conjugately match the two complex impedances directly together with a single L-network and thus reduce the number of interstage matching elements. Therefore, we can use the L-Network Impedance Matching Synthesis Tool, introduced in section 5.6.2, or utilize Equations (5-12) and (5-13) in an Equation Editor to design the interstage matching network. The Equation Editor solution is shown in Figure 8-6.

```
1   %Enter Design Parameters
2   RS=13.773
3   XS=-16.471
4   RL=0.688
5   XL=4.271
6   f=2.35e9
7   %Calculate B3, X3 and Element Values
8   B3=((RS*XL)+sqrt(RS*RL*(RL^2+XL^2-RS*RL)))/(RS*(RL^2+XL^2))
9   X3=(RS*XL-RL*XS)/RL+(RL-RS)/(B3*RL)
10  L2=X3/(2*pi*f)
11  C1=B3/(2*pi*f)
```

Figure 8-6 Calculation of interstage matching network elements

The inter-stage matching network between SHF-0189 and RT243 is shown in Figure 8-7. To display the response of the interstage matching network, set up a new Design with Schematic in Genesys and place the matching elements with the source and load impedances on the schematic, as shown. Simulate the schematic from 1350 MHz to 3350 MHz and display both the return loss, S11, and insertion loss, S21, in dB, as shown in Figure 8-7.

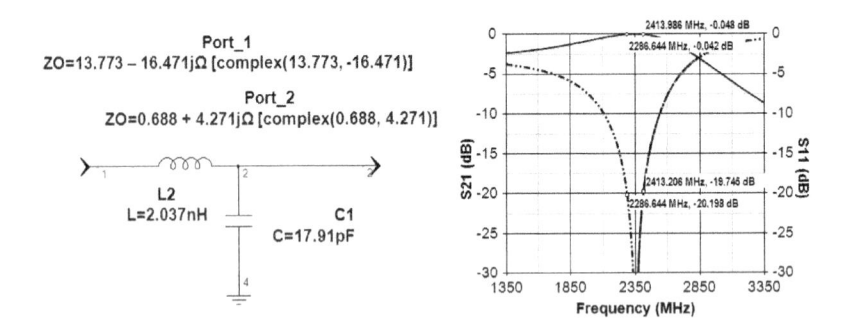

Figure 8-7 Simulated response of the interstage matching network

8.2.5 Second Stage Output Matching Network

Use Equations (5-24) and (5-25) to calculate the element values of the output matching network. The schematic in Figure 8-8 shows that the shunt capacitor C1=6.48 pF and the series inductor L1 = 0.577 nH. Then set up new schematic in Genesys and select the element values from the Equation Editor for the output matching network. Simulate the schematic from 1350 MHz to 3350 MHz and display the insertion loss, S21, in dB.

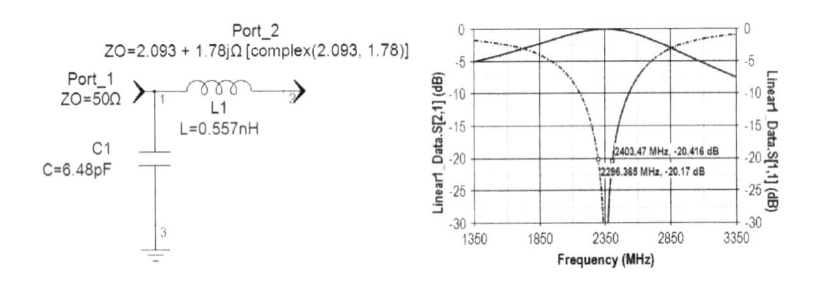

Figure 8-8 Schematic and response of the output matching network

8.3 Two-Stage Amplifier Simulation

The matching networks are attached to the stabilized devices to form a two stage amplifier. The input, inter-stage, and output matching networks are cascaded along with the devices as shown in Figure 8-9. We will only deal with the ideal schematic at this point. The final design, of course, would include the physical models of the lumped element components and the interconnecting microstrip circuit elements.

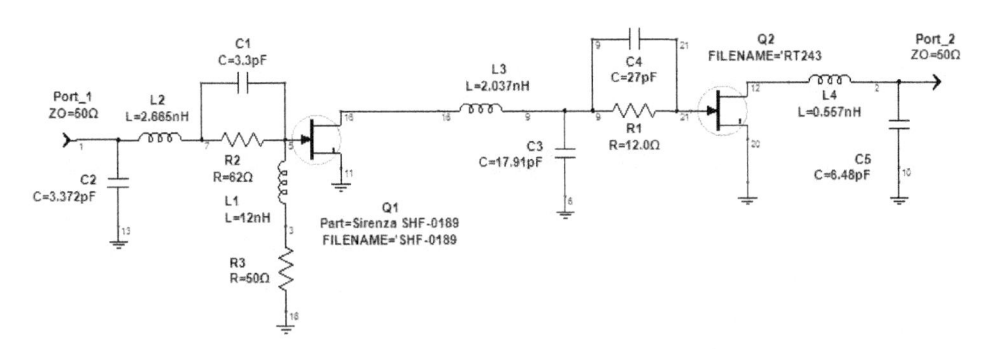

Figure 8-9 Two-stage amplifier schematic with ideal matching networks

The simulated gain and return loss are shown in Figure 8-10. The simulated gain of the two-stage cascade at 2.35 GHz is 34.564 dB which is very close to the direct summation of the GMAX for each stage. The gain of the amplifier is greater than 34 dB from 2260 MHz to 2380 MHz, satisfying one of the design specifications of the amplifier. The input and output return loss is well below -10 dB, thereby satisfying all of the specifications of the amplifier.

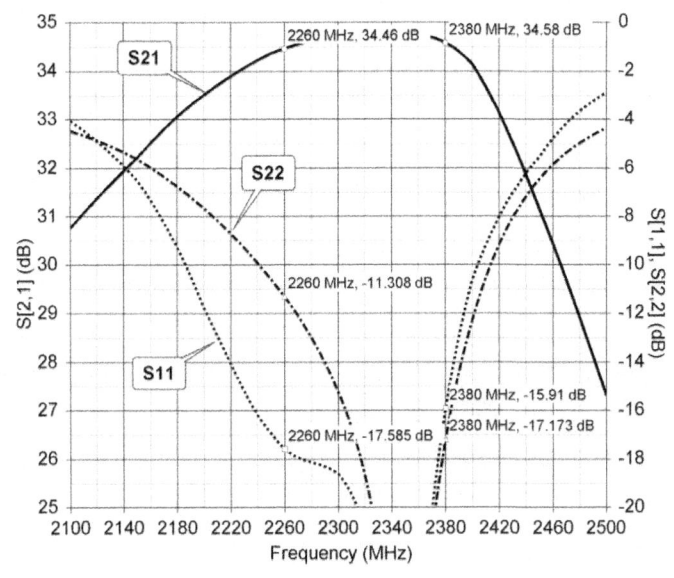

Figure 8-10 Ideal amplifier cascade response

8.4 Parameter Sweeps

In this section we will examine the effect of variation in the nominal value of a component on the overall amplifier response. For simplicity we will continue to work with the ideal amplifier schematic of Figure 8-9. There may be times that we wish to sweep a component's value to determine the change in the circuit's response. This could be useful in analyzing a circuit's response to a component's changing value. A component's value can change due to temperature or simply by the tolerance in the manufacturing process. Multilayer chip capacitors and wire wound chip inductors are manufactured to various tolerances assigned to their nominal values. Chip inductors are often found in tolerances of ± 1% to ± 2.5% while chip capacitors often range from ± 1% to ± 20%. The smaller tolerance values result in components with a higher cost. In volume manufacturing of electronic products it is a necessary trade-off to analyze circuit performance and manufacturing cost. A Parameter Sweep allows the nominal value of any schematic element to be varied in addition to the frequency. As an example consider the input matching network capacitor C2 in Figure 8-9. The nominal value of C2 is entered as 3.372 pF. If we have a 20% tolerance (±10%) we can use a Parameter Sweep to observe the change in gain (S21) as the capacitance value is swept over its tolerance range. Assuming a uniform distribution, the minimum capacitance value would be 3.03 pF and the maximum value would be 3.71 pF. Add a Parameter Sweep to the design through the Evaluations selection. On the Properties window shown in Figure 8-11, enter the analysis and component to sweep. Note that all components that have been made tunable will show up on the drop down box for the parameter to sweep. Then enter the minimum and maximum range of capacitance value. Lastly enter the number of points between the minimum and maximum range to sweep. Sweep the circuit and create a graph to observe the change in the 2-stage amplifier gain. Select the data set associated with the Parameter Sweep and plot S21. A series of plots of the gain will be plotted as shown in Figure 8-12. You can add markers to print out the exact value of gain for each sweep.

Multi-Stage Amplifier Design and Yield Analysis 399

Figure 8-11 Parameter sweep properties tab

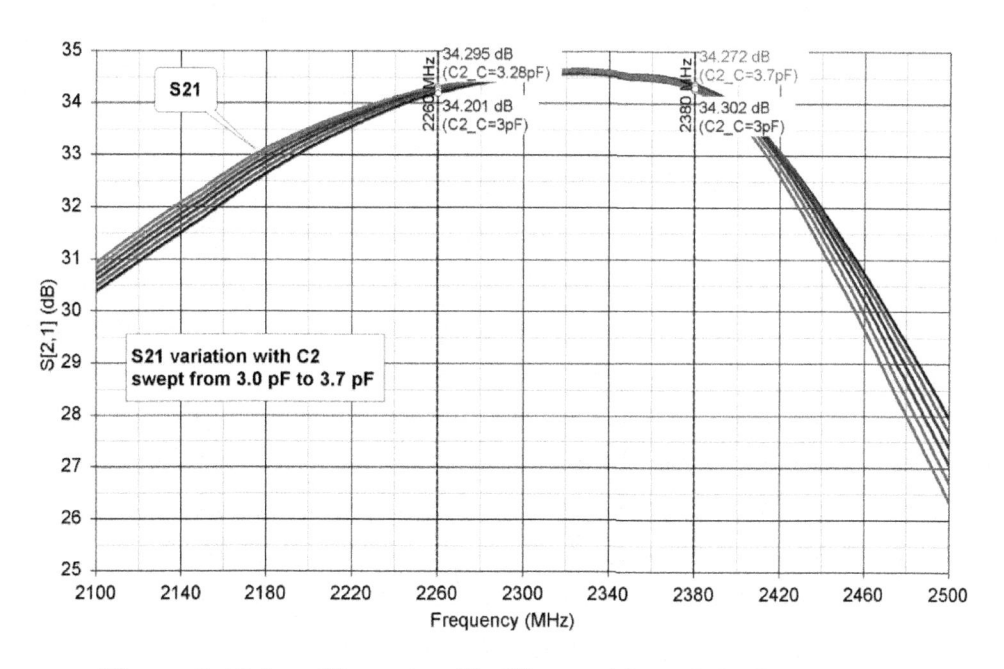

Figure 8-12 Amplifier gain with C2 swept from 3.0 pF to 3.70 pF

8.5 Monte Carlo and Sensitivity Analysis

The Parameter Sweep is useful for observing the amplifier's sensitivity to one or two components but it naturally raises the issue of how the amplifier's response would change to the tolerance of all components in the amplifier design. This type of study is known as Monte Carlo analysis. Set all of the matching elements to be tunable variables on the ideal cascade schematic of Figure 8-9. Keep the value of the stabilization components fixed. In Genesys, the Monte Carlo analysis is added to a workspace through the Evaluations selection. On the General tab, enter 100 samples and select the appropriate Linear Analysis. On the Measurements tab, select S21 and the appropriate data set to observe the gain of the Linear Analysis. Then on the Variables tab, select all the tuned variables. For each inductor, select a uniform distribution with ±2.5% tolerance. For each inductor, select a uniform distribution with a ± 5% tolerance. Run the Monte Carlo analysis and plot S21 on a rectangular graph as shown in Figure 8-14. The input and output return loss is plotted on a separate graph shown in Figure 8-15.

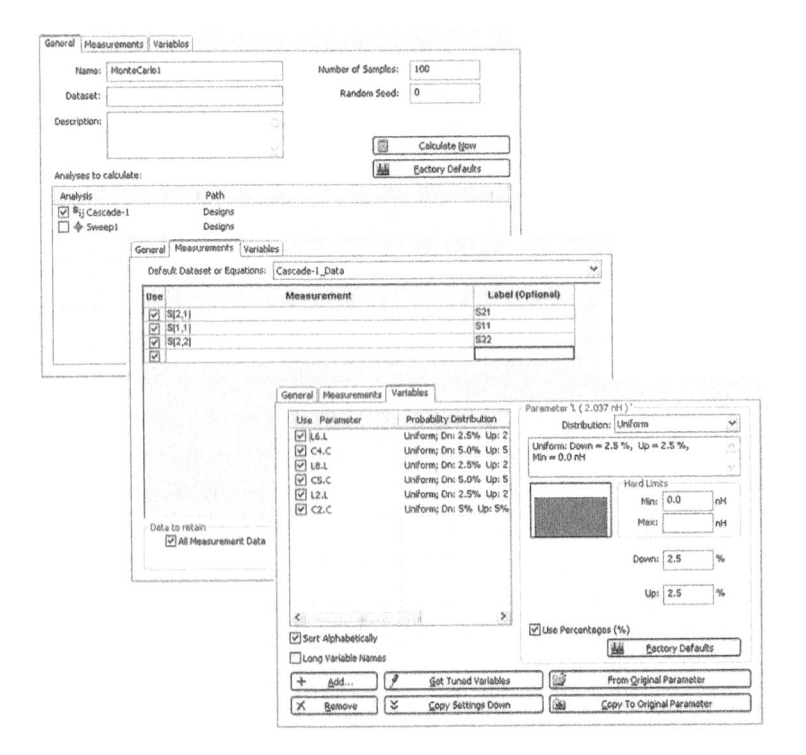

Figure 8-13 Monte Carlo analysis setups

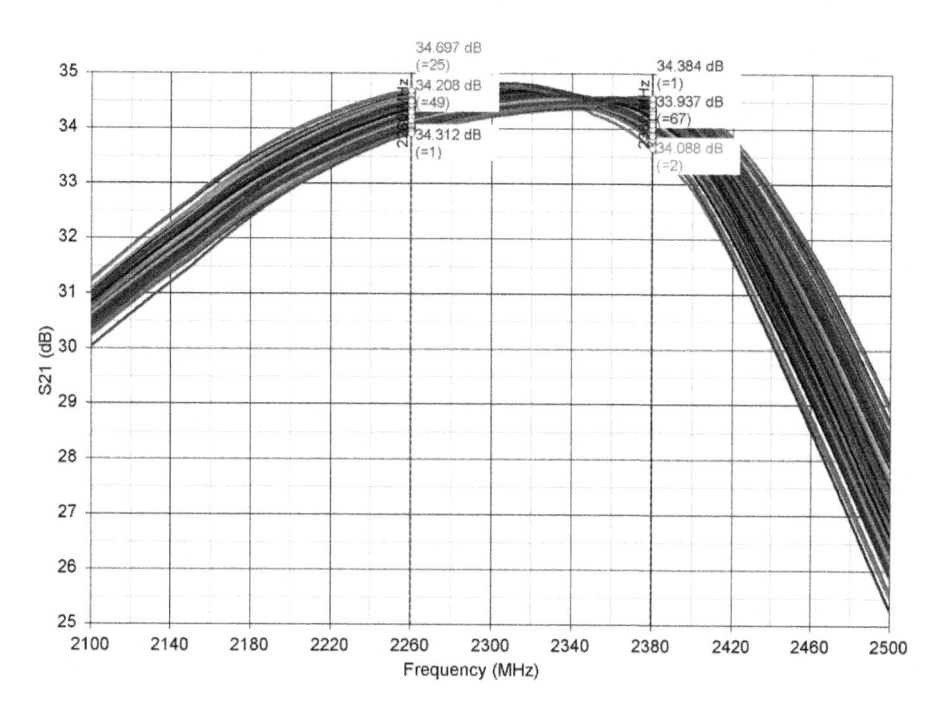

Figure 8-14 Initial Monte Carlo analysis of the 2-stage amplifier, S21, with capacitor tolerance set at ±5%

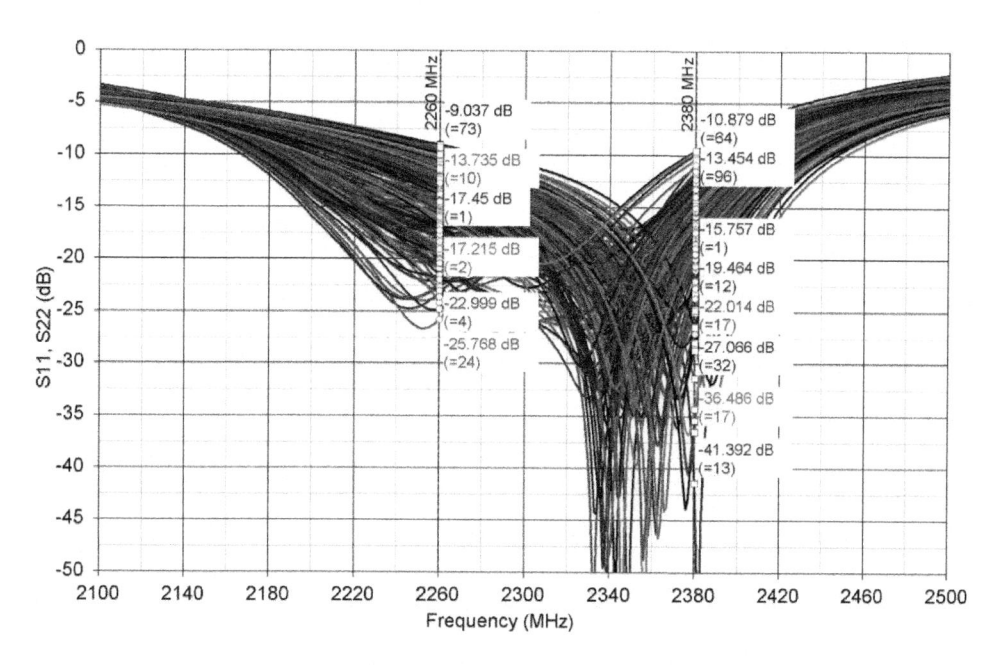

Figure 8-15 Initial Monte Carlo analysis of the 2-stage amplifier, S11 and S22 with capacitor tolerance set at ±5%

Consider that we intend to manufacture this amplifier in volume for a WiFi application. Each amplifier must be capable of achieving 30dB minimum gain from 2260 MHz to 2380 MHz. Line markers can be placed on the plot to better visualize the gain at the design frequency limits. As Figure 8-14 shows the amplifier currently does not meet the design specification as the gain at the high end of the band decreases below 34 dB. In order to keep the cost minimized we would like to use higher tolerance components. Rather than specify lower tolerances for all of the components we could use Parameter Sweeps to see if certain components are more sensitive to achieving the design goal. From the Parameter Sweep performed on C2 in section 8.4 we know that this component is quite sensitive to the overall amplifier gain. On the Monte Carlo variables tab, change the tolerance of C2 to ± 2.5%. Figure 8-16 shows the Monte Carlo analysis of the S21 response is much improved with the high end gain at 34 dB. However an examination of the output return loss of Figure 8-17 shows that the return loss increases to -9.6 dB which fails the design specification of ≤ -10 dB. Next try setting C2 to ± 1% tolerance. This Monte Carlo analysis plot of Figure 8-18 now shows that the amplifier is meeting the design specification with some margin. The minimum gain at the low end is 34.3dB while the high end minimum gain is also 34.3dB. The return loss shown in Figure 8-19 is now -10 dB maximum and satisfies the design specification. We have been able to meet the manufacturing design goals by making a single component have a ±1% tolerance. This will allow us to keep the overall component cost of the amplifier low and meet manufacturing cost targets.

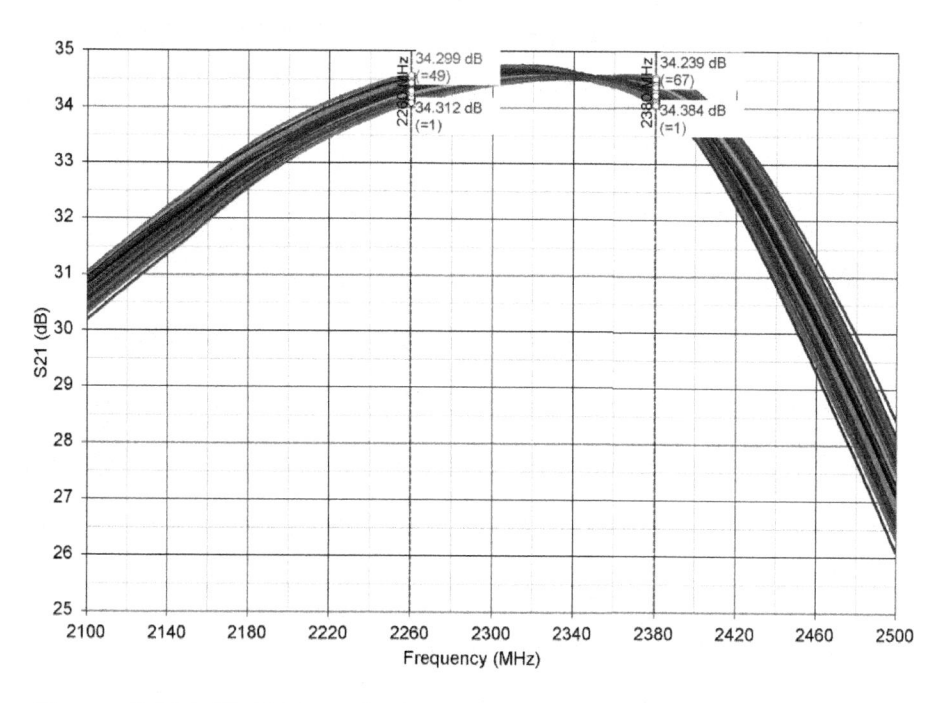

Multi-Stage Amplifier Design and Yield Analysis 403

Figure 8-16 Initial Monte Carlo analysis of the 2-stage amplifier, S21, with capacitor, C2, tolerance set at ±2.5%

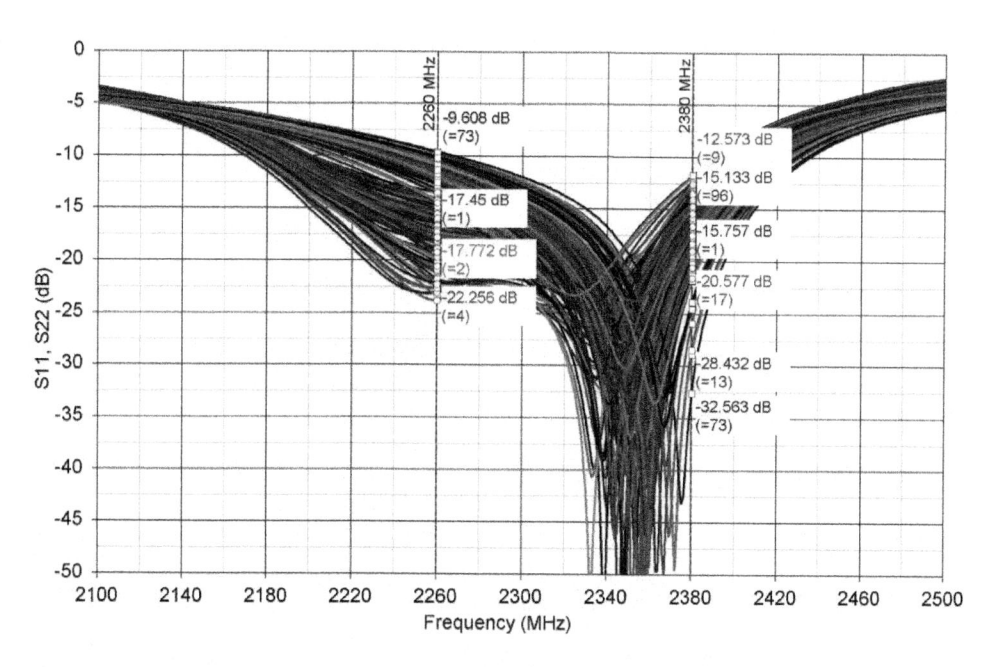

Figure 8-17 Initial Monte Carlo analysis of the 2-stage amplifier, S11 and S22 with capacitor, C2, tolerance set at ±2.5%

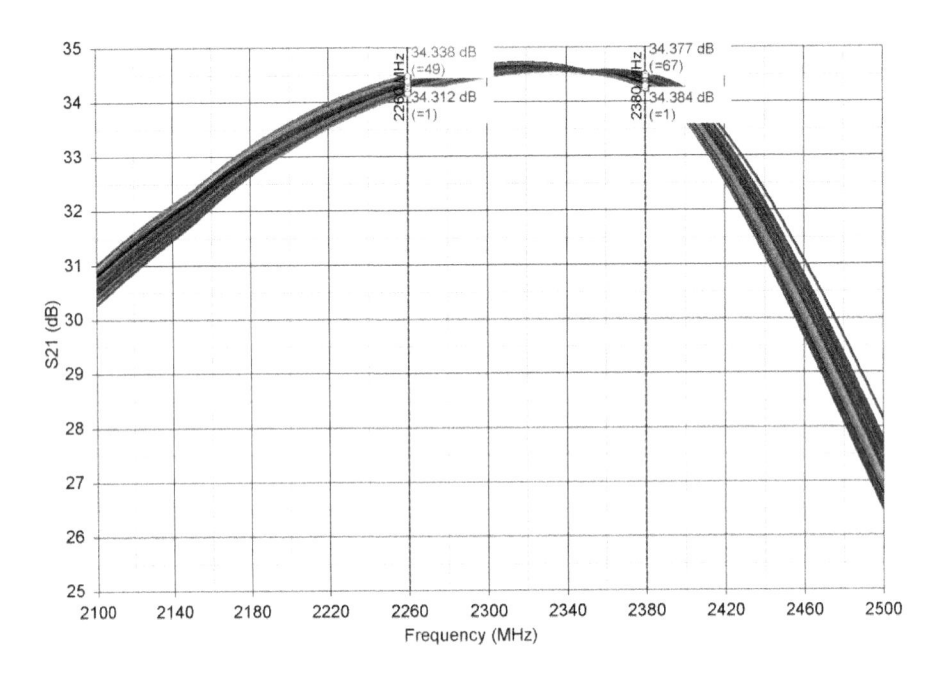

Figure 8-18 Initial Monte Carlo analysis of the 2-stage amplifier, with capacitor, C2, tolerance set at ±1%

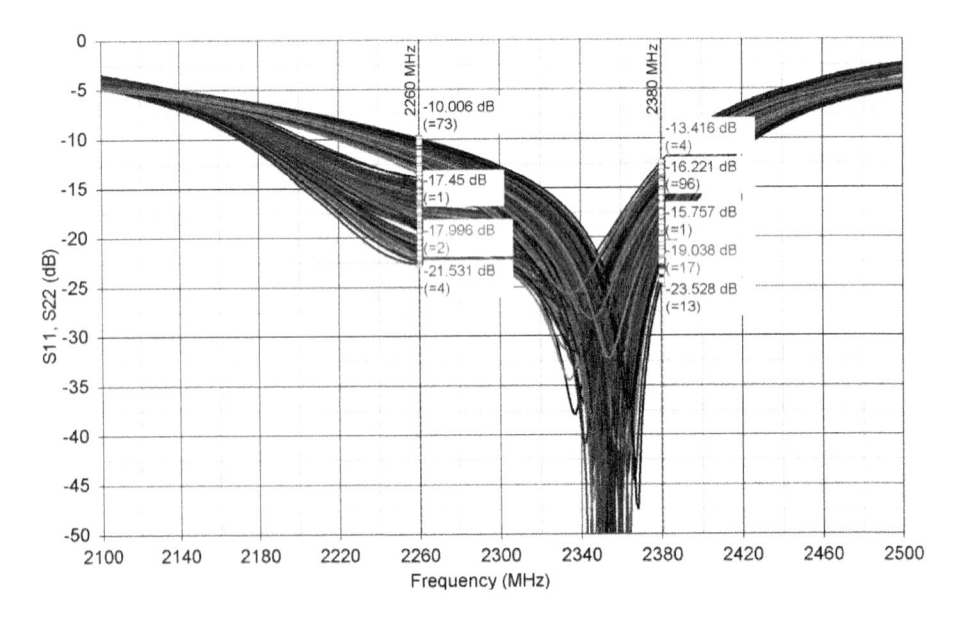

Figure 8-19 Initial Monte Carlo analysis of the 2-stage amplifier, S11 and S22 with capacitor, C2, tolerance set at ±1%

8.6 Yield Analysis

Yield analysis is similar to Monte Carlo analysis in that the tunable elements are statistically varied and the successive sweeps can be observed. As the tunable element values are perturbed, the Yield analysis attempts to achieve a desired target goal similar to Optimization. A given yield is then calculated for the design goal versus the component tolerance values. An observation of the error function can also aid in the determination of the optimal set of design centered values, given the specified tolerances. Yield and Monte Carlo analysis are both probabilistic methods to simulate component variations caused by the production process. These methods can be used to determine which components need to be low tolerance (usually more expensive components) or help create designs that are able to accommodate parameter variation. As the component stepping occurs, the Yield analysis retains a pass-fail value for each round (step) based on the targets. Monte Carlo simply retains the stepped values at each round along with the desired measurements for graphical or tabular comparison. Yield Targets are identical to Optimization Targets, except the "=" and "%" operators are almost never used because the yield would be zero. The ">" and "<" operators are used to specify the range of output parameters which constitute a successful unit. The Optimization Targets are used to find component values which result in the desired nominal responses. The Yield Targets are used to set acceptable limits for definition of what is a successful unit during the analysis. Add a Yield analysis to the design through the Evaluations section. On the General tab, enter 100 samples and select the appropriate Linear Analysis. On the Targets tab set a goal for the gain, S21, > 30 dB from 2260 MHz to 2380 MHz. Finally on the Variables tab, import all variables. Then enter a ±2.5% tolerance for the inductors and a ± 5% tolerance for the capacitors both with a uniform distribution. Figure 8-20 shows the Yield Analysis setup for the design criteria. After simulating the Yield analysis, the computed yield can be displayed either in the Simulation Log or the Yield DataSet. Table 8-1 gives the resulting yield versus capacitor tolerance value. The Yield analysis is helpful for Industrial Engineers to determine the appropriate tradeoffs for cost effective manufacturing. It may not be cost effective to reduce the component tolerance to a level which achieves a 100% yield.

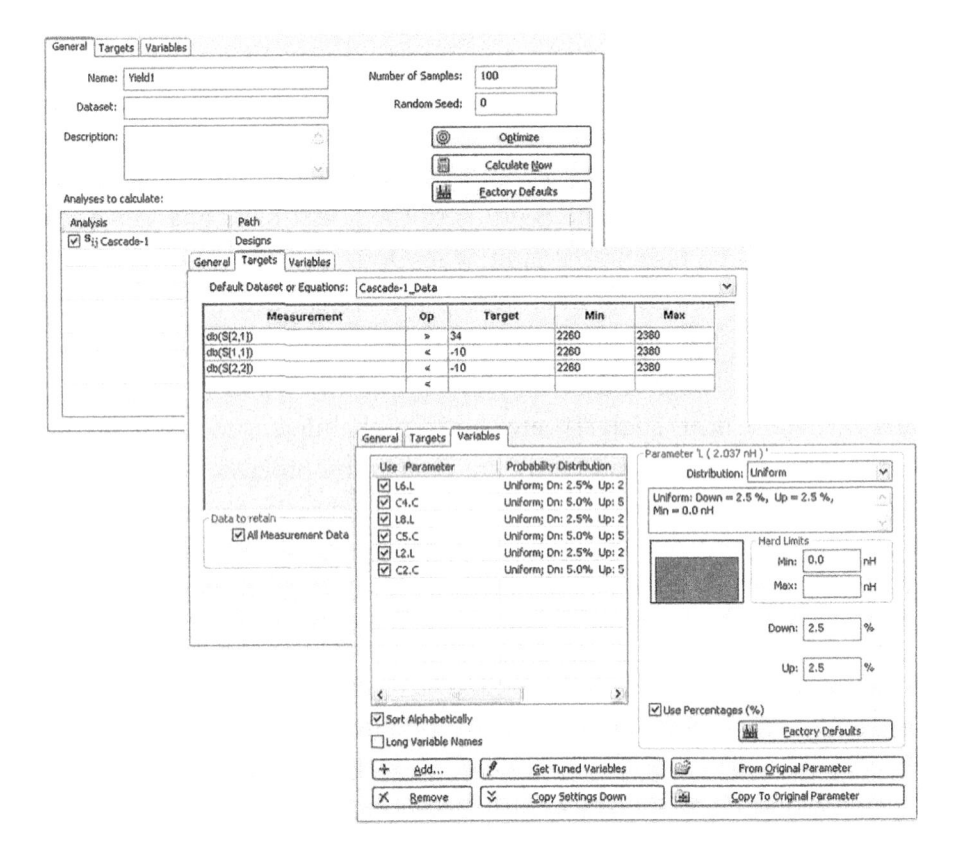

Figure 8-20 Yield analysis setup

Manufacturing Engineers may determine that a 96% yield with a 2.5% capacitor tolerance is a better tradeoff for the manufacturing process. It may be worth accepting 4% fall out for the lower cost of the components used in the design. The ±1% tolerance capacitor may add excessive cost to the amplifier circuit. We may be requested to decrease the cost of the circuit and maintain a 96% yield. The Genesys Yield analysis can prove invaluable for this type of statistical what-if analysis.

Capacitor Tolerance	Calculated Yield
±5%	70%
±2.5%	96%
±1%	99%

Table 8-1 Cascaded amplifier yield to target specifications versus capacitor tolerance

8.6.1 Design Centering

To improve the yield of a design and allow larger tolerance components we need to statistically center the nominal values. We can plot the error function of the Yield analysis from the Yield Dataset. We can also plot the 'Settings' from the Yield Data Set on the same graph as shown in Figure 8-21. The 'Settings' correspond to the tunable element values as defined by the 'VarNames' in the Dataset. To determine the best set of component nominal values look for areas of zero error and set a vertical marker to display the component values. Select an area on the x-axis that results in a zero error function and place a vertical marker. This allows us to easily read the component values at this round. This type of analysis can assist the engineer in properly centering the nominal component value so that the yield can be optimized with higher tolerance components. It can also give insight to whether only one or two components could increase yield of the circuit.

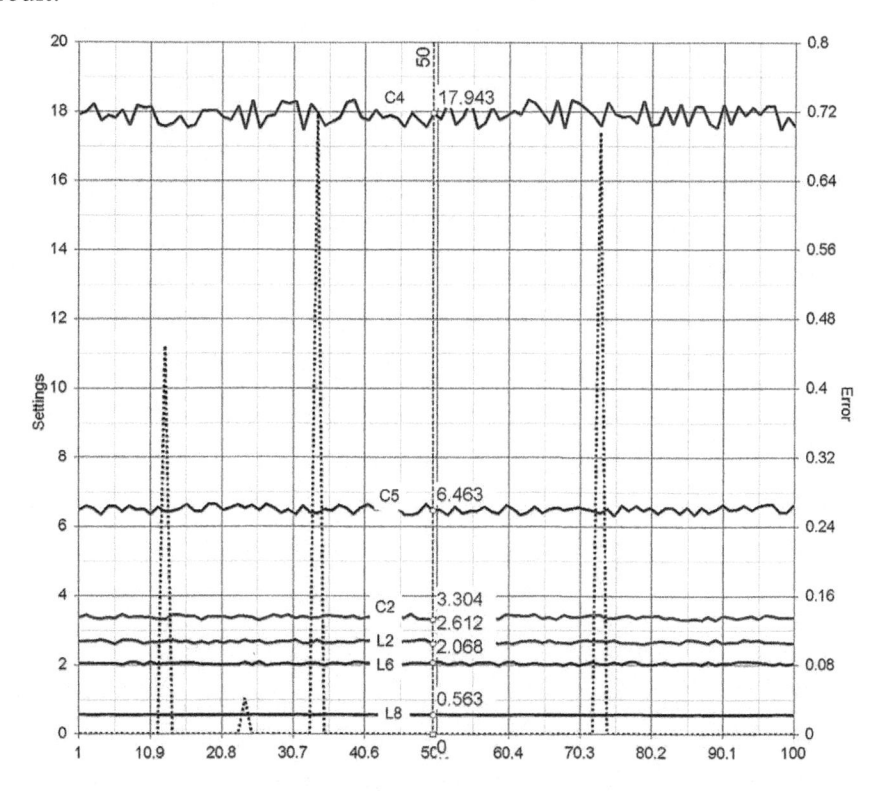

Figure 8-21 Yield analysis error function and perturbed component values

8.7 Low Noise Amplifier Cascade

In practical amplifier applications, a single stage amplifier is rarely adequate to meet the overall gain and output power requirements of an amplifier design. Therefore it is necessary to cascade multiple gain stages to achieve a particular gain specification. In this section the Gain and Noise Figure of the multi-stage cascaded LNA is discussed.

8.7.1 Cascaded Gain and Noise Figure

As we have seen in the example of Section 8.3 the overall gain of a cascaded amplifier is simply the algebraic sum of the gains (or losses) in dB. The gain of an amplifier cascade is then given by Equation (8-7).

$$Cascade\ Gain, G_{dB} = G1_{dB} + G2_{dB} + G3_{dB} + \ldots \quad (8\text{-}7)$$

The overall noise figure is defined by the Friis formula [6]. As the Friis formula of Equation (8-8) shows, the overall noise figure of a cascade is influenced by all stages in the cascade. The noise figure contribution of the second stage is reduced by the gain of the first stage. Therefore to maintain a low noise figure it is important that the first stage have high gain. The noise factor of a given stage is reduced by the gain of the preceding stages throughout the cascade.

$$Noise\ Factor,\ F = F_1 + \frac{F_2 - 1}{G_1} + \frac{F_3 - 1}{G_1 G_2} + \frac{F_4 - 1}{G_1 G_2 G_3} + \ldots \quad (8\text{-}8)$$

where,

$$Noise\ Factor,\ F = 10^{\frac{F_{dB}}{10}} \qquad Gain,\ G = 10^{\frac{G_{dB}}{10}}$$

In the solution of the Friis formula it is important to realize that the noise figure and gain must not be in dB format. This is referred to as the noise factor and related to the noise figure by the 10[log(F)] function. For the three-stage LNA of Figure 8-22, the cascaded gain is easily calculated as 41 dB.

Multi-Stage Amplifier Design and Yield Analysis

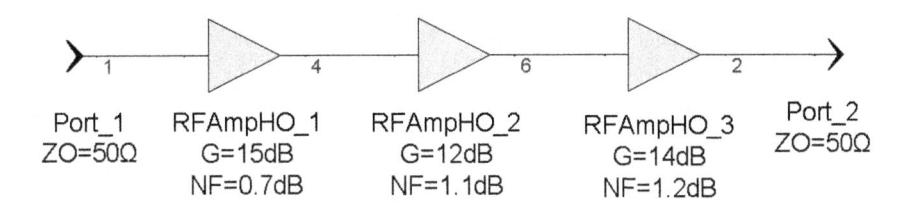

Figure 8-22 Three-stage cascaded low noise amplifier

Using the Friis formula to solve for the resulting noise factor leads to the following result.

$$F = 10^{\frac{0.7}{10}} + \frac{10^{\frac{1.1}{10}}-1}{10^{\frac{15}{10}}} + \frac{10^{\frac{1.2}{10}}-1}{10^{\frac{15}{10}} \cdot 10^{\frac{12}{10}}} = 1.1844$$

Converting the noise factor back to noise figure gives a total cascaded noise figure of 0.736 dB.

$$F_{dB} = 10 \cdot \log(1.1844) = 0.736 \ dB$$

The cascaded gain and noise figure can be calculated in Genesys using the simple block diagram of Figure 8-22. This three stage cascade uses the RFAmpHO system level model but can be used in Linear simulation for gain and noise figure. Setup a Linear simulation for any frequency range as the frequency is independent for this use of the RFAmpHO model. The simulated cascade gain, S21, and noise figure, NF, are shown in Figure 8-23. We can see that the overall noise figure of the cascade is 0.736 dB which correlates with the solution of the Friis formula. The noise figure of the first stage dominates the overall noise figure and is slightly degraded by the 2nd and 3rd stages.

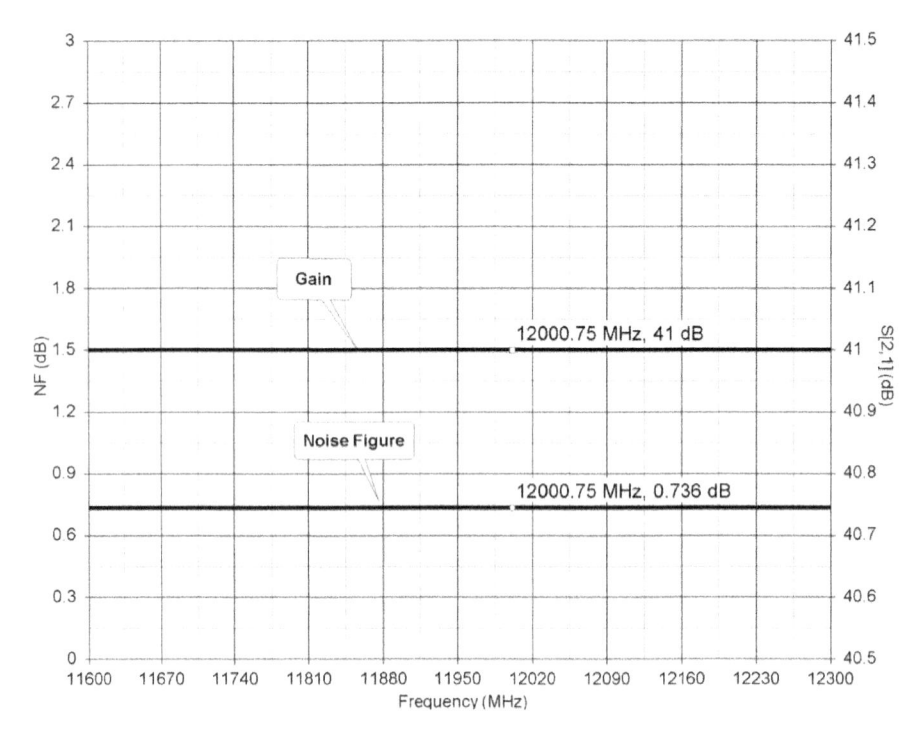

Figure 8-23 Simulated LNA cascade noise figure and gain

8.7.2 Impedance Match and the Friis Formula

A Low Noise Amplifier is typically used as the front end of a radio receiver. Therefore it is often attached to an antenna and filter combination. From Chapter 4 we have seen that the impedance of a filter can vary significantly across its passband as determined by the ripple and return loss. The same can be true of an antenna or the antenna feed network. Because the Noise Figure of an LNA is dependent on its source impedance, large errors can be obtained in the application of the Friis formula when calculating the overall noise figure of such a cascaded network. The Friis formula assumes that there is a perfect impedance match between each stage in the cascade and there are no variations in impedance across the frequency band. As an example consider a typical Low Noise Amplifier used in the Ku Band frequency range of 11.7 to 12.2 GHz. To eliminate any interference from the uplink signal or terrestrial sources a low loss waveguide filter is placed at the input of the LNA. This system level block diagram is shown in Figure 8-24.

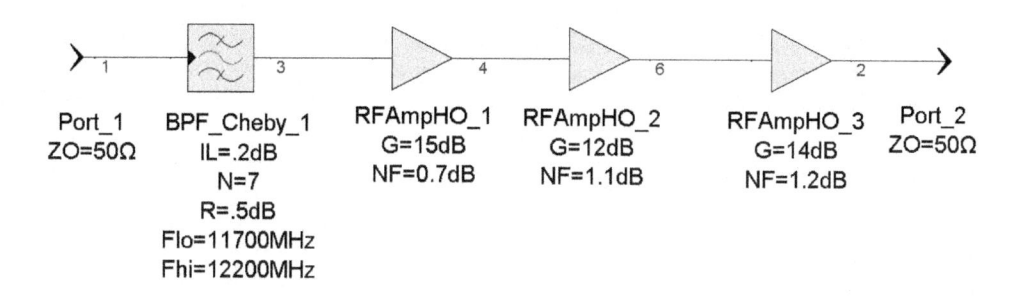

Multi-Stage Amplifier Design and Yield Analysis 411

```
>──1──[BPF_Cheby_1]──3──▷──4──▷──6──▷──2──>
Port_1    BPF_Cheby_1    RFAmpHO_1   RFAmpHO_2   RFAmpHO_3    Port_2
ZO=50Ω    IL=.2dB        G=15dB      G=12dB      G=14dB       ZO=50Ω
          N=7            NF=0.7dB    NF=1.1dB    NF=1.2dB
          R=.5dB
          Flo=11700MHz
          Fhi=12200MHz
```

Figure 8-24 Ku band LNA with bandpass filter

Note that a passive device's insertion loss is entered as a negative gain in dB in the Friis formula. The noise figure then becomes the absolute value of this loss or simply 0.2 dB in the case of the bandpass filter. Using the Friis formula to solve for the resulting noise figure leads to the following result.

$$F = 10^{\frac{0.2}{10}} + \frac{10^{\frac{0.7}{10}} - 1}{10^{\frac{-0.2}{10}}} + \frac{10^{\frac{1.1}{10}} - 1}{10^{\frac{-0.2}{10}} \cdot 10^{\frac{15}{10}}} + \frac{10^{\frac{1.2}{10}} - 1}{10^{\frac{-0.2}{10}} \cdot 10^{\frac{15}{10}} \cdot 10^{\frac{12}{10}}} = 1.2402$$

Converting the noise factor back to noise figure gives a total cascaded noise figure of 0.935 dB.

$$F_{dB} = 10 \cdot \log(1.2402) = 0.935 \; dB$$

Based on this simple application of the Friis formula the engineer would expect the system noise figure to be 0.935 dB. Simulating the block diagram of Figure 8-24 in Genesys reveals a vastly different result. The simulated noise figure of Figure 8-25 shows that there is significant ripple in the noise figure. The worst case noise figure is actually greater than 2 dB across the passband of the amplifier. The 0.5 dB ripple in the filter actually results in greater than 1.5 dB ripple in the noise figure. This is due to the fact that the noise figure of the input device is very sensitive to the impedance that is presented to it. The 0.935 dB noise figure calculation is an erroneous result that is obtained when using the Friss formula or one of the many spreadsheet cascade analysis programs. By modeling the simple cascade in Genesys we can quickly become aware of this condition.

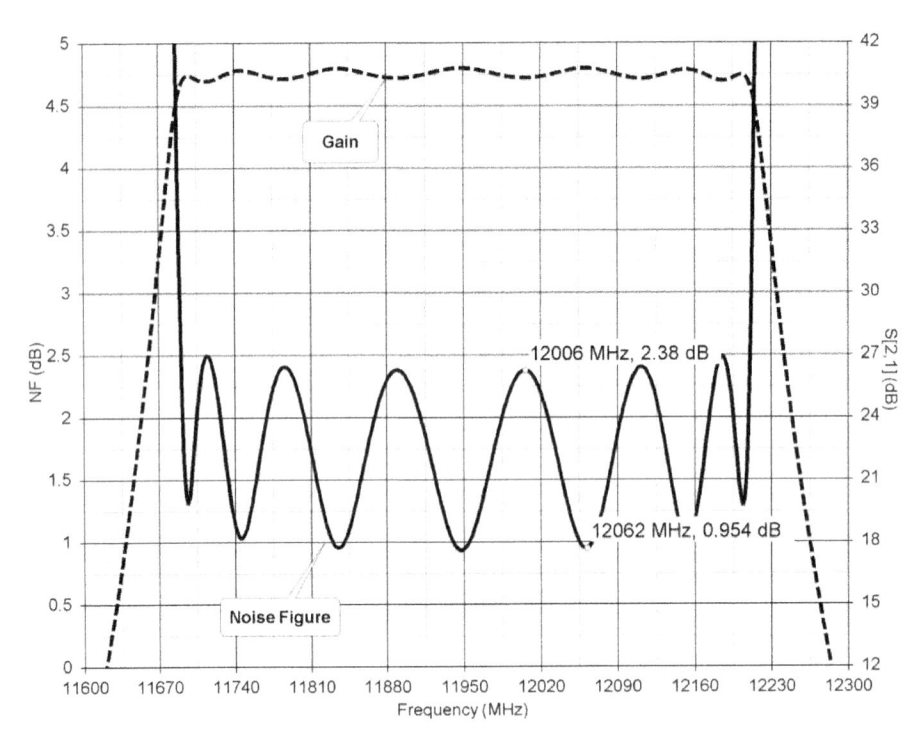

Figure 8-25 Ku band LNA with bandpass filter simulated response

8.7.3 Reducing the Effect of Source Impedance Variation

In practice it is often desirable to place a low-loss isolator at the input of the LNA to buffer the effects of impedance variation due to the filter ripple. It is important that the isolator have very low loss as its insertion loss will also add to the overall noise figure. This can easily be modeled in Genesys by adding an isolator to the block diagram as shown in Figure 8-26. An isolator with 0.1 dB insertion loss reduces the noise figure by 0.8 dB. Perform a Linear Analysis over the range of 11 GHz to 13 GHz with 600 points. This puts the noise figure of the LNA closer to 1.5 dB, which is a better value for the reception of Ku Band satellite signals from space. The resulting cascaded gain and noise figure is shown in Figure 8-27. Note the tremendous decrease in the noise figure due to the addition of the isolator between the filter and the amplifier.

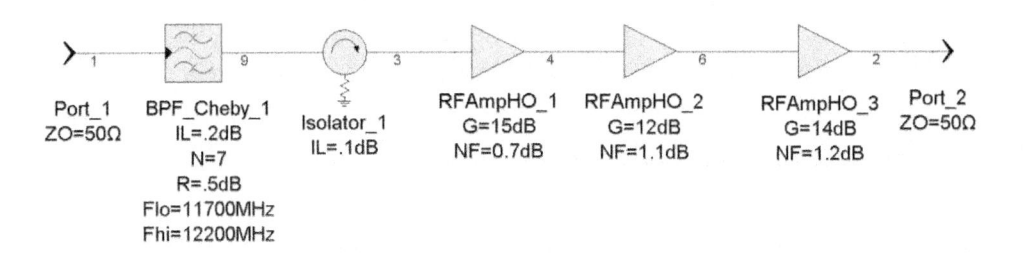

Multi-Stage Amplifier Design and Yield Analysis 413

Figure 8-26 Ku band LNA with isolator and bandpass filter

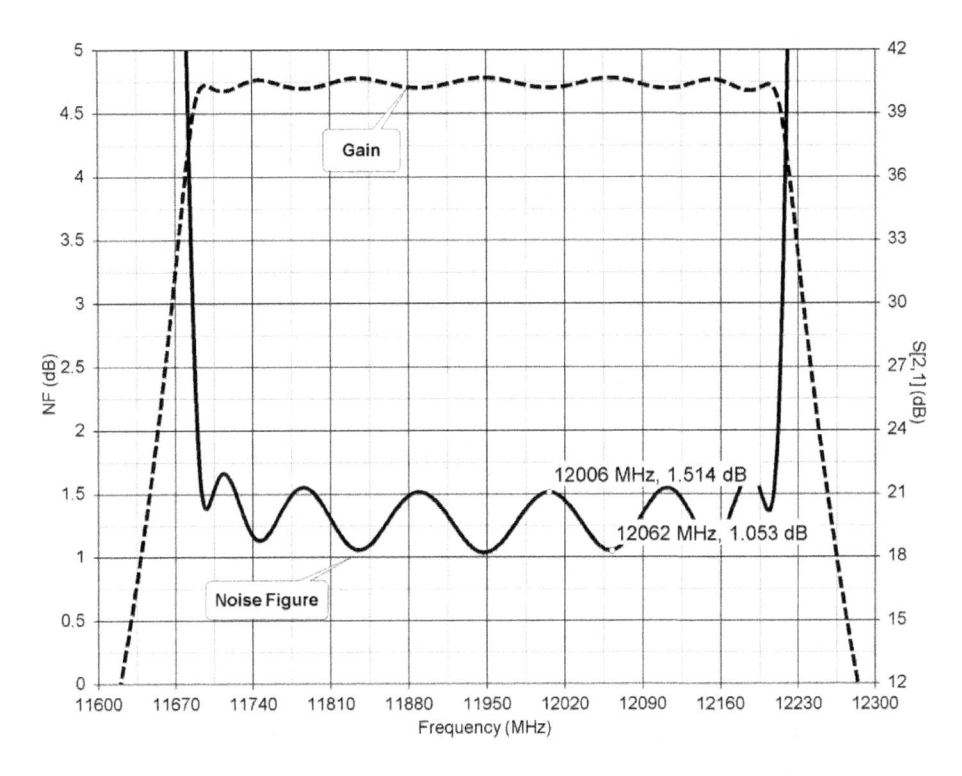

Figure 8-27 LNA cascade noise figure with input matching circuit

8.8 Summary

The example of Section 8.6 gives an introduction to the important subject of system, or block diagram, level simulation. The second edition of this two part volume will begin with a continuation of this subject material extended to the Genesys system simulator which includes nonlinear and frequency translation capability. With this capability we can also evaluate nonlinear parameters such as the cascade output power, intercept points, and intermodulation properties. As the previous example shows there are also system level computations that can be evaluated with Linear simulation techniques. Linear simulation continues to be a very important topic and is the foundation for all RF and microwave CAD work. This volume provides the reader with a thorough coverage of the linear circuit design techniques that can be accomplished with Linear simulation in Genesys. As an applications manual this text forms a bridge between the classic theory and practical engineering problem solving.

References and Further Reading

[1] Guillermo Gonzales, *Microwave Transistor Amplifiers – Analysis and Design,* Second Edition, Prentice Hall Inc., Upper Saddle River, New Jersey

[2] Randy Rhea, *The Yin-Yang of Matching: Part 1 – Basic Matching Concepts*, High Frequency Electronics, March 2006

[3] Steve C. Cripps, *RF Power Amplifiers for Wireless Communications*, Artech House Publishers, Norwood, MA. 1999

[4] David M. Pozar, *Microwave Engineering*, Fourth Edition, John Wiley & Sons, New York, 2012

[5] R. Ludwig, P. Bretchko, *RF Circuit Design*, Theory and Applications, Prentice Hall, Upper Saddle River, NJ, 2000

[6] Ali A. Behagi and Stephen D. Turner, *Microwave and RF Engineering,* BT Microwave LLC, State College, PA, 2011

Problems

8-1. Design a Two–Stage Maximum Gain Amplifier

(a) Use the Agilent HBFP0405 and the Mitsubishi MGF0911a transistors to design a two-stage maximum gain amplifier at 2.2 GHz. All the input, output, and interstage impedance matching networks should be analytically designed using two-element L-networks.

(b) Design the stabilizing network for the first stage using Agilent HBFP0405 transistor

(c) Using the HBFP0405 stabilized transistor create a tabular output of the simultaneous match source and load impedance and Gmax

(d) Design the first stage input matching network by matching the 50 Ohm source impedance to the conjugate impedance looking into the stabilized HBFP0405 device, ZM1.

(e) Design the stabilizing network for the second stage using the Mitsubishi device

(f) Using the MGF0911a stabilized transistor create a tabular output of the simultaneous match source and load impedance and Gmax

(f) Design the second stage output matching network by matching the 50 Ohm load impedance to the conjugate impedance looking into the output of the stabilized MGF0911a device, ZM2

(g) Design the interstage matching network between stage one ZM2 and stage two ZM1.

(h) Cascade the two amplifiers and display the overall response of the cascaded network. The cascaded gain of the amplifier must be the sum of the maximum gains of the two single-stage amplifiers.

8-2. Increase the Bandwidth of the Matching Networks

(a) Increase the bandwidth of the input matching network by selecting a virtual resistor R_M between the input resistor of the matching network, R_S, and the output resistor of the matching network, R_L, such that $R_M = \sqrt{R_S R_L}$

and then conjugately match the input and output impedances to the virtual resistor R_M with L-networks.

(b) Cascade the two matching networks and measure the fractional bandwidth at 20 dB return loss.

(c) Compare the measured fractional bandwidth with the fractional bandwidth of the original matching network at 20 dB return loss and calculate the percentage of the increase in the bandwidth.

(d) Repeat the above procedure for the interstage and output matching networks.

8-3. LNA Cascade Using the Friis Formula

(a) Consider a typical Low Noise Amplifier used in the X Band of 8.7 to 9.2 GHz. Using the Friis formula calculate the overall noise figure of the following LNA cascade.

LNA #1	G1 = 15 dB	F1 = 1.1 dB
LNA #2	G1 = 14 dB	F1 = 1.2 dB
LNA #3	G1 = 13 dB	F1 = 1.3 dB

Appendix A

Straight Wire Parameters for Solid Copper Wire

Wire Size (AWG)	Diameter in Mils	Resistance Ohms/1000 ft.	Area in circular Mils	Suggested Maximum Current Handling, Amperes[1]
0000	460.0	0.049	211600	1000
000	409.6	0.062	167800	839
00	364.8	0.078	133100	665
0	324.9	0.098	105500	527
1	289.3	0.124	83690	418
2	257.6	0.156	66360	332
3	229.4	0.197	52620	263
4	204.3	0.249	41740	208
5	181.9	0.313	33090	165
6	162.0	0.395	26240	131
7	144.3	0.498	20820	104
8	128.5	0.628	16510	83
9	114.4	0.793	13090	65
10	101.9	0.999	10380	52
11	90.7	1.26	8230	41
12	80.8	1.56	6530	32
13	72.0	2.00	5180	26
14	64.1	2.52	4110	20
15	57.1	3.18	3260	16
16	50.8	4.02	2580	13
17	45.3	5.05	2050	10
18	40.3	6.39	1620	8.0
19	35.9	8.05	1290	6.0
20	32.0	10.1	1020	5.0
21	28.5	12.8	812	4.0
22	25.3	16.2	640	3.0
23	22.6	20.3	511	2.5
24	20.1	25.7	404	2.0
25	17.9	32.4	320	1.6
26	15.9	41.0	253	1.2
27	14.2	51.4	202	1.0
28	12.6	65.3	159	0.80
29	11.3	81.2	123	0.61
30	10.0	104.0	100	0.50
31	8.9	131	79.2	0.40
32	8.0	162	64.0	0.32
33	7.1	206	50.4	0.25
34	6.3	261	39.7	0.19
35	5.6	331	31.4	0.16
36	5.0	415	25.0	0.12
37	4.5	512	20.2	0.10
38	4.0	648	16.0	0.08
39	3.5	847	12.2	0.06
40	3.1	1080	9.61	0.05
41	2.8	1320	7.84	0.04
42	2.5	1660	6.25	0.03
43	2.2	2140	4.84	0.024
44	2.0	2590	4.00	0.020
45	1.76	3350	3.10	0.016
46	1.57	4210	2.46	0.012
47	1.40	5290	1.96	0.010
48	1.24	6750	1.54	0.008
49	1.11	8420	1.23	0.006
50	0.99	10600	0.98	0.005

Current Handling based on 1 Amp/200 Circular Mils-no insulation and free air conditions. Insulated and stranded Copper wire must be de-rated from the values in the Table.

Appendix B

Workspace Schematics for the Q_o Measurement Smith Chart Overlay

Appendix B contains the collection of Genesys Workspace schematics that are used to produce the Smith Chart overlays used for resonator Qo measurement of Section 4.6.2.

B.1 Γ_i Line Generation

The Γ_i Line is produced by using the manual S-Parameter model of Figure B1-1. A Parameter Sweep is used to vary the magnitude of S11 from -1 to +1. This will produce a straight line across the Smith Chart. The angle of S11 is made a variable in the Equation Editor. This will allow the 'angle' variable to be shared with the overlays described in Appendix B-2 and B-3. The angle of the Γ_i can be rotated around the Smith Chart using the Tune control. The resulting line on the Smith Chart is shown in Figure B1-3.

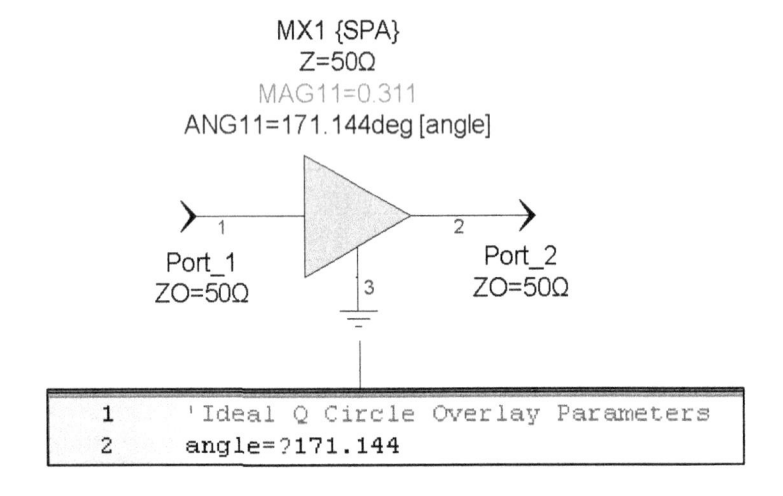

Figure B1-1 Manual S-Parameter Model and Variable Defined in the Equation Editor

Appendix B 419

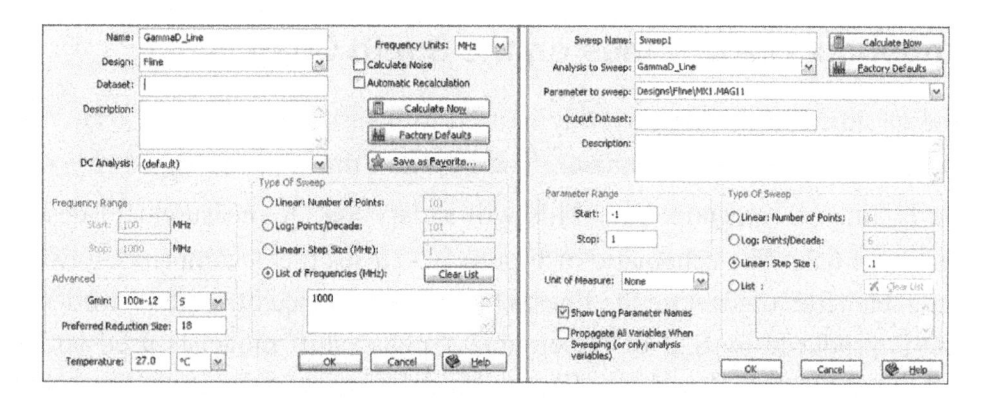

Figure B1-2 Linear Analysis and Parameter Sweep used for the Γ_i Line Generation Schematic

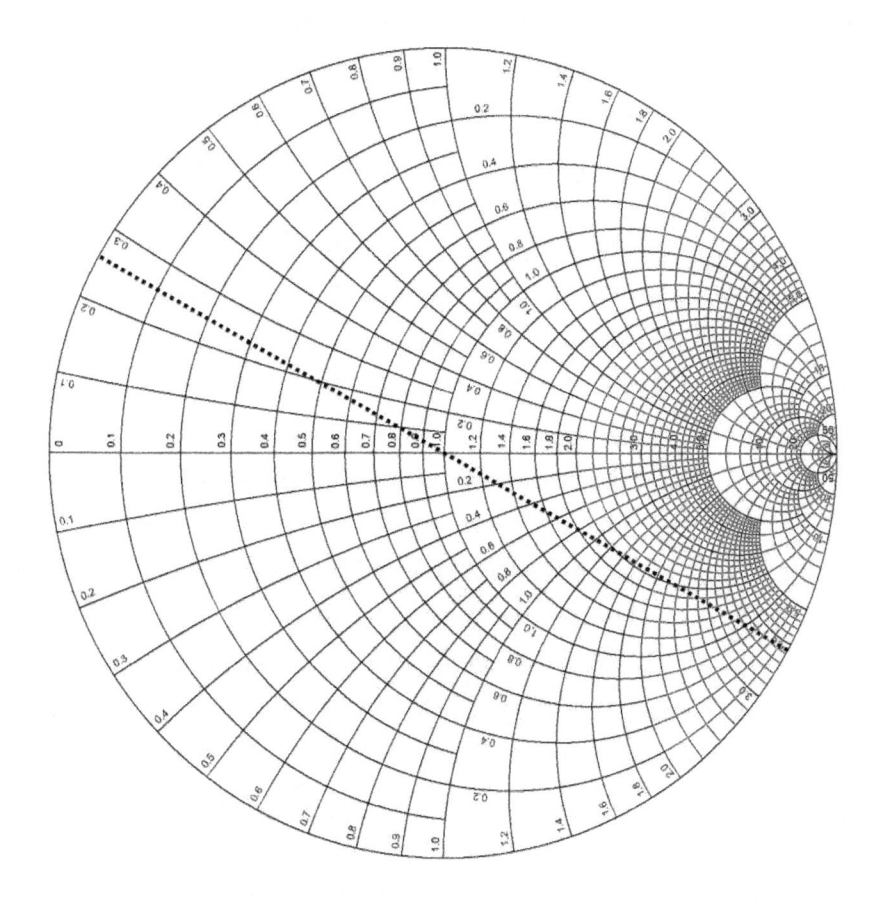

Figure B1-3 Γ_i Line drawn across Smith Chart

Appendix B.2 Q_L Lines on the Smith Chart

The Q_L line Smith Chart overlay is used to measure the loaded Q, Q_L, from the S11 measurement of a resonator. The lines are drawn at an angle of $\pm 45°$ from the Γ_i line of Appendix A.1. This will satisfy the Q_L measurement described in Section 4.6.2. The schematic of Figure B2-1 uses the accompanying two-port S-Parameter file to generate the lines. One line is produced from S11 while the other line is produced by S22 of the circuit. A phase shift model is used on the input and output of the S-Parameter file to rotate the intersection of the lines around the Smith Chart in sync with the Γ_i line. An attenuator element is used on the input and output so that the intersection of the lines can be moved inward to the center of the Smith Chart. This allows proper alignment with Γ_d for lossy coupling cases. The attenuation and phase shift values are defined by the same variable and defined in the Equation Editor.

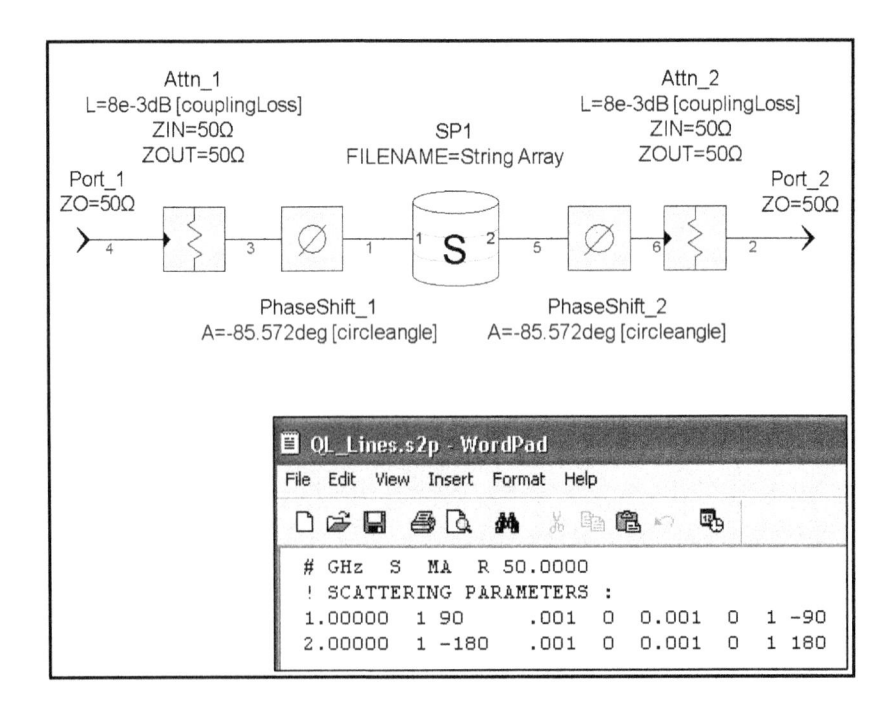

Figure B2-1 Schematic used for Generation of Q_L Lines on the Smith Chart

```
1       'Ideal Q Circle Overlay Parameters
2       angle=?171.144
3       circlediameter=?1988.72
4       couplingLoss=?0.008
5       circleangle=-angle/2
```

Figure B2-2 Equation Editor with Variables required by the Smith Chart Overlay

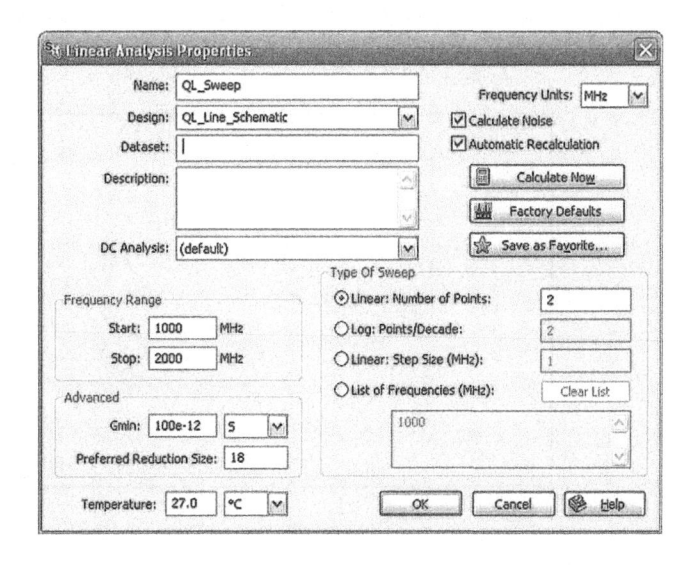

Figure B2-3 Linear Analysis used by the Q_L Line Schematic

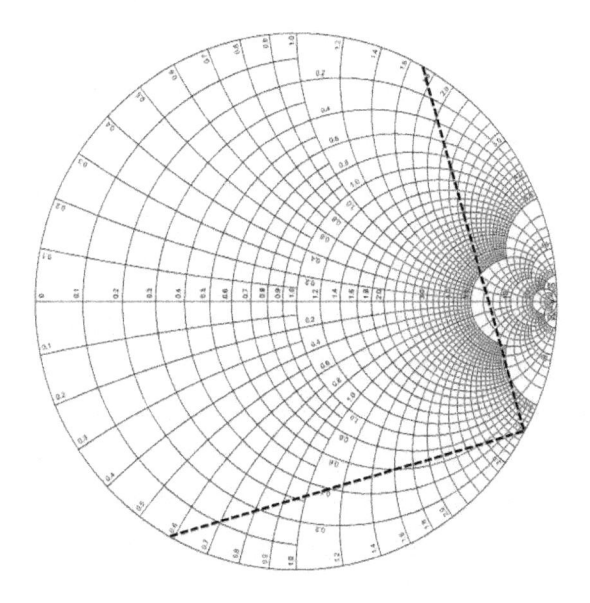

Figure B2-4 Q_L Lines Produced on the Smith Chart

Appendix B.3 Ideal Q Circle on the Smith Chart

The schematic of Figure B3-1 is used to generate the ideal Q circle Smith Chart overlay. A parallel RLC resonator is coupled to the load by an ideal transformer model. The turn ratio of the transformer is defined by variable 'circlediameter' because the coupling determines the size of the Q circle on the Smith Chart. A phase shift element and attenuator are used to move the circle inward toward the center of the chart for lossy coupling cases.

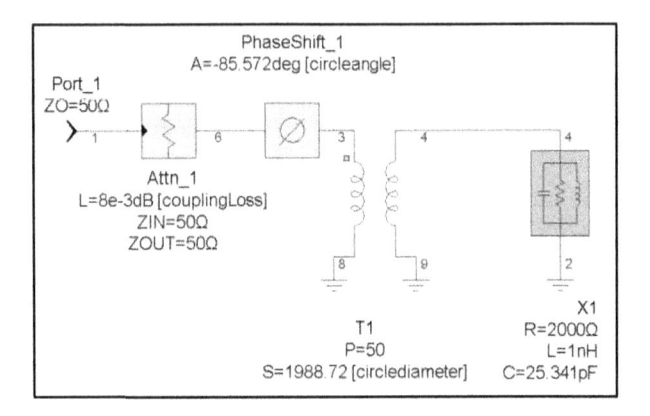

Figure B3-1 Schematic used for Ideal Q Circle Drawn on the Smith Chart

Figure B3-2 Linear Analysis used for the Ideal Q Circle Generation

Appendix B 423

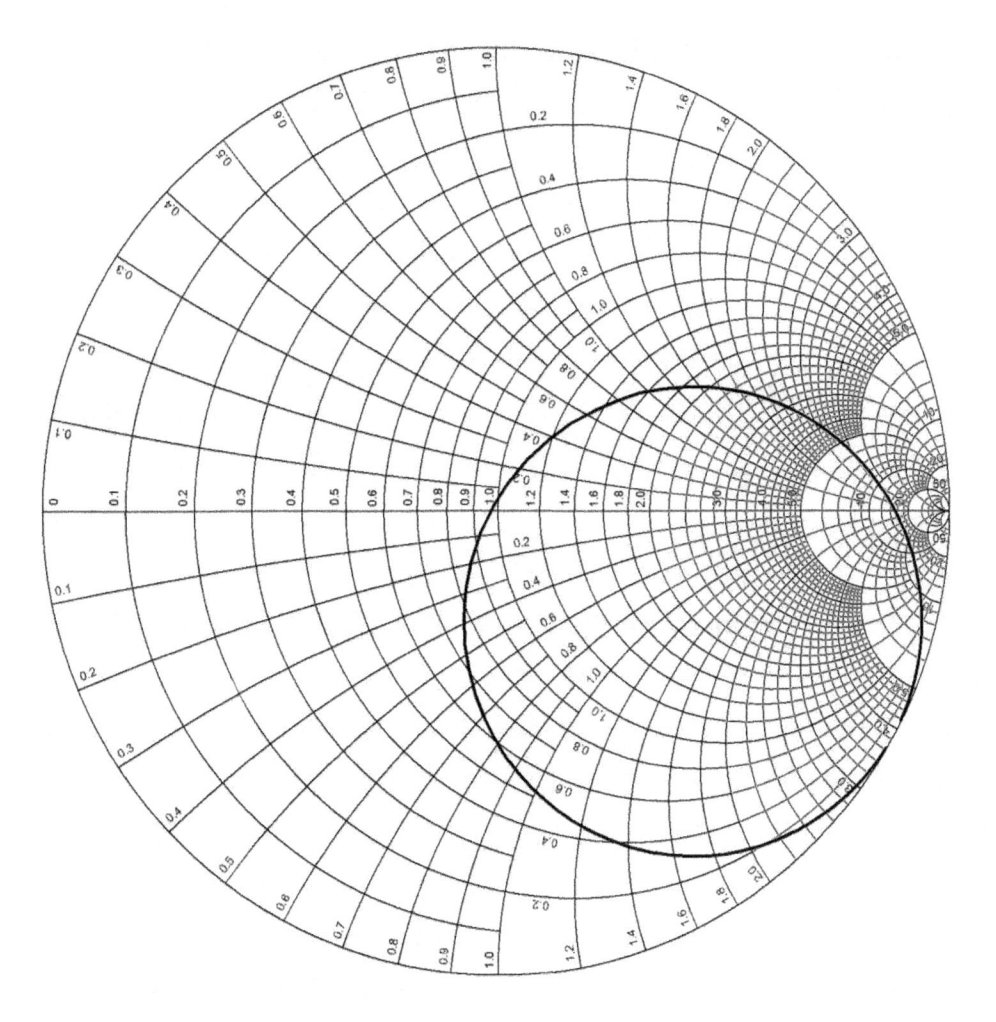

Figure B3-3 Ideal Q Circle Overlay Drawn on the Smith Chart

Appendix B.4 Q_o Measurement on the Smith Chart

The complete set of measurements required for Qo measurement are shown in Figure B4-1. Each output is identified on the Smith Chart of Figure B4-2.

Measurement	Label (Optional)	Complex Format	Hide?	Color
QL_Sweep_Data.S[1,1]		Mag abs+Angle	☐	
QL_Sweep_Data.S[2,2]		Mag abs+Angle	☐	
Sweep1_Data.S[1,1]		Mag abs+Angle	☐	
Linear2_Data.S[1,1]		Mag abs+Angle	☐	
Linear4_Ideal_Data.S[1,1]		Mag abs+Angle	☐	

Figure B4-1 Output of Four Datasets on Single Smith Chart

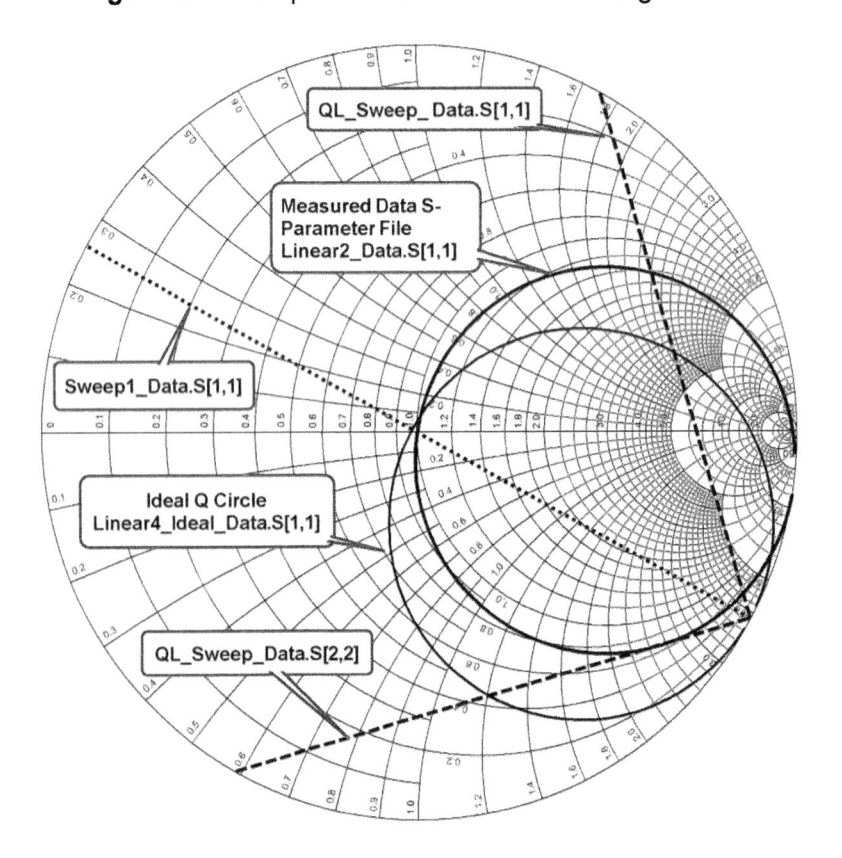

Figure B4-2 Datasets identified on the Smith Chart

Appendix C

VBScript file listing for the Matching Utility of Chapter 5.

```
Dim WsDoc
Dim GL, BL, L2, L1, L3, RS, XS,z, A, A1, B,L, X, C, C1,C2, wo,T
Dim RL, XL, F, complex, pos, i, r, xi, Ls, Cs, Cp, Lp,nsolutions
WsDoc= theApp.GetWorkspaceByIndex(0)
Sim=WsDoc.Designs.Linear1
F = WsDoc.Designs.Sch1.PartList.CWSource_3.ParamSet.F.GetValue()
'PARSE THE REAL AND IMAGINARY PART OF THE IMPEDANCE

complex = WsDoc.Designs.Sch1.PartList.Port_1.ParamSet.ZO.GetData()
z=len(complex)
pos=instr(complex,"j")
If pos=0 then
XS=0
RS = WsDoc.Designs.Sch1.PartList.Port_1.ParamSet.ZO.GetData()
else XS=mid(complex,pos-1)
XS=replace(XS,"j","")
RS=left(complex,z-((z-(pos-1))+1))
end if
```

'PARSE THE REAL AND IMAGINARY PARTS OF THE PORT 2 IMPEDANCE

```
complex = WsDoc.Designs.Sch1.PartList.Port_2.ParamSet.ZO.GetData()
z=len(complex)
pos=instr(complex,"j")
If pos=0 then
XL=0
RL = WsDoc.Designs.Sch1.PartList.Port_2.ParamSet.ZO.GetData()
else XL=mid(complex,pos-1)
XL=replace(XL,"j","")
RL=left(complex,z-((z-(pos-1))+1))
end if
```

'My_L_Network=WsDoc.Designs.Sch1.PartList.L1.ParamSet.L.SetValue(XL)

r=RL/RS

xi=XL/RS

wo=(2*3.14*F)
i=WsDoc.Equation.VarBlock.Solution.GetValue()

'   First Condition

A1=((RS^2)+(XS^2))-(RL*RS)

'   Second Condition

A=((RL^2)+(XL^2))-(RL*RS)

'_____

If A1>0 and A>0 Then
nsolutions=4
WsDoc.Designs.Sch1.Schematic.Page.Annotate2.Max.Set(nsolutions)
Select Case i

case 1
B = ((RL * XS) + Sqr(RS * RL * (RS ^ 2 + XS ^ 2 - RS * RL))) / (RL * (RS ^ 2 + XS ^ 2))
X = (RL * XS - RS * XL) / RS + (RS - RL) / (B * RS)
If B>0 Then   'B is a shunt C
B=B/wo
ShuntC
elseif B<0 then   'B is a shunt L
B=-1/(wo*B)
ShuntL
else
end if

If X>0 then   'X is a series L
X=X/wo
SeriesL_X2
elseif X<0 then   'X is a series C
X=-1/(wo*X)

SeriesC_X2
else
end if
PlotResponse
'
---

case 2

B = ((RL * XS) - Sqr(RS * RL * (RS ^ 2 + XS ^ 2 - RS * RL))) / (RL * (RS ^ 2 + XS ^ 2))
X = (RL * XS - RS * XL) / RS + (RS - RL) / (B * RS)
If B>0 Then   'B is a shunt C
B=B/wo
ShuntC
elseif B<0 then   'B is a shunt L
B=-1/(wo*B)
ShuntL
else
end if

If X>0 then   'X is a series L
X=X/wo
SeriesL_X2
elseif X<0 then   'X is a series C
X=-1/(wo*X)
SeriesC_X2
else
end if
PlotResponse

'
---

case 3

B = ((RS * XL) + Sqr(RS * RL * (RL ^ 2 + XL ^ 2 - RS * RL))) / (RS * (RL ^ 2 + XL ^ 2))
X = (RS * XL - RL * XS) / RL + (RL - RS) / (B * RL)
If B>0 Then   'B is a shunt C
B=B/wo
ShuntC
elseif B<0 then   'B is a shunt L
B=-1/(wo*B)

ShuntL
else
end if

If X>0 then   'X is a series L
X=X/wo
SeriesL_X3
elseif X<0 then   'X is a series C
X=-1/(wo*X)
SeriesC_X3
else
end if
PlotResponse

case 4

B = ((RS * XL) - Sqr(RS * RL * (RL ^ 2 + XL ^ 2 - RS * RL))) / (RS * (RL ^ 2 + XL ^ 2))
X = (RS * XL - RL * XS) / RL + (RL - RS) / (B * RL)
If B>0 Then   'B is a shunt C
B=B/wo
ShuntC
elseif B<0 then   'B is a shunt L
B=-1/(wo*B)
ShuntL
else
end if

If X>0 then   'X is a series L
X=X/wo
SeriesL_X3
elseif X<0 then   'X is a series C
X=-1/(wo*X)
SeriesC_X3
else
end if
PlotResponse
End select
'
_____

ElseIf A1>0 and A<0 then
nsolutions=2

Appendix C 429

WsDoc.Designs.Sch1.Schematic.Page.Annotate2.Max.Set(nsolutions)
Select Case i

case 1

$B = ((RL * XS) + Sqr(RS * RL * (RS \wedge 2 + XS \wedge 2 - RS * RL))) / (RL * (RS \wedge 2 + XS \wedge 2))$
$X = (RL * XS - RS * XL) / RS + (RS - RL) / (B * RS)$
If B>0 Then   'B is a shunt C
B=B/wo
ShuntC
elseif B<0 then   'B is a shunt L
B=-1/(wo*B)
ShuntL
else
end if

If X>0 then   'X is a series L
X=X/wo
SeriesL_X2
elseif X<0 then   'X is a series C
X=-1/(wo*X)
SeriesC_X2
else
end if
PlotResponse

case 2

$B = ((RL * XS) - Sqr(RS * RL * (RS \wedge 2 + XS \wedge 2 - RS * RL))) / (RL * (RS \wedge 2 + XS \wedge 2))$
$X = (RL * XS - RS * XL) / RS + (RS - RL) / (B * RS)$
If B>0 Then   'B is a shunt C
B=B/wo
ShuntC
elseif B<0 then   'B is a shunt L
B=-1/(wo*B)
ShuntL
else
end if

If X>0 then   'X is a series L

```
X=X/wo
SeriesL_X2
elseif X<0 then            'X is a series C
X=-1/(wo*X)
SeriesC_X2
else
end if
PlotResponse
end select

'
_____

ElseIf A1<0 then
nsolutions=2
WsDoc.Designs.Sch1.Schematic.Page.Annotate2.Max.Set(nsolutions)
Select case i

case 1

B = ((RS * XL) + Sqr(RS * RL * (RL ^ 2 + XL ^ 2 - RS * RL))) / (RS * (RL ^ 2 + XL ^ 2))
X = (RS * XL - RL * XS) / RL + (RL - RS) / (B * RL)
If B>0 Then                'B is a shunt C
B=B/wo

ShuntC
elseif B<0 then            'B is a shunt L
B=-1/(wo*B)
ShuntL
else
end if

If X>0 then                'X is a series L
X=X/wo
SeriesL_X3
elseif X<0 then            'X is a series C
X=-1/(wo*X)
SeriesC_X3
else
end if
PlotResponse
```

Appendix C

case 2

B = ((RS * XL) - Sqr(RS * RL * (RL ^ 2 + XL ^ 2 - RS * RL))) / (RS * (RL ^ 2 + XL ^ 2))
X = (RS * XL - RL * XS) / RL + (RL - RS) / (B * RL)
If B>0 Then   'B is a shunt C
B=B/wo
ShuntC
elseif B<0 then   'B is a shunt L
B=-1/(wo*B)
ShuntL
else
end if

If X>0 then   'X is a series L
X=X/wo
SeriesL_X3
elseif X<0 then   'X is a series C
X=-1/(wo*X)
SeriesC_X3
else
end if

PlotResponse

'_____

end select
Else
End if

'_____Define Network Configurations_____
Sub ShuntC
My_L_Network=WsDoc.Designs.Sch1.PartList.X1.ChangeSymbol("CAPACITOR")
My_L_Network=WsDoc.Designs.Sch1.PartList.X1.ChangeModel("CAP")
My_L_Network=WsDoc.Designs.Sch1.PartList.X1.ParamSet.C.SetValue(B)
end sub

Sub ShuntL
My_L_Network=WsDoc.Designs.Sch1.PartList.X1.ChangeSymbol("INDUCTOR")

```
My_L_Network=WsDoc.Designs.Sch1.PartList.X1.ChangeModel("IND")
My_L_Network=WsDoc.Designs.Sch1.PartList.X1.ParamSet.L.SetValue(B)
end sub

Sub SeriesL_X2
L2=.0001*10^-20
My_L_Network=WsDoc.Designs.Sch1.PartList.X3.ChangeSymbol("Ferrite")
My_L_Network=WsDoc.Designs.Sch1.PartList.X3.ChangeModel("IND")
My_L_Network=WsDoc.Designs.Sch1.PartList.X3.ParamSet.L.SetValue(L2)
My_L_Network=WsDoc.Designs.Sch1.PartList.X2.ChangeSymbol("INDUCTOR")
My_L_Network=WsDoc.Designs.Sch1.PartList.X2.ChangeModel("IND")
My_L_Network=WsDoc.Designs.Sch1.PartList.X2.ParamSet.L.SetValue(X)
end sub

Sub SeriesC_X2
L2=.0001*10^-20
My_L_Network=WsDoc.Designs.Sch1.PartList.X3.ChangeSymbol("Ferrite")
My_L_Network=WsDoc.Designs.Sch1.PartList.X3.ChangeModel("IND")
My_L_Network=WsDoc.Designs.Sch1.PartList.X3.ParamSet.L.SetValue(L2)
My_L_Network=WsDoc.Designs.Sch1.PartList.X2.ChangeSymbol("CAPACITOR")
My_L_Network=WsDoc.Designs.Sch1.PartList.X2.ChangeModel("CAP")
My_L_Network=WsDoc.Designs.Sch1.PartList.X2.ParamSet.C.SetValue(X)
end sub

sub SeriesL_X3
L3=.0001*10^-20
My_L_Network=WsDoc.Designs.Sch1.PartList.X3.ChangeSymbol("INDUCTOR")
My_L_Network=WsDoc.Designs.Sch1.PartList.X3.ChangeModel("IND")
My_L_Network=WsDoc.Designs.Sch1.PartList.X3.ParamSet.L.SetValue(X)
My_L_Network=WsDoc.Designs.Sch1.PartList.X2.ChangeSymbol("Ferrite")
My_L_Network=WsDoc.Designs.Sch1.PartList.X2.ChangeModel("IND")
My_L_Network=WsDoc.Designs.Sch1.PartList.X2.ParamSet.L.SetValue(L3)
end sub

sub SeriesC_X3
L1=.0001*10^-20
My_L_Network=WsDoc.Designs.Sch1.PartList.X3.ChangeSymbol("CAPACITOR")
My_L_Network=WsDoc.Designs.Sch1.PartList.X3.ChangeModel("CAP")
```

Appendix C

My_L_Network=WsDoc.Designs.Sch1.PartList.X3.ParamSet.C.SetValue(X)
My_L_Network=WsDoc.Designs.Sch1.PartList.X2.ChangeSymbol("Ferrite")
My_L_Network=WsDoc.Designs.Sch1.PartList.X2.ChangeModel("IND")
My_L_Network=WsDoc.Designs.Sch1.PartList.X2.ParamSet.L.SetValue(L1)
end sub

Sub PlotResponse
Sim.ClearModelCache
Sim.RunAnalysis
Graph=WsDoc.Designs.Graph1
Graph.OpenWindow
End Sub.

Appendix D

VBScript file listing for the Line and Stub Matching Utility of Chapter 6.

```
Dim WsDoc
Dim RS, d1,d2,pi, serieslinelength, B1, B2,so1,so2,ss1,ss2,
OpenShuntStub,ShortedShuntStub
Dim RL, XL, F, complex, pos, i, r, x, t1, t2 ,nsolutions, z1,Length,e
pi=3.141592653589793
WsDoc= theApp.GetWorkspaceByIndex(0)

F = WsDoc.Designs.Sch1.PartList.CWSource_3.ParamSet.F.GetValue()
My_LineStub=WsDoc.Designs.Sch1.PartList.TL1.ParamSet.F.SetValue(F)
My_LineStub=WsDoc.Designs.Sch1.PartList.TL2.ParamSet.F.SetValue(F)
RS = WsDoc.Designs.Sch1.PartList.Port_1.ParamSet.ZO.GetData()
complex = WsDoc.Designs.Sch1.PartList.Port_2.ParamSet.ZO.GetData()
pos=instr(complex,"j")
complex=left(complex,pos-1)
z=len(complex)

' Find the 1st space position

space_position=instr(complex," ")
RL=left(complex,(space_position))
XL=Mid(complex,space_position+1,((z-space_position)+1))

' Convert image string to number

s=len(XL)
sign=mid(XL,1,1)
XL=mid(XL,3,(s-2))
If sign="+" then
XL=XL
else
XL=XL*-1
end if
```

' Calculation of Series Line Length
r=RL/RS
x=XL/RS
t1=(x+sqr(r*(r^2+x^2-(2*r)+1)))/(r-1)
t2=(x-sqr(r*(r^2+x^2-(2*r)+1)))/(r-1)

' Calculate the electrical lengths

If t1>0 then
d1=(360/(2*pi))*atn(t1)
else
d1=360*(pi+atn(t1))/(2*pi)
end if
If t2>0 then
d2=(360/(2*pi))*atn(t2)
else
d2=360*(pi+atn(t2))/(2*pi)
end if
If d1<d2 then
SeriesLineLength=d1
else
SeriesLineLength=d2
end if

' Calculation of Shunt Line Length

B1=(x*t1^2+(r^2+x^2-1)*t1-x)/(RS*(r^2+x^2+t1^2+2*x*t1))
B2=(x*t2^2+(r^2+x^2-1)*t2-x)/(RS*(r^2+x^2+t2^2+2*x*t2))

' Open circuit stubs

so1=-360*(atn(RS*B1))
so2=-360*atn(RS*B2)/(2*pi)

If so1<0 then
OpenShuntStub=so2
else
OpenShuntStub=so1
end if

' Shorted shunt stubs

ss1=(360*atn(1/(RS*B1)))/(2*pi)
ss2=(360*atn(1/(RS*B2)))/(2*pi)

If ss1<0 then
ShortedShuntStub=ss2
else
ShortedShuntStub=ss1
end if

' Determine valid solutions

nsolutions=2
WsDoc.Designs.Sch1.Schematic.Page.Annotate2.Max.Set(nsolutions)
i=WsDoc.Equation.VarBlock.Solution.GetValue()
Select Case i

case 1
If ss1>0 then
serieslinelength=d1
SeriesLine
ShortedShuntStub=ss1
ShortedCircuitStub
elseif ss1<0 then
serieslinelength=d2
SeriesLine
ShortedShuntStub=ss2
ShortedCircuitStub
end if
Plotresponse

case 2

If so1>0 then
serieslinelength=d1
SeriesLine
OpenShuntStub=so1
OpenCircuitStub
elseif so1<0 then

```
serieslinelength=d2
SeriesLine
OpenShuntStub=so2
OpenCircuitStub
end if
Plotresponse
End Select

' Define Transmission Line Configurations
Sub OpenCircuitStub
Length=so2
Length=Length*(pi/180)
My_LineStub=WsDoc.Designs.Sch1.PartList.TL2.ParamSet.L.SetValue(Length)
z1=100e6
My_LineStub=WsDoc.Designs.Sch1.PartList.Port_4.ParamSet.ZO.SetValue(z1)
My_LineStub=WsDoc.Designs.Sch1.PartList.Port_4.ChangeSymbol("TRL_END
")
End sub

Sub ShortedCircuitStub ShortedShuntStub=ShortedShuntStub*(pi/180)
My_LineStub=WsDoc.Designs.Sch1.PartList.TL2.ParamSet.L.SetValue(Shorted
ShuntStub)
z1=.0001
My_LineStub=WsDoc.Designs.Sch1.PartList.Port_4.ParamSet.ZO.SetValue(z1)
My_LineStub=WsDoc.Designs.Sch1.PartList.Port_4.ChangeSymbol("GROUND
")
End sub

Sub SeriesLine serieslinelength=serieslinelength*(pi/180)
My_LineStub=WsDoc.Designs.Sch1.PartList.TL1.ParamSet.L.SetValue(serieslin
elength)
End sub

' Simulate and Plot Response of Matching Network

Sub PlotResponse
Sim=WsDoc.Designs.Linear1
Sim.RunAnalysis
Sim.ClearModelCache
```

```
Graph=WsDoc.Designs.Graph1
Graph.OpenWindow
End Sub.
```

Index

A

Admittance circles, 128–129, 142, 301
Amplifier design, 217, 234, 321–389, 391–416
    low noise, 321–322, 348–350, 360–362, 365–366, 371–375, 385, 389, 391, 408–413
    maximum gain, 321–322, 324, 334–336, 347, 353, 360, 366, 372, 385, 387–389, 391–392, 414–415
    power, 184, 213–214, 321–322, 376–385, 391
    specific gain, 321, 347–360, 366, 385, 388, 391
Amplifier noise figure, 371
Amplifier noise temperature, 373–375
Amplifier types, 321–389
Attenuation constant, 53, 56, 59, 62

B

Band pass filter, 205–207
Bandwidth limitation of matching networks, 264–266
Broadband matching networks, 217, 247–259, 307–308, 311–318

C

Capacitors, aluminum, 41
    ceramic, 41, 43–44
    chip, 41, 43–48, 51, 80, 185–186, 188–189, 191–192, 341, 398
    multilayer, 41, 43–48
    polystyrene, 41
    silver mica, 41
    single layer, 41–43, 51
    tantalum, 41
    thin film, 43
Capacitor tolerance, 401, 405–406
Cascaded gain, 408–410, 412–413, 415
Cascaded L-networks, 248, 250–251, 253–259
Cascaded matching networks, 253, 307, 310
Coaxial transmission line. *See* Transmission lines

Conductor Q factor, 163
Coupled line filter, 205–207
Coupling coefficient, 168–171, 175
Critically coupled resonator, 173–174

D

Design centering, 407
Dielectric loss, 80, 161, 163
Directional coupler, 53, 80, 91, 105–109, 112, 377 coupling,
    105 directivity,
    105 insertion loss,
    105 isolation, 105
Distributed bias feed, 100–102, 111

E

Edge coupled filter, 207–211, 215
Effective area, 11
Effective capacitance, 47–48
Effective dielectric constant, 76, 78, 81, 93, 97, 161–162, 196
Electric field, 1, 41, 53, 67–68, 77, 83, 103
Electromagnetic spectrum, 1–3
Electromagnetic waves, 1, 49, 53
Enclosure effects, 200, 210–211
Equal-Q L-networks, 248, 253–254, 257–259, 276
External Q, 151–152, 168, 260, 263, 281

F

Filter attenuation, 177–178, 184-186, 188, 210, 213
Filter order, 177–178, 181–182, 184, 188, 215
Filter return loss, 180–183
Filter ripple, 182, 412
Filter synthesis utility, 184–185, 205, 270
Filter topology, 176
Filter type, 176–181, 183–184
Flat ribbon inductance, 11–12
Fractional bandwidth, 152, 205, 230–233, 239–242, 246–247, 251–253, 256–257, 276, 279–282, 284, 286, 289–292, 296, 304, 307–308, 314,

318–320, 336, 416. *See also* Quarter-wave matching network
Free space, 1, 4, 9, 54–55, 67–69, 82, 89, 200, 207
Friis formula, 408–411, 416

G

Genesys artwork replacement, 337
GMAX, 325–328, 335, 348–349, 351, 353, 359, 365–366, 387–388, 394, 397, 415

H

High pass filter, 84, 87, 177, 184, 187–191, 195, 214, 266
High pass matching L-networks, 267

I

Impedance matching,
  complex load to complex source, 227–234
  complex load to real source, 234–242
  real load to real source, 242–247
Incident power. *See* Normalized waveforms
Inductors,
  air core, 17–28, 32, 36, 50
  chip, 17, 28–31, 51, 80, 186, 398
  coil, 16, 18–24, 32–33, 50, 186
  cylindrical, 8, 16, 18, 33
  magnetic core, 18, 32–40
  rectangular, 16
  rod, 32–33
  toroidal, 17, 32–34, 36, 37–40, 51
Input reflection coefficient, 64, 110–111, 120, 166, 169, 279–280, 283, 285–286, 290, 367
Insertion loss, 74, 86–87, 101, 105, 108, 111, 124, 153, 155, 164, 182, 199, 213, 222, 225, 231–233, 238, 240, 249–250, 270, 273, 330, 341, 394, 396, 411–412
Intermediate resistors, 248–249, 252, 254, 264, 305, 308–309, 315
Interstage matching network, 225, 392, 394–396, 415

L

L-network matching utility, 217
L-Network Synthesis Utility, 329
Loaded Q, 151–154, 164, 168–169, 175, 257, 260, 264–265, 276, 280–285
Lossless transmission line, 61–63, 67, 293
  characteristic impedance, 61–63, 293
  parameters, 61–62
  propagation constant, 61–62
Low pass Chebyshev filter, 182–184
Low pass filter, 177, 181–188, 195–196, 198–199, 204–205, 214
Low pass matching L-networks, 266

M

Magnetic fields, 1, 3, 16, 32–33, 53, 55, 67–68, 76–77, 83–84
Magnetic flux, 9, 16, 21, 32–33
Magnetic flux density, 32
Magnetization intensity, 32
Maximum gain amplifier. *See* Amplifier design
Maximum power transfer, 217–225, 227, 234–235, 242–243, 275, 278
Microstrip bias feed, 99
Microstrip resonator, 102, 159, 162–167, 171–175, 214
Microstrip stepped impedance, 195
Microstrip transmission lines, 70, 76–80
Microwave resonators, 147, 158–159, 167–175. *See also* One port microwave resonators
Mismatch loss, 65–66, 75, 182
Monte Carlo analysis, 400–405

N

Network parameters, 113–146
  ABCD parameters, 113, 116–117, 120
  h parameters, 113, 115, 122
  Y parameters, 113–115, 122
  Z parameters, 113–114, 122
Normalized impedance, 125–128, 137, 378
Normalized waveforms, 118 incident power, 61–62, 106, 118 reflected power, 65–66, 105–106, 118

Index

Number of L-networks, 253, 257–259

O

One port microwave resonators, 167–170
Over coupled resonators, 166, 174–175

P

Parallel resonant circuits, 26, 94, 139, 149–151, 158–159, 164, 175
Parameter sweeps, 23, 65, 91, 94, 134–136, 261, 287, 398–400, 402
Parasitic capacitance, 13, 16, 25
Permeability, 9–10, 32–36, 54, 56, 67–68
   free space, 9
   relative, 9
Permittivity, 54, 67–69
Plane waves, 53–56
   equations, 53–56
   in a good conductor, 55–56
   in a lossless medium, 53–55
   phase velocity, 54
Power gain circles, 353–354, 376, 388

Q

Q factor, 17–18, 27–29, 31, 33–39, 44–47, 51, 147, 150–154, 158–159, 161–163, 168, 185, 211, 247, 251–254, 257–260, 263, 272–273, 276–277, 280–285, 308, 310–311, 319–320. *See also* Radiation Q factor
Q_0 measurement on the Smith Chart, 171–175
Q-curves, 259–264
Quarter-wave matching networks, 277–292, 307–320
   characteristic impedance, 278–279, 282, 284, 286, 288
   fractional bandwidth, 279–282, 284, 286, 289–292
   loaded Q factor, 280–282, 284
   matching bandwidth, 277, 280, 286–292
   power loss, 277, 285–286, 290–292, 319

Quarter-wave transformer, 63, 110–111, 277, 281, 288, 307–308, 311–312, 314, 316

R

Radiation Q factor, 161, 163
Reactance power of capacitor, 26, 125, 149, 224
Reactance power of inductor, 148
Rectangular waveguide, 82–85, 111
   characteristic impedance, 84
   cutoff frequency, 84
   field pattern, 83 intrinsic impedance, 84
   propagation mode, 82–84
   TE mode, 83
   TM mode, 83
   wavelength, 84
Reentrant modes, 204–205, 214
Reflected power. *See* Normalized waveforms
Resistivity, 5, 9–10, 12, 17, 71, 80, 86
Resistors,
   chip, 14–16, 28–29, 44, 50
   leaded, 12–14
   thick film, 14–15
Resonant circuits, 26, 147–215. *See also* Parallel resonant circuits, Series resonant circuits
Resonator decoupling, 155–158
Resonator Q circle, 170, 172, 174
Resonator series reactance, 166–175
Return loss, 53, 61, 64–66, 75, 101, 105–106, 110–111, 124, 145, 166, 180–183, 185, 214, 230–233, 238–242, 246–251, 253, 256, 264–266, 270–273, 280–286, 289–292, 296, 304, 307–308, 310–311, 314, 318–320, 322–323, 330, 334–336, 343–345, 359, 363, 365–366, 372, 384–385, 389, 391, 394, 396–397, 400, 402, 410, 416. *See also* Transmission lines

S

Scalar S-Parameters, 123–124
   input return loss, 124
   insertion gain, 124
   insertion loss, 124

isolation, 124
output return loss, 124
reverse gain, 124
Scattering matrix, 117, 119
Self inductance, 3
Self resonant frequency, 24, 26, 29, 31, 36, 39, 45, 47–48, 50–51
Sensitivity analysis, 400–404
Series resonant circuits, 48, 96, 147–148, 150–151
Single-stub matching networks, 277, 292–307, 311–314, 316–318, 367, 370–371
　analytical design, 292–298
　calculation of line length, 297–299
　calculation of stub length, 297–299
　graphical design, 301–304
Single-stub matching utility, 299–301
Skin depth, 3, 8–11, 50, 56, 160
Skin effects, 8, 10, 17, 36, 38, 71
S parameters, 28–29, 31, 44, 117–124, 127, 134–135, 145, 147, 153, 167–168, 171, 185–186, 188, 200, 222, 272, 325, 335, 339–341, 351, 359, 362–363, 365, 376, 380, 382–384, 387–389
　file tuning, 190–195
　forward transmission, 120
　input reflection coefficient, 120
　output reflection coefficient, 120
　reverse transmission, 120
Stability circle, 350–354, 357, 363–366, 369–370, 388–389
　input, 350
　output, 350
Stability factor, 324–326, 351, 353, 387
Stability measure, 324–326, 387
Stability parameters, 325–327, 365, 384, 387, 389
Step discontinuities, 98, 199
Straight wire inductance, 3–8
Stripline transmission lines, 70, 80–82
Surface roughness, 160

T

Tapped capacitor resonator, 156–157
Tapped inductor resonator, 157–158
TEM wave propagation, 68–69
Transistor stability, 323–328
Transmission line components, 91–102
　distributed capacitive, 69, 96–98
　distributed inductive, 69, 96–98
　half wavelength, 63, 295
　open circuited, 92, 94–96, 101–102, 141–143, 159, 292, 301–302
　quarter wavelength, 63, 92
　short circuited, 62, 91–94, 139–143, 159, 266, 292, 300, 303–304
Transmission line resonator, 159–166
Transmission lines, 10, 53–112, 137–143, 145, 147, 158–162, 169, 195–196, 199, 203–204, 214, 234, 260, 277, 281, 284, 288, 292–296, 298, 300–304, 319. *See also* Lossless transmission line
　attenuation constant, 53, 56, 59, 62
　broadside coupled, 102–103
　characteristic impedance, 53, 59, 61–63, 67–68, 71, 73, 77, 81, 84, 104
　coaxial, 50, 53, 68, 70–75, 80, 85–86, 89, 111, 159, 186, 189
　coupled, 53, 102–109, 159, 167, 169, 205–207
　distributed capacitive, 69, 96–98
　distributed inductive, 69, 96–98
　edge coupled, 102–104, 107–109, 205–207
　end coupled, 102–103
　equations, 57–66
　equivalent model, 53
　group delay, 53, 89–90
　microstrip, 53, 68, 70, 76–80, 85, 89, 91–102, 104–105, 107–108, 111, 159–163, 167, 169, 186, 195–198, 205–207, 214, 292, 344, 372. *See also* Microstrip transmission lines
　parameters, 53, 57–66
　phase constant, 53, 59, 62
　reflection coefficient, 53, 59–61, 64–66, 110–111
　return loss, 53, 61, 64–66, 75, 101, 105–106, 110–111
　stripline, 68, 70, 80–82. *See also* Stripline transmission lines
　terminations, 62–63, 94
　two parallel wire, 56
　voltage standing wave ratio, 53, 60–61

Index

U

Under-coupled resonator, 172–173
Unloaded Q, 151, 162–163, 168–169, 175, 181, 214, 264–265

V

VBScript code, 191–194, 268–269, 299–300
   listing, 191–194
   slider control, 192–194, 268–269, 299
Velocity of light, 1, 55
Voltage standing wave ratio (VSWR), 53, 60–61, 64–66, 75, 105–106, 109–112, 134–137, 144–145, 182, 225, 265

W

Wavelength, 1–2, 49, 53, 55, 57, 63, 68–69, 84, 92–93, 97, 111, 117, 138–139, 141, 143, 145, 160–162, 217, 277, 295–296, 337

Y

Yield analysis, 391–416

About the Authors

Ali Behagi and Stephen Turner bring a unique perspective to this subject material. Their long association of over 25 years blends a strong academic and industrial background to the text.

Ali A. Behagi received the Ph.D. degree in electrical engineering from the University of Southern California and the MS degree in electrical engineering from the University of Michigan. He has several years of industrial experience with Hughes Aircraft and Beckman Instruments. Dr. Behagi joined Penn State University as an associate professor of electrical engineering in 1986. He has devoted over 20 years to teaching microwave engineering courses and directing university research projects. While at Penn State he has introduced Keysight/Agilent ADS software in teaching RF and microwave circuit design courses and laboratory experiments. After retirement from Penn State he has been active as an educational consultant. Dr. Behagi is a Keysight Certified Expert, a senior member of the Institute of Electrical and Electronics Engineers (IEEE) and the Microwave Theory and Techniques Society.

Stephen D. Turner has over 30 years of experience in the microwave industry with focus on amplifier, oscillator, and system design. He is an avid microwave CAD enthusiast going back to Touchstone v1.0. Having familiarity with all of the major commercial microwave CAD products, these days he is a proponent and user of the Keysight/Agilent Advanced Design System (ADS) and Genesys. He received the Master of Engineering degree from Penn State University and a BS degree from the University of Pittsburgh. He is a registered professional engineer in the state of Pennsylvania and member of the Institute of Electrical and Electronics Engineers and the Microwave Theory and Techniques Society.

CPSIA information can be obtained
at www.ICGtesting.com
Printed in the USA
BVOW10*2001210616
452932BV00010B/26/P